RNA Biochemistry and Biotechnology

NATO Science Series

A Series presenting the results of activities sponsored by the NATO Science Committee. The Series is published by IOS Press and Kluwer Academic Publishers, in conjunction with the NATO Scientific Affairs Division.

A. Life Sciences	IOS Press
B. Physics	Kluwer Academic Publishers
C. Mathematical and Physical Sciences	Kluwer Academic Publishers
D. Behavioural and Social Sciences	Kluwer Academic Publishers
E. Applied Sciences	Kluwer Academic Publishers
F. Computer and Systems Sciences	IOS Press

1. Disarmament Technologies	Kluwer Academic Publishers
2. Environmental Security	Kluwer Academic Publishers
3. High Technology	Kluwer Academic Publishers
4. Science and Technology Policy	IOS Press
5. Computer Networking	IOS Press

NATO-PCO-DATA BASE

The NATO Science Series continues the series of books published formerly in the NATO ASI Series. An electronic index to the NATO ASI Series provides full bibliographical references (with keywords and/or abstracts) to more than 50000 contributions from internatonal scientists published in all sections of the NATO ASI Series.
Access to the NATO-PCO-DATA BASE is possible via CD-ROM "NATO-PCO-DATA BASE" with user-friendly retrieval software in English, French and German (WTV GmbH and DATAWARE Technologies Inc. 1989).

The CD-ROM of the NATO ASI Series can be ordered from: PCO, Overijse, Belgium

Series 3. High Technology – Vol. 70

RNA Biochemistry and Biotechnology

edited by

Jan Barciszewski
Institute of Bioorganic Chemistry
of the Polish Academy of Sciences,
Poznan, Poland

and

Brian F.C. Clark
Institute of Molecular and Structural Biology,
University of Aarhus, Denmark

Springer-Science+Business Media, B.V.

Proceedings of the NATO Advanced Research Workshop on
RNA Biochemistry and Biotechnology
Poznan, Poland
October 10–17, 1998

A C.I.P. Catalogue record for this book is available from the Library of Congress.

ISBN 978-0-7923-5862-6 ISBN 978-94-011-4485-8 (eBook)
DOI 10.1007/978-94-011-4485-8

Printed on acid-free paper

Originally published by Kluwer Academic Publishers in 1999
Softcover reprint of the hardcover 1st edition 1999

TABLE OF CONTENTS

PREFACE

This volume contains contributions from the speakers at the NATO Advanced Research Workshop on "RNA: biochemistry and biotechnology" which was held in Poznań, Poland, 10 - 17 October, 1998.

RNA plays many roles in biological processes and our knowledge of its importance is expanding very rapidly. By understanding the three dimensional structure of RNA we can significantly extend our understanding of its biological functions. This can be achieved only in close cooperation among molecular biologist, chemists and biophysicists. The discovery of catalytic activity in RNAs has prompted various attempts to solve the mechanisms of their assembly into functional native states starting from linear strands. The folding of RNA was thought until now to take place in two steps. The RNA first folds into a secondary structure of stems, loops, bulges and mismatches, and in the second step a tertiary structure is formed due to long range interactions of single stranded portions of the sequence. Recently it has been shown (I. Tinoco) that RNA folding causes a secondary structure rearrangement. The ground rules for the formation of secondary structure have been derived from physical studies of oligoribonucleotides. Powerful NMR techniques and X-ray analysis have revealed more details of RNA structure including novel conformations. A wealth of information has been obtained by studying the relatively small RNA molecules (tRNA, ribozymes, aptamers). A few of these have been crystallized, enabling determination of their three dimensional structures. Independent evidence for three dimensional folding stems from high resolution proton NMR studies. Molecular dynamics calculations promise to provide us with a detailed knowledge of the atomic motions in these molecules. Details of the structures and the interaction with ligands are also derived from data obtained by a variety of spectroscopic and genetic techniques.
The structural features of RNA that are important for RNA-protein specific recognition have only recently come under investigation. Several strategies for recognizing specific RNA sites can be discerned.

An unsolved problem in biology is the origin and evolution of tRNA aminoacylation systems that dictate correct expression of the genetic code at the translational level. The aminoacylation of RNA structures generally is considered the starting point for the emergence of the theatre of proteins from the RNA World. Transfer RNAs and aminoacyl-tRNA synthetases are the center of attention for various biochemical, genetic and phylogenetic studies. There is, with some exceptions, one aminoacyl-tRNA synthetases for each amino acid and, because of the degeneracy of the genetic code, one or more tRNAs for each amino acid. The aminoacyl-tRNA synthetases have long been upheld as a paradigm of molecular evolution, because their products aminoacyl-tRNAs are essentially the same in all living organisms. The role of aminoacyl-tRNA synthetases is to interpret, to decode amino acids, providing the essential link between RNA and protein without which translation of the genetic information would be impossible. In the aminoacyl-tRNA reactions, each amino acid is joined to the tRNA that harbours the anticodon triplet of the code for that amino acid. Thus, the question of the origin of the code has to deal at some point with tRNA and aminoacyl-tRNA synthetases. Once an aminoacyl ester linkage is established with an RNA acceptor, peptide bond formation is thermodynamically favoured and such complex with elongation factor in the GTP form is transported the A-site on a ribosome.

This volume contains scientific research reports covering the foregoing description of RNA biochemistry and biotechnology.

The NATO Scientific Affairs Division is gratefully acknowledged for granting an award that made the organisation of the workshop and the preparation of this book possible.

Jan Barciszewski
Brian F. C. Clark

LIST OF CONTRIBUTORS

Adamiak Ryszard W., Institute of Bioorganic Chemistry, Polish Academy of Sciences, Noskowskiego 12, 61-704 Poznan, Poland
E-mail: adamiakr@ibch.poznan.pl

Barciszewska Miroslawa, Institute of Bioorganic Chemistry, Polish Academy of Sciences, Noskowskiego 12, 61-704 Poznan, Poland
E-mail: mbarcisz@ibch.poznan.pl

Barciszewski Jan , Institute of Bioorganic Chemistry, Polish Academy of Sciences, Noskowskiego 12, 61-704 Poznan, Poland
E-mail: jbarcisz@ibch.poznan.pl

Bayer Peter, MRC Laboratory of Molecular Biology, Hills Road, Cambridge CB2 2QH, UK

Berkhout Ben, Department of Human Retrovirology, Academic Medical Center, University of Amsterdam, Meibergdreef 15, 1105 AZ Amsterdam, The Netherlands
E-mail: b.berkhout@amc.uva.nl

Betzel Christian, Universitätskrankenhaus Eppendorf der Universität Hamburg, Inst. f. Physiol. Chemie, AG Makro. Strukturanalyse, c/o DESY, D-22603 Hamburg, Germany
E-mail: Betzel@unisgi1.desy.de

Bielecki Lukasz, Institute of Bioorganic Chemistry, Polish Academy of Sciences, Noskowskiego 12, ,61-704 Poznan, Poland
E-mail: bielecki@ibch.poznan.pl

Blechschmidt Bernd, Universität Bayreuth, Lab. für Biochemie, 95440 Bayreuth, Germany

Boczkowska M., Polish Academy of Sciences, Centre of Molecular and Macromolecules Studies, Department of Bioorganic Chemistry, Sienkiewicza 112, 90-363 Lodz, Poland

Brock S., Universität Bayreuth, Lab. für Biochemie, 95440 Bayreuth, Germany

Bunjun S., Department of Molecular Biophysics and Biochemistry, Yale University, New Haven, CT 06520-8114, USA

Ciesiolka Jerzy, Institute of Bioorganic Chemistry, Polish Academy of Sciences, Noskowskiego 12, 61-704 Poznan, Poland
E-mail: ciesiolk@ibch.poznan.pl

Clark Brian F.C., Institute of Molecular and Structural Biology, University of Aarhus, Gustav Wieds Vej 10, 8000 Aarhus C, Denmark
E-mail: clark@biobase.dk

Cole Paul, MRC, Laboratory of Molecular Biology, Hills Road, Cambridge CB2 2QH, UK

Das Atze T., Department of Human Retrovirology, Academic Medical Center, University of Amsterdam, Meibergdreef 15, 1105 AZ Amsterdam, The Netherlands

Dragon Francois, Friedrich Miescher Institute, P.O. Box 2543,, Maulbeerstrasse 66, 4002 Basel, Switzerland

Drozdz M. , Institute of Bioorganic Chemistry, Polish Academy of Sciences, Noskowskiego 12, 61-704 Poznan, Poland

Erdmann Volker A., Freie University Berlin, Institute of Biochemistry, Thielallee 63, 14195 Berlin, Germany
E-mail: erdmann@chemie.fu-berlin.de

Filipowicz Witold, Friedrich Miescher Institute, P.O. Box 2543,, Maulbeerstrasse 66, 4002 Basel, Switzerland
E-mail: Filipowi@fmi.ch

Fürste Jens Peter, NOXXON, Pharma AG, Gustav-Meyer-Allee 25, D-13355 Berlin, Germany
E-mail: jfuerste@noxxon.net

Gdaniec Zofia, Institute of Bioorganic Chemistry, Polish Academy of Sciences, Noskowskiego 12, , 61-704 Poznan, Poland
E-mail: zgdan@ibch.poznan.pl

Giel-Pietraszuk Malgorzata, Institute of Bioorganic Chemistry, Polish Academy of Sciences, Noskowskiego 12, 61-704 Poznan, Poland
E-mail: giel@ibch.poznan.pl

Gustafson E. Hilary, The Rockefeller University, , 1230 York Avenue, New York, NY 10021, USA

Ibba Michael, Department of Molecular Biophysics and Biochemistry, Yale University, New Haven, CT 06520-8114, USA
E-mail: mibba@trna.chem.yale.edu

Ismaili Naima,The Rockefeller University, 1230 York Avenue, , New York, NY 10021, USA

Ke Y., Children's Hospital Oakland , Research Institute, 747 52nd Street, Oakland, CA 94609, USA

Kirsebom L., Uppsala University, Department of Microbiology, BMC, Box 581, 75123 Uppsala, Sweden

Knudsen Charlotte R., Institute of Molecular and Structural Biology, University of Aarhus, Gustav Wieds Vej 10, 8000 Aarhus C, Denmark
E-mail: crk@imsb.au.dk

Kois Pavol, The Rockefeller University, 1230 York Avenue, New York, NY 10021, USA

Konarska Maria M., The Rockefeller University, 1230 York Avenue, New York, NY 10021, USA
E-mail: konarsk@rockvax.rockfeller.edu

Konno Michiiko, Deparptment of Chemistry, Faculty of Science, Ochanomizu University, 2-1-1- Otsuka, Bunkyo-ku, Tokkyo 112-8610, Japan

Koziolkiewicz Maria, Polish Academy of Sciences, Centre of Molecular and Macromolecules Studies, Department of Bioorganic Chemistry, Sienkiewicza 112, 90-363 Lodz, Poland
E-mail: mkoziol@bio.cbmm.lodz.pl

Krakowiak A., Polish Academy of Sciences, Centre of Molecular and Macromolecules Studies, Department of Bioorganic Chemistry, Sienkiewicza 112, 90-363 Lodz, Poland

Krzyzosiak Wlodzimierz, Institute of Bioorganic Chemistry, Polish Academy of Sciences, Noskowskiego 12, 61-704 Poznan, Poland
E-mail: wlodkrzy@ibch.poznan.pl

Kulinski Tadeusz, Institute of Bioorganic Chemistry, Polish Academy of Sciences, Noskowskiego 12, 61-704 Poznan, Poland
E-mail: tadkul@ibch.poznan.pl

Kupryszewski Gerard, Faculty of Chemistry, University of Gdansk, Sobieskiego 18, 80-952 Gdansk, Poland

Leonitz N.B., Chemistry Department, Bowling Green State University, Bowling Green, OH 43 403 , USA

Losey H., Department of Molecular Biophysics and Biochemistry, Yale University, New Haven, CT 06520-8114, USA

Mani Peter, Gentechnologie & Gesellschaft, ETH Zentrum, Rämistrasse 101, 8092 Zürich, Switzerland
E-mail: mani@huwi.ethz.ch

Mansilla F., Institute of Molecular and Structural Biology, University of Aarhus, Gustav Wieds Vej 10, 8000 Aarhus C, Denmark

McCloskey Jeffrey, The Rockefeller University, 1230 York Avenue, New York, NY 10021, USA

Min B., Department of Molecular Biophysics and Biochemistry, Yale University, New Haven, CT 06520-8114, USA

Morikawa Kosuke, Biomolecular Engineering Research Institute (BERI), 6-2-3 Furuedai, Suita, Osaka 565, Japan

Mucha Piotr , Faculty of Chemistry, University of Gdansk, Sobieskiego 18, 80-952 Gdansk, Poland
E-mail: fly@chemik.chem.univ.gda.pl

Napierala Marek, Institute of Bioorganic Chemistry, Polish Academy of Sciences, Noskowskiego 12, 61-704 Poznan, Poland
E-mail: napmarek@ibch.poznan.pl

Nawrot Barbara, Polish Academy of Sciences, Centre of Molecular and Macromolecules Studies, Department of Bioorganic Chemistry, Sienkiewicza 112, 90-363 Lodz, Poland

Okruszek Andrzej, Polish Academy of Sciences, Centre of Molecular and Macromolecules Studies, Sienkiewicza 112, 90-363 Lodz, Poland
E-mail: okruszek@bio.cbmm.lodz.pl

Owczarek A., Polish Academy of Sciences, Centre of Molecular and Macromolecules Studies, Department of Bioorganic Chemistry, Sienkiewicza 112, 90-363 Lodz, Poland

Pedersen G.N., Institute of Molecular and Structural Biology, University of Aarhus, Gustav Wieds Vej 10, 8000 Aarhus C, Denmark

Pelczar Pawel, Friedrich Miescher Institute, P.O. Box 2543, Maulbeerstrasse 66, 4002 Basel, Switzerland

Perbrandt M., Institut für Biochemie, Freie Universität Berlin, Theilallee 63, 14195 Berlin, Germany

Pogacic Vanda, Friedrich Miescher Institute, P.O. Box 2543, Maulbeerstrasse 66, 4002 Basel, Switzerland

Ramos Andres, MRC Laboratory of Molecular Biology, Hills Road, Cambridge CB2 2QH, UK

Rekowski Piotr, Faculty of Chemistry, University of Gdansk, Sobieskiego 18, 80-952 Gdansk, Poland

Rhodes Daniela , Medical Research Council, Lab. of Molecular Biology, Hills Road, Cambridge CB2 2QH, UK
E-mail: Rhodes@mrc-lmb.cam.ac.uk

Schlosser Andreas, Universität Bayreuth, Lab. für Biochemie, 95440 Bayreuth, Germany

Sha Ma, The Rockefeller University, 1230 York Avenue, New York, NY 10021, USA

Sierzputowska-Gracz, Department of Biochemistry, North Carolina State University , Box 7622, 128 Poll Hall, Raleigh N.C. 27695-7622, USA

Söll Dieter, Department of Molecular Biophysics and Biochemistry, Yale University, New Haven, CT 06520-8114, USA

Sprinzl Mathias , Universität Bayreuth, Lab. für Biochemie, 95440 Bayreuth, Germany

Szymanski Maciej, Institute of Bioorganic Chemistry, Polish Academy of Sciences, Noskowskiego 12, 61-704 Poznan, Poland
E-mail: mszyman@ibch.poznan.pl

Theil Elizabeth, Department of Biochemistry, North Carolina State University , Box 7622, 128 Poll Hall, Raleigh N.C. 27695-7622, USA
E-mail: theil@unity.ncsu.edu

Varani Gabriele, Medical Research Council, Lab. of Molecular Biology, Hills Road, Cambridge CB2 2QH, UK
E-mail: gv1@mrc-lmb.cam.ac.uk

Varani Luca, MRC Laboratory of Molecular Biology, Hills Road, Cambridge CB2 2QH, UK

Westhof Eric UPR 9002 du CNRS, IBMC, 15 rue 12 Descartes, 67084 Strasbourg Cedex, France
E-mail: westhof@ibmc.u-strasbg.fr

Wyszko Eliza, Institute of Bioorganic Chemistry, Polish Academy of Sciences, Noskowskiego 12, 61-704 Poznan, Poland E-mail: wyszkoe@ibch.poznan.pl

Yokoyama Shigeyuki, Dept. of Biophysics and Biochemistry, Graduate School of Science, University of Tokyo, 7-3-1 Hongo, Bunkyo-ku, Tokyo 113-0033, Japan
E-mail: yokoyama@y-sun.biochem.s.u-tokyo.ac.jp

Zuker Michael , Institute for Biomedical Computing, Washington University, Campus Box 8036, 700 S. Euclid Avenue, St.Louis, MO 63110, USA
E-mail: zuker@snark.wustl.edu; zuker@ibc.wustl.edu

WHY RNA?

J. BARCISZEWSKI[1)] and B.F.C. CLARK[2)]
[1)] Institute of Bioorganic Chemistry of the Polish Academy of Sciences, Noskowskiego 12, 61704 Poznań, Poland and [2)] Institute of Molecular and Structural Biology, Aarhus University, Gustav Wieds Wej 10C, DK-8000 Aarhus C, Denmark

Every era in biological and chemical sciences has had a current state of the art philosophy. Early in the century, it was sufficient to describe an extract with interesting properties. Soon, however, science required increasing levels of purification starting with simple precipitation methods and ranging through sophisticated chromatography and HPLC. Thus, in modern times, it is rare to see a biochemical paper that does not boast of having achieved a "single band" on gel electrophoresis. With the advent of protein and nucleic acid sequencing methods a number of sequence data banks were established so that various comparisons of proteins and nucleic acids were available. Now we have the structural era, where the most crucial information is the 3-dimensional structure of the key protein or nucleic acid. The data bases of structures are expanding rapidly as X-ray and NMR developing technologies allow structural analysis of more and more proteins, nucleic acids and their complexes. Recently, complex structures have involved various ribonucleic acid molecules.

The history of ribonucleic acids started in 1868, when Friedrich Miescher first isolated nucleic acids, but only in 1909 was ribose identified in them. Identification of deoxyribose as a component of DNA in 1929 helped finally to differentiate both nucleic acid species. The participation of ribonucleic acid (RNA) in protein synthesis seemed likely by 1940, several years before it was shown that DNA is the genetic material. As James D. Watson has recently confessed, RNA first became known to him during the fall of 1947 when he took Luria's course on viruses. It was clear that a given virus had either RNA or DNA in contrast to cells which contained both. But whether RNA carried a genetic specificity was not clear at that time [1]. The distribution of RNA and DNA in single eukaryotic cells has been shown by light microscopy, making use of the fact that both RNA and DNA strongly absorb 260 nm ultraviolet light, whereas only DNA found in the nucleus is intensely stained by the Feulgen reagent. RNA has been found in the cytoplasmic fractions, in small particles in association with protein. These particles consisting of RNA and protein have been shown to be sites of a protein synthesis. The 3'-5' linkage had been established for both DNA and RNA in 1953, however the question of whether nature used the 2' hydroxyl of RNA as a way of building a chain was undecided at that time (Fig. 1). Watson and Crick had not anticipated that the RNA double helix could be formed by adopting a slightly different conformation than proposed for the DNA double helix. The research of RNA changed dramatically with

J. Barciszewski and B.F.C. Clark (eds.), RNA Biochemistry and Biotechnology, 1–9.

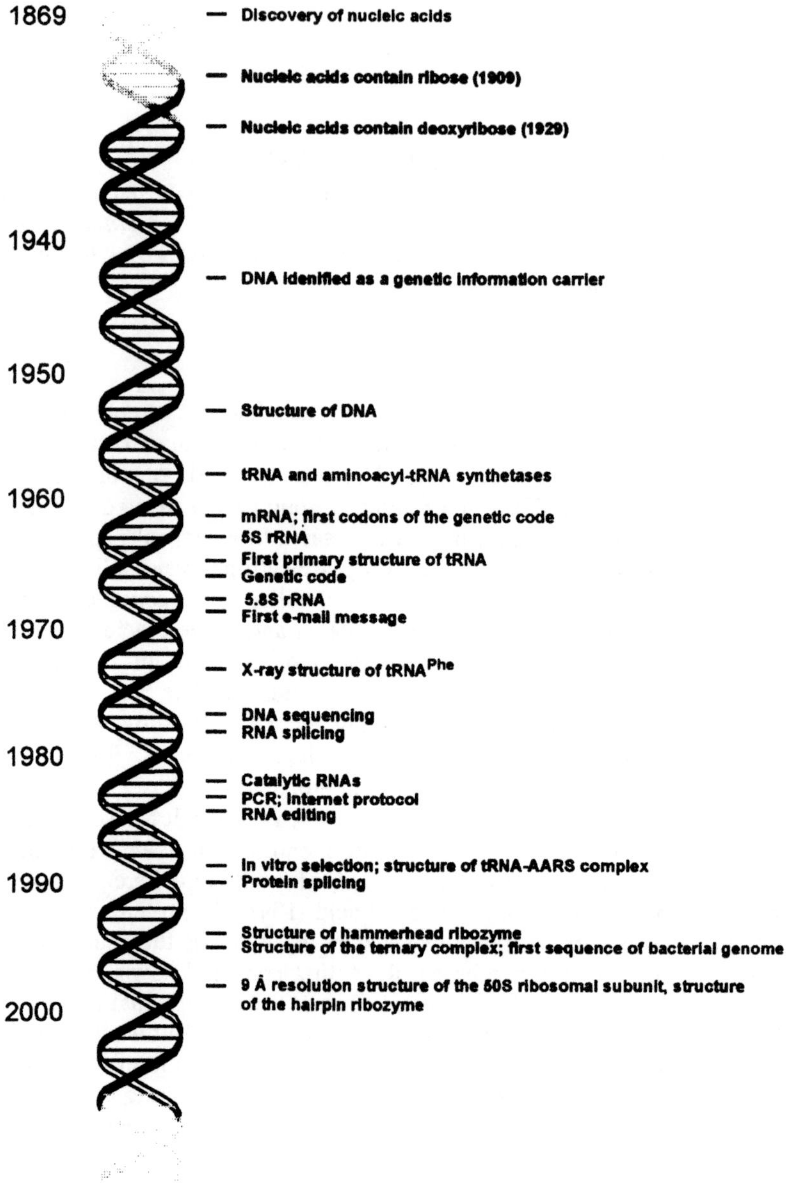

Fig. 1. Milestones in molecular biology projected on B-form of DNA where 1 bp corresponds to 1 year. The selected most important achievements in the field are marked. Two important dates concerning the Internet are also included, because at present it is difficult to imagine molecular biology without the Internet.

the discovery of the polynucleotide phosphorylase, which allowed the synthesis of long polyribonucleotides. When two polynucleotides were dissolved together, there was rapid increase in viscosity. This has been interpreted in terms of a double helix in which adenine paired with uracil and guanine with cytidine in a manner similar to that in DNA. The driving force of this process is the stability of the duplex molecule, compared with the unstructured form of the components. That was in fact the first hybridisation experiment where two nucleic acid molecules form a double helix. Shortly after there was shown a formation of DNA-DNA duplexes, DNA-RNA heteroduplexes and three stranded nucleic acids. A DNA-RNA hybrid is a stable structure, which transmits information from DNA to RNA. The first hybridisation of a DNA and an RNA strand is one that is still widely used today in the isolation of messenger RNA. This was also basis for the biological production of RNA by a mechanism in which a single strand of DNA serves as the template for the production of a complementary RNA strand. At that time it seemed reasonable to believe that a primitive polynucleotide chain could act as a template for synthesis of its complement, to build a two-stranded molecule. In this proposal, these molecules would act rather inefficiently for assembling their complements and facilitating their polymerisation into double-stranded RNA molecules. This observation might be regarded as a very early statement on what is known today as the "RNA world". In about 1961, the concept of messenger RNA had been developed. Clearly, if RNA copies were made of both complementary chains of DNA, one of them would be involved in coding for proteins, while the other might be a component of a control or regulatory system, that is now known as "anti-sense" RNA [2].

In 1955, F.H.C. Crick using pure deductive reasoning anticipated the existence of an RNA adaptor molecule capable of linking the sequence of DNA to encoded amino acid sequences [3]. This idea was not then known to M. Hoagland, M. Stephenson and P. Zamecnik who found a soluble RNA fraction to which the activated amino acids are transferred prior to protein synthesis and in fact confirmed the hypothesis (Fig. 2).

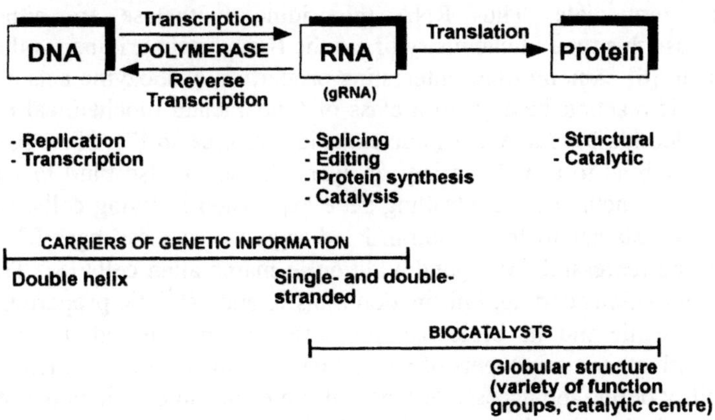

Fig. 2. Central dogma of molecular biology. An involvement of DNA, RNA and protein in various processes is summarized. One can easily notice that RNA is a molecule of dual properties. It is the carrier of genetic information as well as having biocatalytic properties. gRNA is guide RNA for the editing mechanism.

Ten years later R. Holley's group solved the nucleotide sequence of such a molecular adaptor, the yeast alanine specific transfer ribonucleic acid (in those days called soluble RNA). Using enzymatically synthesised RNA as messengers for protein synthesis, the correct assignments for all of the triplet codon were determined by early 1966. The genetic code is an algorithm that relates triplets of nucleotides called codons in messenger RNA with amino acid components of proteins. A further relationship was established by aminoacylation reactions in which specific amino acids are joined enzymically to their cognate transfer RNAs (tRNAs), each of which contains anticodon triplets of the genetic code. A triplet and a amino acid attachment site are separated by 75 angstroms. In 1968 F. Crick argued that RNA must have been the first genetic molecule, further suggesting that RNA, besides acting as a template, might also act as an enzyme and, in doing so, catalyse its own self-replication [1]. This very probably first hypothesis of an RNA World without protein was largely forgotten but has now become fashionable again because the remarkable discoveries by S. Altman and T. Cech of RNA molecules that indeed have catalytic activities of one sort or another. The idea of an "RNA world" containing RNA (ribozyme) rather than protein catalysts has been published 12 years ago [4]. A real surprise of the past decade is the realisation that RNA readily acquire specific structures that are adapted, in their shapes and chemical properties, for molecular recognition and catalysis. A number of ribozymes have been described as well as short RNA aptamers with high affinity and selectivity for small molecules which have been selected from pools of random sequence RNA. Developments in *in vitro* selection and amplification methodologies have enabled identification of unique RNA folds from random RNA libraries that can potentially target any ligand of interest which affinities in the 10^{-6}-10^{-9} molar range. There are many examples of RNA aptamers that target cofactors, antibiotics, amino acids, nucleotides, saccharides, peptides, proteins and other nucleic acids [5]. The ribozyme catalyses formation the glycosylic bond that joins a sugar to a base to make a nucleoside. The mechanism of RNA thiouridine synthetase action is not known. So far a protein uridine synthetase acts by an S_{N1} reaction mechanism involving a stable carbocation intermediate. Thus RNA thiouridine synthetase are either the first ribozymes to use this type of chemistry or are the first protein uridine synthetases to use S_{N2} mechanism [6]. Among other interesting properties, a ribozyme acts as thiouridine synthetase. This reaction belongs to a class of fundamental biochemical processes, by which the nucleotide constituents of nucleic acids are created [7]. Short RNA aptamers that specifically bind to a wide variety of ligands *in vitro*, also bind *in vivo*, enabling development of a method for controlling gene expression in living cells. Insertion of a small molecule aptamer to the 5' untranslated region of a messenger RNA allows its translation to be repressible by ligand addition in mammalian cells [8]. The versatility of RNA is that exhibits coding, information transfer and catalytic properties. RNA is an interactive molecule and its biological properties are manifested in its multifaceted interactions with other components of the cellular machinery. RNA structure, variability and adaptability define the accessible range of molecular interactions that contribute to RNAs finctional diversity. RNA structural diversity now appears to be comparable to that of proteins (Fig. 3).

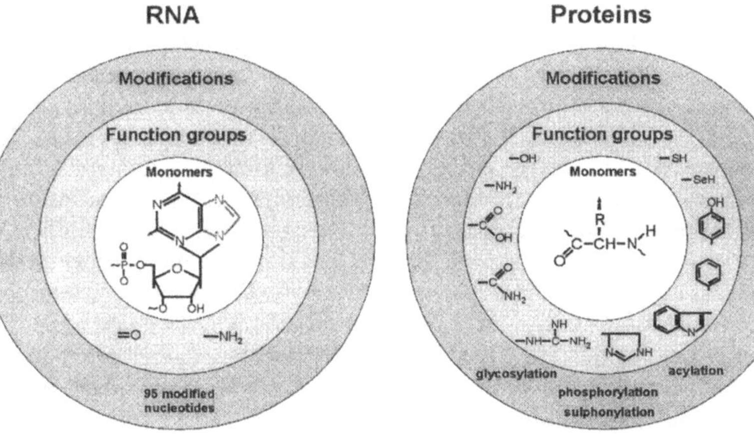

Fig. 3. Schematic presentation of differences and potentials of RNA and proteins for mutual interactions. There are only four basic units of RNA which have small numbers of functional groups. On the other hand, there are 21 amino acids (monomers) which can be further modified.

Although it possesses regular and stable helical secondary structure from canonical base pairing, additional non-canonical and tertiary interactions can give RNA various shapes including protein-like which accounts for the diversity of RNA structure-function relationships. Those additional interactions such as unusual hydrogen bonding within hairpin or internal loops, uridine turns within hairpin loops, cross strand stacks, adenosine platforms and base triplets contribute to stable tertiary folds which can be achieved by several different strategies of crosslinking individual segments of RNA structure [9]. An RNA can form stable structures with only a small number of nucleotides. A tertiary structure such as that of the hammerhead ribozyme can be constructed with as few as ca 40 nucleotides, and that of transfer RNA requires ca. 70 nucleotides. Thus larger RNAs of more than 100 nucleotides are expected to be able to form several structures, some of which may form only transiently. To make a protein of 100 amino acids, a messenger RNA needs to be longer than 300 nucleotides, so it is not surprising that nature takes advantage of the enormous structure-forming potential of RNA. Small assemblies of secondary structure elements are known to form autonomous and functional entities, which on the next level of organisation are associated with other molecules through numerous contacts and interactions. The term tectonics was coined for this process [10]. Most ribozymes are considered to be metalloenzymes, in which divalent metal ions play critical roles. A strategically positioned metal ion is directly (innersphere) coordinated to a substrate atom, to assist in deprotonation of the attacking nucleophile, stabilisation of the leaving group, or stabilisation of the transition state. There is also a possibility that neither direct binding of metal ions to phosphate oxygens nor nucleophilic activation of metal-bound water is required for activity. Instead the critical metal ions required for the structure formation or for catalysis are likely to interact with the RNA through outer-sphere coordination [11]. The structure problem of the hammerhead activity does not arrise from the size of the ribozyme, but is due to a scarcity of structural interconnections within the folded RNA structure relative to proteins (Fig. 3). The diversity in size, shape and polarity of the amino acid side chains allows proteins to fold via formation of a closely packed hydrophobic core and networks of hydrogen bonding interactions. The resulting extensive structural interconnections prevent an overall collapse of the structure upon removal of individual side chains. In contrast RNA side chains are limited in number and diversity. This could be a reason why a ribozyme is unable to position efficiently catalytic groups in the ground state. Rather it has to undergo an unfavourable folding transition to adopt its active conformation. Once folded, a small RNA enzyme may further lack the interconnections required to provide precise positioning and optimal catalyses. In contrast protein enzymes of similar size can achieve folding and efficient positioning through formation of extensive networks of interactions. These observations termed structural redundancy underscore the basic interrelationship, common to both RNA and protein systems between folding and catalysis [12]. How are different RNA species organised. At the present time, there is no consensus on definition of what really constitutes a tertiary structure in RNA. The definition is suggested by thermodynamic behaviour of RNA structure, because tertiary structures correspond to cooperatively folding domains of an RNA. The major stabilising force of RNA is base stacking, which is associated with large enthalpy. RNA molecules have several hydrogen bond donors and acceptors on each base, ribose and phosphate and charge on each phosphate. The bases tend to stack, but they do not need to be exposed to the solvent. On the other hand proteins tend to bury their hydrophobic amino acids and leave their polar and ionic side chains on the surface, bonding to water and salts in the solvent (Fig. 3).

A general feature of RNA tertiary structure is its preferential stabilisation by multivalent ions. It seems that the metal ion core in RNA is the equivalent of hydrophobic core in proteins. Therefore RNA secondary structure is formed first and then secondary structure motifs fold into the biologically relevant tertiary structure. The secondary structure may be modified somehow to accommodate the formation of the tertiary fold, but most of the secondary structure elements still persists in the final folded form. It seems that complex functional RNA molecules may be composed of relatively small number of RNA building blocks which form the "language" of these structural motifs which can be the basis for prediction of the tertiary structures from sequence information alone [13]. The rules are rather simple: Watson-Crick base pairs lower the free energy of the secondary structure, but loops and bulges raise the free energy [14]. In contrast to proteins, where the total free energy of folding rarely exceeds 15 kcal/mol even small RNA hairpins can have folding free energies of this magnitude, and the total free energy of folding a large RNA can be more than 100 kcal/mol [15].

Tertiary structure analysis of P4-P6 domain of the *Tetrahymena thermophila* group I intron at high resolution by NMR spectroscopy and X-ray has showed that an RNA subdomain in solution changes when the subdomain becomes part of its parent intron and then it is part of a domain of the intron in a crystal. Mg^{++} induced folding monitored by NMR clearly indicates secondary structure rearrangement to a folded conformation consistnt with the crystal structure [16]. This observation is in contrast to the standard assumption that RNA folds by first forming base-paired helices and loops, the loop regions are than available for the tertiary interactions that hold the helices in their biologically required structure [16].

When the first structure of a large RNA molecule, phenylalanine tRNA, was determined at atomic resolution, it became immediately clear that the 2'-hydroxyls play a key role in the stabilisation of RNA tertiary structure. tRNAs are the key actors in protein synthesis, in which their ultimate fate is to facilitate mRNA dependent peptide bond formation on the ribosome. Prior to protein synthesis, each tRNA precursor has to be recognized by ribonuclease P, CCA nucleotidyltransferase and its cognate aminoacyl-tRNA synthetase. During protein synthesis, each aminoacylated tRNA interacts specifically with messenger RNA and the ribosome through its anticodon and with elongation factor-Tu and the ribosome at its amino acid-acceptor end. The tRNAs thus have the universal properties required both for processing and for binding to the ribosome and EF-Tu. Each tRNA also has a special anticodon for reading the message and must also have special identity elements that are recognised by each aminoacylating enzyme. A knowledge of the universal and unique structural features of tRNAs has allowed both the recognition of naturally occurring tRNA mimics in other molecules, and the synthesis of functional tRNA mimics [17]. It is well established that RNA structure is essential for the function of ribosomal, splicesomal and catalytic RNAs. Although studying the structures of these stable RNAs is exciting area of research, most molecular biologists deal only with messenger RNAs, usually considered simply to be the intermediary in genetic information transfer from DNA to proteins [3]. Being the only functional group in the sugar phosphate backbone capable of acting as both an acceptor and a donor for hydrogen bonds, the RNA 2' hydroxyl group can interact with backbone and base sites in many other ways besides the ribose zipper motif. The 2'-hydroxyl groups mediate intricate inter residue contacts through forming numerous hydrogen bonds to bases, phosphate oxygens, and other 2'-hydroxyl groups, thus leading to a dramatic extension of the usually encountered pairing schemes between bases [18]. Moreover, specific 2'-hydroxyl groups were shown to be important for protein-RNA and RNA-RNA recognition. Through conferring a large number of water molecules on the RNA backbones and grooves, the 2'-hydroxyl groups are furthermore responsible for the higher thermodynamic stability of the RNA duplex compared with DNA. The problem of understanding RNA-protein interactions is important because RNA is more involved in a variety of functions than is DNA. However the larger structural variations manifested in RNA compared to DNA make its study more difficult. The complexity of the problem resembles the case of protein-protein interactions. In some ways the difference between DNA and RNA binding sites corresponds to a difference in dimensionality; the DNA double helical structure being nearly one dimensional. On the other hand, RNA presents highly variable surfaces for interaction, more similar to protein structures. Interaction between RNA and protein ought to provide a variety of interactions similar to that between pairs of proteins. For proteins there are potentially 20 types of residues to interact compared to the four nucleotides bases (Fig. 3). How does the RNA-protein interface achieve a comparative level of diversity for recognition? The search for protein binding motifs has proved to be almost pointless, since a wide variety of protein structure elements are now known to interact with DNA.

The much greater diversity of RNA structures would seem to make the dominance of only a few structural motifs even less likely. However this does not preclude the occurrence of dominant motifs of atomic interactions. Therefore for RNA-protein binding, the present strategy is to look at frequencies of base-amino acid interactions, without consideration of their structural context. How do the structural differences between RNA and DNA affect protein binding? In RNA there are additional features in comparison to DNA, that facilitate specific recognition by proteins. The additional G-U base pair type, beyond the canonical Watson-Crick types of base pairs, adds to the diversity of possible interactions. Furthermore, the available RNA structures already show remarkable variety of other ways in which bases can hydrogen bond to one another as triplets, bulges, loops and non-canonical base pairs. It means that the bases themselves do offer a rich diversity for binding to the 21 types of amino acids (including SeCys) (Fig. 2). Also some amino acids are capable of interacting simultaneously with several base pairs so this provides a further variety on the RNA surface for protein interactions.

In summary, this array of potential interacting RNA surface features affords a sufficient number of ways to achieve their specific recognition by proteins. For RNA the dominant major groove interaction is arginine-guanine but for the minor groove bonding arginine-uracil, lysine-cytosine, aspartic acid-cytosine, glutamic acid-guanine, alanine-guanine, tyrosine-cytosine and serine or threonine-adenine interactions are possible [19]. RNA and its modified counterparts can be used as antisense RNA. The search for oligonucleotides with sufficient metabolic stability for in vivo applications has provided strong support for modified RNA, but chemical modification should not impair RNA affinity and pairing selectivity. In the course of one of the attempts to use antisense RNA to block gene expression in the maternal germ line, it has been found that antisense and sense RNA preparations induced remarkably precise phenocopies of the targeted gene. Furthermore it was shown that the double stranded RNA (dsRNA) was an order of magnitude more potent at inducing interference than are preparations of either single strand. The properties of the interference mechanism prompted abandonment of the term "antisense" and reference to the process as "RNA interference" [20]. This and other applications of RNA are covered by the term modern biotechnology, which brings achievements of molecular biology to the marketplace (Fig. 4).

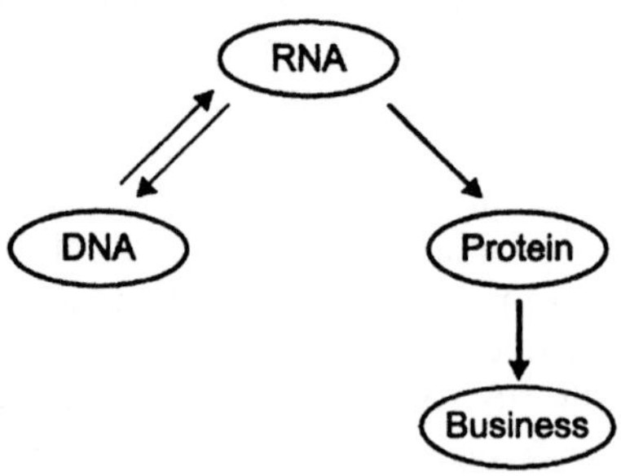

Fig. 4. The general rule of modern biotechnology. Various outcomes of the central dogma of molecular biology are the basis for further business.

The book is unique in that it not only describes structure-function relationships of different RNA species but also it conveys the message that RNA structure is relevant for anyone interested in fully understanding the pathway that leads from DNA to RNA to protein (Fig. 2). In this book one can find answers to question in the title of the text.

How is that RNA with only four chemical building blocks, each composed of an inflexible base and an invariant sugar phosphate can find more than one solution to the problem of forming unique structures and specific binding sites for small molecules (Fig. 3). As one can see from the articles in this book the RNA molecules are capable of carrying out unexpected gymnastic feats.

REFERENCES

1. Watson, J.D. (1993) Early speculations and facts about RNA templates, in The RNA World, Gesteland, R.F. and Atkins, J. F. Eds., Cold Spring Harbor Laboratory Press.
2. Rich, A. (1995) The nucleic acids. A backward glance, Ann. New York Acad. Sci. 758, 78-142.
3. Thiefry, D. and Sorkar, S. (1998) Forty years under the central dogma, Trends in Biochem.Sci. 23, 312-316.
4. The RNA World, Gesteland, R.F. and Atkins, J.F. Eds. Cold Spring Harbor Laboratory Press 1993.
5. Patel, D.J. (1997) Structural analysis of nucleic acid aptamers, Curr. Opinion Chem. Bol. 1, 32-46.
6. Robertson, M.P. and Ellington, A.D. (1998) How to make a nucleotide, Nature 395, 223-225.
7. Unrau, P.J. and Bartel, D.P. (1998) RNA-catalysed nucleotide synthesis, Nature 395, 260-263.
8. Werstuch, G. and Green, M.R. (1998) Controlling gene expression in living cells through small molecule - RNA interactions, Science 282, 296-298.
9. Draper, D.E.(1996) Strategies for RNA folding, Trends Biochem. Sci. 21, 145-149.
10. Westhof, E., Masquida, B. and Jaeger, L. (1996) RNA tectonics: towards RNA design, Folding and Design 1, R78-R88.
11. Suga, H., Cowan, J.A. and Szostack, J. W. (1998) Unusual metal ion catalysis in acetyl-transferase ribozyme, Biochemistry 37, 10118-10125.
12. Peraoclu, A., Karpeisky, A., Maloney, L., Beigelman, L. and Herschlag, D. (1998) Folding model for catalysis by the hammerhead ribozyme accounts for its extraordinary sensitivity to classic mutations, Biochemistry 37, 14765-14775.
13. Conn, G. L. and Draper, D.E. (1988) RNA structure, Curr. Opinion Struct. Biol. 8, 278-285.
14. Tinoco, I. Jr. and Kieft, J.S. (1997) The ion core in RNA folding, Nature Struct. Biol. 4, 509-512.
15. Turner, D. H., Sugimoto, N. and Freier, S. M. (1988) RNA structure prediction, Ann. Rev. Biophys. Biophys. Chem. 17, 167-192.
16. Wu, M. and Tinocco, I. Jr (1998) RNA folding causes secondary structure rearrangement, Proc. Natl. Acad. Sci. USA 95, 11555-11560.
17. Hagerman, P. and Tinoco, I. Jr. (1998), Nucleic acids - understanding structure, Curr. Opinion Struct. Biol. 8, 275-277.
18. Berger, I. and Egli, M. (1997) The role of backbone oxygen atoms in the organisation of nucleic acids tertiary structure: zippers, networks, clamps, and C-H---O hydrogen bonds, Chem. Eur. J., 3, 1400-1404.
19. Lustig, B., Arora, S. and Jernigan, R.L. (1997) RNA base-amino acid interaction strengths derived from structures and sequences, Nucleic Acids Res. 25, 2562-2565.
20. Fire, A., Xu, S., Montgomery, M.K., Kostas, S.A., Driver, S.E. and Mello, C.C. (1998) Potent and specific genetic interference by double stranded RNA in *Caenorhabditis elegans*, Nature 391, 806-811.

ALGORITHMS AND THERMODYNAMICS FOR RNA SECONDARY STRUCTURE PREDICTION: A PRACTICAL GUIDE.

M. ZUKER
Institute for Biomedical Computing
Washington University
St. Louis, MO 63110

AND

D.H. MATHEWS & D.H. TURNER
Department of Chemistry
University of Rochester
Rochester, NY 14627-0216

Abstract

This article is about the current status of the *mfold* package for RNA and DNA secondary structure prediction using nearest neighbor thermodynamic rules. The details of the free energy rules and of the latest version 3.0 software are described. Future plans are also discussed.

The *mfold* software now runs on a variety of Unix platforms; specifically SGI Irix, Sun Solaris, Dec alpha Ultrix and on Linux. While the older interactive programs of version 2 still exist, they are now run by a variety of scripts that make for much easier handling. There is both a command line interface for *mfold* and an HTML interface that runs in a Unix environment but can be accessed by anyone with a web browser.

The thermodynamic basis for the folding model is presented in detail, with references given for both specific free energy parameters and to overview articles that have summarized the state of these nearest neighbor parameters over the past dozen years. Both RNA and DNA rules are discussed, with some mention of parameters for RNA/DNA hybridization. Although the thermodynamic model has grown in complexity to accommodate new types of information, the folding algorithm has not yet incorporated some features, such as coaxial stacking of adjacent helices, and other features will probably remain too difficult or computationally expensive to implement. For this reason, a new energy calculation program has been introduced to recompute the free energies of predicted foldings to reflect the best of our knowledge.

The most significant improvements in the *mfold* software are:

1. Folding times have been greatly reduced in recent years, partly because of

J. Barciszewski and B.F.C. Clark (eds.), RNA Biochemistry and Biotechnology, 11–43.

faster computers and partly because of improvements in the algorithm.

2. Folding constraints have been expanded and are now implemented without the use of ***bonus energies*** that distort the results.

3. The output is significantly improved. Clear and enlargeable images of ***dot plots*** and of predicted foldings are now produced in PostScript and ***gif*** formats. Bases in structures may be annotated using different colors that reflect how ***well-determined*** they are in terms of their tendency to pair with other bases or to be single-stranded. Similarly, base pair probabilities from partition function calculations may be used for annotation. A detailed decomposition of each predicted folding into stacked pairs and loops with associated free energies is now provided.

The ***mfold*** software has a variety of parameters that may be adjusted to improve the predictions. Several examples are presented to illustrate how to interpret folding results and how to adjust these parameters to obtain better results.

1 Introduction

RNA is ubiquitous in the cell and is important for many processes. The activity of RNA is determined by its structure, the way it is folded back on itself. Secondary structure modeling of RNA predicts, or otherwise determines, the pattern of Watson-Crick (WC), wobble and other, non-canonical pairings that occur when the RNA is folded. For the purposes of this article, a non-canonical base pair is defined as non Watson-Crick (non W-C) and non-wobble (not GU or UG). There are many different kinds of RNA. Ribosomal RNA (rRNA) is a crucial part of the ribosome which is found in all living cells and in organelles such as mitochondria and chloroplasts. Small nuclear RNAs (snRNA) form a vital part of sliceosomes that process mRNAs in eukaryotes. These are 2 examples of ***structural*** RNAs.

Messenger RNAs (mRNA) do more than just carry information. Secondary structures can be used in part to explain translational controls in mRNA [1, 2], and replication controls in single-stranded RNA viruses [3]. Although the vast majority of known mRNAs code for proteins or structural RNAs, some do not [4, 5]; and it is likely that the secondary structures of these transcripts play an important role in their regulatory function in the cell. It is to be expected that many more such functional RNAs will be discovered in the future.

RNA is not just a passive structural element or a regulator. It is also an active component in many situations. Thus RNA acting alone is able to catalyze RNA processing [6, 7]. In a protein-RNA complex, the RNA component of ribonuclease P is an active component of tRNA processing [8].

The function of an RNA can only be understood in terms of its secondary or tertiary structure. For the understanding of catalytic activity, knowledge of secondary structure alone is insufficient. However, few large structures have been determined by crystallography [9, 10, 11] and the need for modeling is great. Secondary structure modeling can reasonably be viewed as a first step towards three dimensional modeling. For example, in small and large subunit rRNA, all tertiary

TABLE 1: *mfold* runs on a variety of Unix platforms. The ones shown have been tested by the first author.

Operating System	Hardware
Irix	SGI
Solaris	Sun SPARC
Ultrix	Dec Alpha
Solaris	Intel
Linux	Intel

interactions, including base triples, involve only 3% and 2% of the nucleotides, respectively [12]. In contrast, nucleotides in secondary structure comprise 60% and 58% of these rRNAs.

Secondary structure modeling is therefore a significant first step to the far more difficult process of three dimensional atomic resolution modeling. Knowledge of secondary structure, together with additional information on structural constraints or tertiary interactions, can be used to construct atomic resolution structural models [13, 14, 15].

2 Software platforms and environment

The *mfold* package [16, 17, 18, 19] consists of a group of programs written in Fortran or C, and a group of Bourne shell or Perl scripts. All of these programs and scripts run in a Unix environment. The *mfold* software is currently running on a number of different platforms, as indicated in Table 1.

In addition, some of the core programs of *mfold* have been translated into C++ and the resulting software is running on Windows 98/Windows NT (Intel hardware) with a point and click interface that uses proprietary Microsoft windowing tools. This program is called *RNAstructure*[20].

Up to version 2.2, the programs in *mfold* were interactive and had to be run individually. Versions 2.3 and 3.0 provide a simple command line interface, as described below, although it is still possible to run individual programs interactively or, for that matter, to write one's own scripts to link programs together in novel ways. The RNA folding web server (`www.ibc.wustl.edu/~zuker/rna/form1.cgi`) provides an HTML interface to *mfold* that uses a similar, but not identical script. This server will be described elsewhere. It's main advantage is that anyone with a web browser can use it.

An early version of *mfold* (2.0) was ported to run on Mac computers. It lacked the ***energy dot plot*** and could only fold about 350 bases. We no longer recommend its further use, especially since the energy parameters it uses have not been updated. The current *mfold* version 3.0 runs in command line mode under Windows, but this implementation requires a type of Unix emulation program to be run.

Two Unix "environment variables" must be defined for *mfold* to function. The first, MFOLDBIN, defines the directory where all the executable files for *mfold* are stored. The second, MFOLDLIB, defines the directory containing all the free energy and other data files. Dynamic memory allocation is not yet available for *mfold* , and the largest fragment size that can be folded is defined by the value of the MAXN parameter in the "Makefile" file.

3 Loops and Nearest neighbor rules

The *mfold* software uses what are called *nearest neighbor* energy rules. That is, free energies are assigned to loops rather than to base pairs. These have also been called loop dependent energy rules. In an effort to keep this article as self-contained as possible, we are including some well-known definitions that may be found elsewhere [21, 22, 23].

A secondary structure, **S**, on an RNA sequence, $\mathbf{R} = r_1, r_2, r_3, \ldots, r_n$, is a set of *base pairs*. A base pair between nucleotides r_i and r_j $(i < j)$ is denoted by $i.j$. A few constraints are imposed.

- Two base pairs, $i.j$ and $i'.j'$ are either identical, or else $i \neq i'$ and $j \neq j'$. Thus base triples are deliberately excluded from the definition of secondary structure.
- Sharp U-turns are prohibited. A U-turn, called a hairpin loop, must contain at least 3 bases.
- Pseudoknots are prohibited. That is, if $i.j$ and $i'.j' \in \mathbf{S}$, then, assuming $i < i'$, either $i < i' < j' < j$ or $i < j < i' < j'$.

Pseudoknots [24, 25, 26, 27, 28, 29] and base triples are not excluded for frivolous reasons. When pseudoknots are included, the loop decomposition of a secondary structure breaks down and the energy rules break down. Although we can assign reasonable free energies to the helices in a pseudoknot, and even to possible coaxial stacking between them, it is not possible to estimate the effects of the new kinds of loops that are created. Base triples pose an even greater challenge, because the exact nature of the triple cannot be predicted in advance, and even if it could, we have no data for assigning free energies.

A base $r_{i'}$ or a base pair $i'.j'$ is called accessible from a base pair $i.j$ if $i < i'(< j') < j$ and if there is not other base pair, $k.l$ such that $i < k < i'(< j') < l < j$. The collection of bases and base pairs accessible from a given base pair, $i.j$, but **not** including that base pair, is called the loop closed by $i.j$. We denote it by $\mathbf{L}(i.j)$. The collection of bases and base pairs not accessible from any base pair is called the exterior (or external) loop, and will be denoted by $\mathbf{L}_e$ here. It is worth noting that if we imagine adding a 0th and an $(n+1)^{st}$ base to the RNA, and a base pair $0.(n+1)$, then the exterior loop becomes the loop closed by this imaginary base pair. We call this the *universal closing base pair* of an RNA structure. If **S** is a secondary structure, then **S'** denotes the same secondary structure with the addition of the universal closing base pair. The exterior loop exists only in linear RNA. It is treated differently than other loops because we assume as a first approximation that there are no conformational constraints, and therefore no associated entropic costs.

Any secondary structure, $\mathbf{S}$, decomposes an RNA uniquely into loops. We can write this as:

$$\mathbf{R} = \bigcup_{i.j \in \mathbf{S}'} \mathbf{L}(i.j)$$

Loops may contain 0, 1 or more base pairs. The term k-loop denotes a loop containing $k-1$ base pairs, making a total of k base pairs by including the closing base pair. We introduce the terms $l_s(\mathbf{L})$ and $l_d(\mathbf{L})$ to denote the number of single-stranded bases and base pairs in a loop, respectively. The size of a 1 or 2-loop is defined as $l_s(\mathbf{L})$.

A 1-loop is called a hairpin loop. Polymer theory predicts that the free energy increment, $\delta\delta G$, for such a loop is given by

$$\delta\delta G = 1.75 \times RT \times \ln(l_s), \qquad (1)$$

where T is absolute temperature and R is the universal gas constant (1.9872 cal/mol/K). The factor 1.75 would be 2 if the chain were not self-avoiding in space. In reality, we use tabulated values for $\delta\delta G$ for l_s from 3 to 30. These values are based on measurements and interpolations of measurements, and are stored in an file named loop.dg, or loop.TC, where TC is a temperature (integral) in °C. We use the latter only when departing from our temperature standard of 37 degrees. Thus loop.dg and loop.37 refer to the same file. The same convention holds for other files defined below. Equation 1 is used to extrapolate beyond size 30. Thus, for $l_s > 30$,

$$\delta\delta G = \delta\delta G_{30} + 1.75 \times RT \times \ln(l_s/30). \qquad (2)$$

Figure 1 shows the information stored in the loop file.

In addition, the effects of *terminal mismatched pairs* are taken into account for hairpin loops of size greater than 3. For loops of size 4 and greater closed by a base pair $i.j$, an extra $\delta\delta G$ is applied. This is referred to as the *terminal mismatch* free energy for hairpin loops. These parameters are stored in a file named tstackh.dg or tstackh.TC, as above. The data are arranged in 4×4 tables that each comprise 4 rows and columns. Figure 2 illustrates how the parameters are stored.

Both the loop and tstackh files treat hairpin loops in a generic way, and assume no special structure for the bases in the loop. We know that this is not true in general. For example, the anti-codon loop of tRNA is certainly not unstructured. For certain small hairpin loops, special rules apply. Hairpin loops of size 3 are called triloops and those of size 4 are called tetraloops. Files of *distinguished* triloops and tetraloops have been created to store the free energy bonus assigned to those loops. These parameters are stored in files triloop.dg and tloop.dg, respectively (or triloop.TC and tloop.TC for a specific temperature, TC). Some typical entries are given in Figure 3

Finally, there are some special hairpin loop rules derived from experiments that will be defined explicitly here. A hairpin loop closed by r_i and r_j ($i < j$) called a "GGG" loop if $r_{i-2} = r_{i-1} = r_i = G$ and $r_j = U$. Such a loop receives a free energy bonus that is stored in the miscloop.dg or miscloop.TC file, which contains

```
DESTABILIZING ENERGIES BY SIZE OF LOOP (INTERPOLATE WHERE NEEDED)
hp3 ave calc no tmm;hp4 ave calc with tmm; ave all bulges
SIZE          INTERNAL          BULGE          HAIRPIN
-------------------------------------------------------
1                .               3.8               .
2                .               2.8               .
3                .               3.2              5.6
4               1.7              3.6              5.5
5               1.8              4.0              5.6
6               2.0              4.4              5.3
7               2.2              4.6              5.8
8               2.3              4.7              5.4
 ...
30              3.7              6.1              7.7
```

Figure 1: The loop.dg or loop.TC contains size based free energy increments for hairpin, bulge and interior loops up to size 30. Entries with '.' are undefined.

a variety of miscellaneous, or extra free energy parameters. Another special case is the "poly-C" hairpin loop, where all the single stranded bases are C. If the loop has size 3, it is given a free energy penalty of $c3$. Otherwise, the penalty is $c_2 + c_1 \times l_s$. The constants c_1, c_2 and c_3 are all stored in the miscloop file.

To summarize, we can write the free energy, $\delta\delta G_H$ of a hairpin loop as:

$$\delta\delta G_H = \delta\delta G_H^1 + \delta\delta G_H^2 + \delta\delta G_H^3 + \delta\delta G_H^4, \tag{3}$$

where

1. $\delta\delta G_H^1$ is the size dependent contribution from the loop file, or from equation 2 for sizes > 30,

```
              5' --> 3'                        5' --> 3'
                 WX                               CX
                 ZY                               GY
              3' <-- 5'                        3' <-- 5'
          Y: A     C     G     U          A     C     G     U
            ----------------------       ----------------------
     X:A | aAA   aAC   aAG   aAU        -1.5 -1.5 -1.4 -1.8
       C | aCA   aCC   aCG   aCU        -1.0 -0.9 -2.9 -0.8
       G | aGA   aGC   aGG   aGU        -2.2 -2.0 -1.6 -1.2
       U | aUA   aUC   aUG   aUU        -1.7 -1.4 -1.9 -2.0
```

Figure 2: On the left, a typical 4×4 table. The pairs WX and YZ are covalently linked. WZ is assumed to be the closing base pair of a hairpin loop, and XY is the mismatched pair. 'X' refers to row , and 'Y' to column, in order A, C, G and U. Thus 'aGU' is the same as 'a34' and is the mismatch free energy for a GU mismatch (X=G and Y=U). On the right is a sample table for W=C and Z=G.

Seq	Energy
GGGGAC	-3.0
...	
CGAAGG	-2.5
CUACGG	-2.5
...	
GUGAAC	-1.5
UGGAAA	-1.5

Figure 3: Sample ***distinguished*** tetraloops together with the free energy bonuses, in kcal/mole, attached to them. These entries include the closing base pair of the loop. Triloops are not shown since they are not currently in use for RNA folding.

2. $\delta\delta G_H^2$ is the terminal mismatch stacking free energy, taken from the tstackh file (0 for hairpin loops of size 3),

3. $\delta\delta G_H^3$ is the bonus free energy for triloops or tetraloops listed in the TRILOOP or TLOOP files. This value is 0 for loops not listed in the TRILOOP or TLOOP files and for loop sizes > 4,

4. $\delta\delta G_H^4$ is the bonus or penalty free energy for special cases not covered by the above.

A 2-loop, $\mathbf{L}$, is closed by a base pair $i.j$ and contains a single base pair, $i'.j'$, satisfying $i < i' < j' < j$. In this case, the loop size, $l_s(\mathbf{L})$, can be written as:

$$l_s(\mathbf{L}) = l_s^1(\mathbf{L}) + l_s^2(\mathbf{L}),$$

where $l_s^1(\mathbf{L}) = i' - i - 1$ and $l_s^2(\mathbf{L}) = j - j' - 1$.

A 2-loop of size 0 is called a *stacked pair*. This refers to the stacking between the $i.j$ and immediately adjacent $i+1.j-1$ base pair contained in the loop. Free energies for these loops are stored in a file named stack.dg, or stack.TC, where TC is a temperature, as defined above. The layout is the same as for the tstackh file. A portion of such a file is given in Figure 4. A group of 2 or more consecutive base pairs is called a *helix*. The first and last are the closing base pairs of the helix. They may be written as $i.j$ and $i'.j'$, where $i < i' < j' < j$. Then $i.j$ is called the external closing base pair and $i'.j'$ is called the internal closing base pair. This nomenclature is used for circular RNA as well, even though it depends on the choice of origin.

Only Watson-Crick and wobble GU pairs are allowed as *bona fide* base pairs, even though the software is written to allow for any base pairs. The reason is that nearest neighbor rules break down for non-canonical, even GU base pairs, and that mismatches must instead be treated as small, symmetric interior loops.

```
            5' --> 3'
               CX
               GY
            3' <-- 5'
      Y:  A    C    G    U
     -----------------------
X:A |     .    .    .  -2.1
  C |     .    .  -3.3   .
  G |     .  -2.4   .  -1.4
  U |  -2.1    .  -2.1   .
```

Figure 4: Sample free energies in kcal/mole for CG base pairs stacked over all possible base pairs, XY. X refers to row and Y refers to column, in the order A, C, G and U respectively. Entries denoted by an isolated period, '.', are undefined, and may be considered as $+\infty$.

Note that the stacks $\begin{array}{ccc} 5' & --> & 3' \\ & WX & \\ & ZY & \\ 3' & <-- & 5' \end{array}$ *and* $\begin{array}{ccc} 5' & --> & 3' \\ & YZ & \\ & XW & \\ 3' & <-- & 5' \end{array}$ are identical, and yet formally different for $W \neq Y$ and $X \neq Z$. These stacked pairs are stored twice in the file, and the *mfold* software checks for symmetry. This is an example of built in redundancy as a check on precision.

A 2-loop, **L**, of size > 0 is called a ***bulge loop*** if $l_s^1(\mathbf{L}) = 0$ or $l_s^2(\mathbf{L}) = 0$, and an interior loop if **both** $l_s^1(\mathbf{L}) > 0$ and $l_s^2(\mathbf{L}) > 0$.

Bulge loops up to size 30 are assigned free energies from the loop file (See Figure 1). For larger bulge loops, equation 2 is used. When a bulge loop has size 1, the stacking free energy for base pairs $i.j$ and $i'.j'$ are used (from the stack file).

Interior loops have size ≥ 2. If $l_s^1(\mathbf{L}) = l_s^2(\mathbf{L})$, the loop is called ***symmetric***; otherwise, it is ***asymmetric***, or lopsided. The asymmetry of an interior loop, a(**L**) is defined by:

$$a(\mathbf{L}) = |l_s^1(\mathbf{L}) - l_s^2(\mathbf{L})|. \tag{4}$$

The free energy, $\delta\delta G_I$, of an interior loop is the sum of 4 components:

$$\delta\delta G_I = \delta\delta G_I^1 + \delta\delta G_I^2 + \delta\delta G_I^3 + \delta\delta G_I^4. \tag{5}$$

1. $\delta\delta G_I^1$ is the size dependent contribution from the loop file, or from equation 2 for sizes > 30.

2. $\delta\delta G_I^2$ and $\delta\delta G_I^3$ are terminal mismatch stacking free energies, taken from the tstacki file. The format of this file is identical to the format of the tstackh

```
5' --> 3'
    X
   C A
   G T
    Y
3' <-- 5'
```

Y:	A	C	G	T
X:A	1.1	2.1	0.8	1.0
C	1.7	1.8	1.0	1.4
G	0.5	1.0	0.3	2.0
T	1.0	1.4	2.0	0.6

```
5' --> 3'
    X
   C  A
   G  U
    YA
3' <-- 5'
```

Y:	A	C	G	U
X:A	3.2	3.0	2.4	4.8
C	3.1	3.0	4.8	3.0
G	2.5	4.8	1.6	4.8
U	4.8	4.8	4.8	4.8

Figure 5: Left: Free energies for all 1×1 interior loops in DNA closed by a CG and an AT base pair. Right: Free energies for all 1×2 interior loops in RNA closed by a CG and an AU base pair, with a single stranded U 3' to the double stranded U. As in similar Figures, X refers to row and Y to column.

file. There are 2 terms because of the terminal stacking of both r_{i+1} and r_{j-1} on the $i.j$ base pair, and of both $r_{i'-1}$ and $r_{j'+1}$ on the $i'.j'$ base pair. This may be visualized as

$$\begin{array}{ccccc} 5'- & r_i & - & r_{i+1} & -3' \\ & \bullet & & \circ & \\ 3'- & r_j & - & r_{j-1} & -5' \end{array} \quad \text{and} \quad \begin{array}{ccccc} 5'- & r_{j'} & - & r_{j'+1} & -3' \\ & \bullet & & \circ & \\ 3'- & r_{i'} & - & r_{i'-1} & -5', \end{array}$$

where $\bullet$ denotes a base pair and $\circ$ denotes a mismatched pair.

3. $\delta\delta G_I^4$ is the asymmetry penalty, and is a function of a(**L**) defined in equation 4. The penalty is 0 for symmetric interior loops. The asymmetric penalty free energies come from the miscloop.dg or miscloop.TC file.

Equation 5 is now used only for loops of size > 4 or of asymmetry > 1. This means that special rules apply to 1×1, 1×2 and 2×2 interior loops. Free energies for these symmetric and almost symmetric interior loops are stored in files sint2.dg, asint1x2.dg and sint4.dg, respectively. As above, the suffix TC is used in place of dg when explicit attention is paid to temperature. These files list all possible values of the single stranded bases, and all possible Watson-Crick and GU base pair closings. The sint2 file comprises a 6×6 array of 4×4 tables. There is a table for all possible 6×6 closing base pairs. The free energy values for each choice of closing base pairs are arranged in 4×4 tables. The term "closing base pairs" refers to the closing base pair of the loop and the contained base pair of the loop, as in the strict definition of a loop. An example of such a table is given in Figure 5.

The asint1x2 file comprises a 24 row by 6 column array of 4×4 tables. There is a 4×4 table for all possible $6 \times 6 \times 4$ closing base pairs and choice of one of the

```
             5' ------> 3'
               C \/ \_/ A
               G /\  |  U
             3' <------ 5'
Y:    A    A    A    C    C    C    G    G    G    U    U    U
      A    C    G    A    C    U    A    G    U    C    G    U
--------------------------------------------------------------
   AA 2.0  1.6  1.0  2.0  2.6  2.6  1.0  1.4  0.2  2.3  1.5  2.2
   AC 2.4  1.9  1.3  2.4  2.4  2.4  1.3  1.7 -0.4  2.1  0.8  1.5
   AG 0.9  0.4 -0.1  0.9  1.9  1.9 -0.1  0.2 -0.1  1.6  1.2  1.8
   CA 1.9  1.5  0.9  1.9  1.9  1.9  0.9  1.3 -0.9  1.6  0.4  1.1
   CC 2.8  1.8  2.2  2.2  2.2  2.2  2.2  2.2  0.4  1.9  1.7  1.4
 X CU 2.7  1.6  2.0  2.1  2.1  2.1  2.0  2.0  0.3  1.8  1.5  1.2
   GA 1.0  0.6  0.0  1.0  2.0  2.0  0.0  0.4  0.0  1.7  1.3  2.0
   GG 1.8  1.3  0.7  1.8  2.4  2.4  0.7  1.1  0.0  2.1  1.2  1.9
   GU 1.8  0.4  1.6  0.8  1.8  1.8  1.6  1.2 -2.0  1.5 -0.7  1.8
   UC 2.7  1.6  2.0  2.1  2.1  2.1  2.0  2.0  0.3  1.8  1.5  1.2
   UG 0.3 -1.1  0.1  0.7  0.3  0.3  0.1  0.3 -3.5  0.0 -2.2  0.3
   UU 2.2  0.7  1.9  1.2  1.2  1.2  1.9  1.5  0.2  0.9  1.5  0.3
```

Figure 6: Free energies for all interior loops in RNA closed by a CG and an AU base pair. Values of 'X' or 'Y' that correspond to bases that could form Watson-Crick pairs have been removed for brevity.

single stranded bases. The free energy values for each choice of closing base pairs and a single stranded base are arranged in 4×4 tables. An example of these tables is given in Figure 5.

Finally, the sint4 file contains 36 16×16 tables, 1 for each pair of closing base pairs. A 2×2 interior loop can have 4^4 combinations of single stranded bases. If, for example, the loop is closed by a GC base pair and an AU base pair, we can write it as:

```
5' ------> 3'
  G \/ \_/ A
  C /\  |  U
3' <------ 5'
```

Both the large 'X' and large 'Y' refer to an unmatched pair of bases that are juxtaposed. They can each take on 16 different values, from 'AA','AC', ..., to 'UU', or 1 to 16, respectively. The number in row 'X' and column 'Y' of the table is the free energy of the 2×2 interior loop with the indicated single stranded bases. Figure 6 shows the full table for the CG and AU closing base pairs.

Some special rules apply to 2-loops. A stacked pair that occurs at the end of a helix has a different free energy than if it were in the middle of a helix. Because of the availablility and precision of data, we distinguish between GC closing and non-GC closing base pairs. In particular, a penalty (terminal AU penalty) is assigned to each non-GC closing base pair in a helix. The value of this penalty is stored in the MISCLOOP file.

Because free energies are assigned to loops, and not to helices, there is no ***a priori*** way of knowing whether or not a stacked pair will be terminal or not. For this reason, the terminal AU penalty is built into the TSTACKH and TSTACKI tables. For bulge, multi-branch and exterior loops, the penalty is applied explicitly. In all of these cases, the penalty is ***formally*** assigned to the adjacent loop, although it really belongs to the helix.

A "Grossly Asymmetric Interior Loop (GAIL)" is an interior loop that is $1 \times n$, where $n > 2$. The special "GAIL" rule that is used in this case substitutes AA mismatches next to both closing base pairs of the loop for use in assigning terminal stacking free energies from the TSTACKI file.

A k-loop, $\mathbf{L}$, where $k > 2$, is called a ***multi-branch*** loop. It contains $k - 1$ base pairs, and is closed by a kth base pair. Thus there are k stems radiating out from this loop. Because so little is known about the effects of multi-branch loops on RNA stability, we assign free energies in a way that makes the computations easy. This is the justification for the use of an ***affine*** free energy penalty for multi-branch loops. The free energy, $\delta\delta G(\mathbf{L})$, is given by:

$$\delta\delta G(\mathbf{L}) = a + b \times l_s(\mathbf{L}) + c \times l_d(\mathbf{L}) + \delta\delta G_{stack}, \tag{6}$$

where a, b and c are constants that are stored in the miscloop file and $\delta\delta G_{stack}$ includes stacking interactions that will be explained below. This simple energy function allows the dynamic programming algorithm used by ***mfold*** to find optimal multi-branch loops in time proportional to n^3. It would take exponentially increasing time (with sequence length) to use a more appropriate energy function derived from Jacobson-Stockmeyer theory [30] that grows logarithmically with $l_s(\mathbf{L})$. In the ***efn2*** program that recalculates folding free energies using more realistic rules (defined below), equation 6 is replaced by:

$$\delta\delta G(\mathbf{L}) = a + 6b + 1.75 \times RT \times \ln(l_s(\mathbf{L})/6) + c \times l_d(\mathbf{L}) + \delta\delta G_{stack}. \tag{7}$$

That is, the linear dependence on l_s changes to a logarithmic dependence for more than 6 single stranded bases in a multi-branch loop.

Stacking free energies, $\delta\delta G_{stack}$ are computed for multi-branch and exterior loops. In the folding algorithm these are single strand stacking free energies, also known as ***dangling base*** free energies, because they are applied to single stranded bases adjacent to a base pair that is either in the loop, or closes the loop. This single stranded base may "dangle" from the $5'$ or $3'$ end of the base pair. These parameters are stored in a file named dangle.dg or dangle.TC, as above.

Figure 7 shows some single strand stacking free energies.

If $i.j$ and $j+2.k$ are 2 base pairs, then r_{j+1} can interact with both of them. In this case, the stacking is assigned to only 1 of the 2 base pairs, whichever has a lower free energy (usually the $3'$ stack). If $k.l$ is a base pair and both r_{k-1} and r_{l+1} are single stranded, then both the $5'$ and $3'$ stacking are permitted. The value of $\delta\delta G_{stack}$ is then the sum of all the single base stacking free energies associated with the base pairs and closing base pair of the loop.

It has been evident for some time that to make the free energy rules more realistic for multi-branch and exterior loops, and to improve folding predictions, we would

```
         X                            X
-----------------            -----------------
A    C    G    U             A    C    G    U
-----------------            -----------------
     5' --> 3'                    5' --> 3'
        CX                           C
        G                            GX
     3' <-- 5'                    3' <-- 5'
-1.7 -0.8 -1.7 -1.2          -0.2 -0.3  0.0  0.0
```

Figure 7: Free energies for all possible single stranded bases that are adjacent to a CG base pair. 'X' refers to column. Note that the $3'$ dangling free energies are larger in magnitude than the $5'$ dangling free energies.

be compelled to take into account the stacking interactions between adjacent helices. Two helices, $\mathbf{H_1}$ and $\mathbf{H_2}$ in a multi-branch or exterior loop are adjacent if there are 2 base pairs $i.j$ and $j+1.k$, $i.j$ and $i+1.k$ or $i.j$ and $k.j-1$ that close $\mathbf{H_1}$ and $\mathbf{H_2}$, respectively. The last 2 cases can only occur in a multi-branch loop. In addition, we define ***almost adjacent*** helices as 2 helices where the addition of a single base pair (usually non-canonical), results in an adjacent pair. The concept of adjacent helices is important, since they are often coaxial in 3 dimensions, with a stacking interaction between the adjacent closing base pairs. The concept of almost adjacent comes from tRNA where, in many cases, the addition of a GA base pair at the base of the anti-codon stem creates a helix that is adjacent to, and stacks on, the D-loop stem.

Mfold does not yet take into account coaxial stacking of adjacent or almost adjacent helices. The *efn2* program that re-evaluates folding energies based on our best estimates does take this into account. It is not a trivial matter to decide which combination of coaxial stacking and single base stacking gives the lowest free energy in a multi-branch or external loop, and a recursive algorithm is employed to find this optimal combination. For example, coaxial stacking excludes single base stacking adjacent to the stacked helices. Free energies for the stacking of adjacent helices are stored in a file called coaxial.dg. The format is the same as for stack.dg. When 2 helices are almost adjacent, then 2 files, named coaxstack.dg and tstackcoax.dg are used. The format is the same as for stacking free energies. The use of these 2 files is explained with the aid of Figure 8. Thus, in the *efn2* program, $\delta\delta G_{stack}$ is a combination of single base stacking and coaxial stacking, depending on the loop.

In the case of circular RNA, the choice of origin is arbitrary. However, once it is made, what would be the exterior loop in linear RNA becomes equivalent to a hairpin, bulge, interior or multi-branch loop, or a stacked pair.

The Turner parameters for RNA folding have been published and summarized a number of times. The most significant older publications are [31, 32, 33], and *mfold* was originally used these results alone. Version 1 of *mfold* had no tloop.dg file, and there was a single terminal stacking free energy file, tstack.dg. Tetraloop

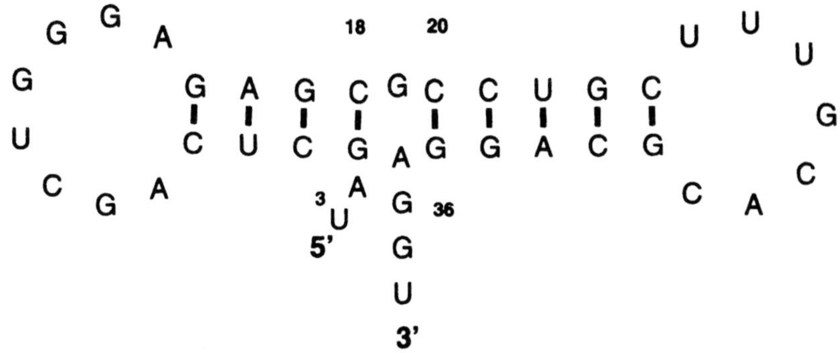

Figure 8: The helices closed by G^{3}-C^{18} and C^{20}-G^{36} are almost adjacent. Their stacking is mediated by a non-canonical G^{19}-A^{37} base pair. The free energy for the G^{19}-A^{37} to C^{20}-G^{36} comes from the **tstackcoax.dg** file. This is used where the phosphate backbone is unbroken, since there are 2 covalent links. The C^{18}-G^{3} to G^{19}-A^{37} stacking free energy comes from the **coaxstack.dg**, which is used for stacking where the backbone is broken. In this case, G^{3} and A^{37} are not linked.

bonus free energies were added in version 2.0. The tstack file was split into 2 files in version 2.2. Version 3.0 introduces triloop bonus energies, and both tetraloop and triloop bonus energies now depend on the closing base pair. Special rules for small interior loops are also new. For example, the 2×2 interior loop rules have evolved from [34]. Coaxial stacking was also introduced in version 3.0, although it's importance was realized eariler [35].

A complete set of DNA folding parameters have recently become available [36]. These are based on measuremments for stacking and mismatches, and on the literature for loop and other effects. Parameters are also available for predicting the formation of RNA/DNA duplexes [37].

4 Constrained folding

In addition to the free energy rules, specific constraints may be used to force or prohibit base pairs. A special file containing commands to constrain folding is used. The command syntax is rigid. The various commands and syntax are given below.

1. Forcing a string of consecutive bases to pair.
   Syntax: F i 0 k
   Ribonucleotides $r_i, r_{i+1}, \ldots, r_{i+k-1}$ are forced to be double stranded. The partners for these bases are chosen by the program. As an example, the command:
   F 23 0 5
   would cause bases 23, 24, 25, 26 and 27 to pair.

TABLE 2: Unsupported ambiguous codes for RNA/DNA. ***mfold*** **does not currently support the convention for ambiguous codes. Unrecognized bases will not be allowed to pair.**

Ambiguity	A,G	C,U/T	A,U/T	C,G	A,C	G,U/T
Code letter	R	Y	W	S	M	K
Ambiguity	C,G,U/T	A,G,U/T	A,C,U/T	A,C,G	A,C,G,U/T	
Code letter	B	D	H	V	N	

2. Forcing a string of consecutive base pairs.
   Syntax: F i j k
   Base pairs $r_i - r_j, r_{i+1} - r_{j-1}, r_{i+2} - r_{j-2}, \ldots, r_{i+k-1} . r_{j-k+1}$ are forced to occur. This is the same thing as forcing a helix to form. The helix is designated by its (external) closing base pair, $i.j$. As an example, the command:
   F 2 110 3
   would force base pairs 2.110, 3.109 and 4.108. Note that these base pairs **must be able to form!** Be aware also that *mfold* filters out isolated base pairs.

3. Prohibiting a string of consecutive bases from pairing.
   Syntax: P i 0 k
   Ribonucleotides $r_i, r_{i+1}, \ldots, r_{i+k-1}$ are prevented from pairing.

4. Prohibiting a string of consecutive base pairs
   Syntax: P i j k
   Base pairs $r_i - r_j, r_{i+1} - r_{j-1}, r_{i+2} - r_{j-2}, \ldots, r_{i+k-1} . r_{j-k+1}$ are not allowed to form. This is equivalent to prohibiting a helix.

5. Prohibiting 1 segment of a sequence from pairing with another
   Syntax: P i-j k-l
   where $i \leq j$ and $k \leq l$. In this case, no base pairs are allowed between $r_i, r_{i+1}, \ldots, r_j$ and $r_k, r_{k+1}, \ldots, r_l$. Note that the 2 segments need not be distinct. For example, the command:
   P i-j i-j
   will not allow $r_i, r_{i+1}, \ldots, r_j$ to pair with itself.

6. Annotated bases.

   (a) *mfold* recognizes A, C, G, U and T. In RNA folding, a 'T' will be treated as a 'U'; and *vice versa* for DNA folding. In addition, B, D, H and V are recognized as A, C, G and U/T, respectively. Bases marked in this way are regarded as susceptible to nuclease cleavage. They are allowed to pair only if their 3′ neighbor is unpaired. This is an old feature of *mfold* .

   (b) *mfold* also recognizes W, X, Y and Z as A, C, G and U/T, respectively. These bases are regarded as "modified" and are allowed to pair only at the ends of helices. At this time, the commonly used ambiguous codes shown in Table 2 are not supported by *mfold* .

5 Running the programs

The main program in the *mfold* package is named *mfold* . It is really a (Unix shell) script that calls a number of Fortran and C programs and puts together a reasonable output. Calling the *mfold* script without any command line parameters will cause the following output:

```
Usage is
mfold SEQ='file\_name' with optional parameters:
[ AUX='auxfile\_name' ] [ RUN\_TYPE=text (default) or html ]
[ NA=RNA (default) or DNA ] [ LC=sequence type (default = linear) ]
[ T=temperature (default = 37) ] [ P=percent (default = 5) ]
[ W=window parameter (default - set by sequence length) ]
[ MAXBP=max base pair distance (default - no limit) ]
[ MAX=maximum number of foldings to be computed (default 100) ]
[ ANN=structure annotation (default = none, p-num or ss-count) ]
[ START=5' base # (default = 1)] [ STOP=3' base # (default = end) ]
[ REUSE=NO/YES (default=NO) reuse existing .sav file ]
```

The meaning and use of these parameters are discussed below.

5.1 COMMAND LINE PARAMETERS

These command line parameters can only be fully understood when the meaning of the different type of output files, described in the next subsection, is known.

SEQ : The user must supply the name of a sequence file, called 'file_name' here. If 'file_name' ends with a suffix, that is, a period ('.') followed by other characters, then the suffix is removed and the result is called 'fold_name'. If no periods exist in 'file_name', then 'fold_name' = 'file_name'. For example, if the sequence is stored in 'trna.seq', then 'file_name' becomes 'trna'. If, on the other hand, the sequence file is named 'trna-file', then 'file_name' becomes 'trna-file'. The 'file_name', which may contain periods, becomes the 'prefix' for all the output files, such as 'file_name.out', 'file_name.det' and others.

Accepted sequence file formats are *GenBank*, *EMBL*, *FASTA* and *IntelliGenetics*. The sequence file may contain multiple sequences. At present, the *mfold* script will fold the first sequence by default. A new command line variable, NUM='#' may be added that directs the script to fold the '#'th sequence in the input file.

AUX : This is the name of an auxiliary input file of folding constraints. If this parameter is not used, *mfold* looks for a file named 'fold_name.aux'. If this file exists and is not empty, then it is interpreted as a constraint file. Thus constraints may be used without the use of this command line parameter.

RUN_TYPE : This parameter takes on 2 values; 'text', by default, and 'html' otherwise. The text option creates plain text files for the 'fold_name.out' and 'fold_name.det' files described below. The html option creates HTML versions of these files for display with a web browser.

TABLE 3: If 'W' is not specified, *mfold* will choose its value from this table based on sequence length. The user is encouraged to experiment with this parameter.

Sequence length	Default window size
1-29	0
30-49	2
50-119	3
120-299	5
300-399	7
400-499	8
500-599	10
600-699	11
700-799	12
800-1199	15
1200-1999	20
$\geq$ 2000	25

NA : This parameter takes on 2 values; 'RNA' by default, and 'DNA' otherwise. It tells *mfold* what type of nucleic acid is being folded.

LC : This parameter takes on 2 values; 'linear' by default, and 'circular' otherwise. It indicates to *mfold* whether a linear or circular nucleic acid is being folded.

T : This is the temperature, in °C. By default, it is 37°. Non-integral values will be rounded down to the nearest integer. Values should be in the range $0 \leq T \leq 100$.

P : This is the percent suboptimality for computing the *energy dot plot* and suboptimal foldings. The default value is 5%. This parameter controls the value of the free energy increment, $\Delta\Delta G$. $\Delta\Delta G$ is set to P% of ΔG, the computed minimum free energy. The *energy dot plot* shows only those base pairs that are in foldings with free energy $\leq \Delta G + \Delta\Delta G$. Similarly, the free energies of computed foldings are in the range from ΔG to $\Delta G + \Delta\Delta G$. No matter the value of P, *mfold* currently keeps $\Delta\Delta G$ in the range $1 \leq \Delta\Delta G \leq 12$ (kcal/mole).

W : This is the window parameter that controls the number of foldings that are automatically computed by *mfold* . 'W' may be thought of as a distance parameter. The distance between 2 base pairs, $i.j$ and $i'.j'$ may be defined as $\max\{|i-i'|, |j-j'|\}$. Then if $k-1$ foldings have already been predicted by *mfold* , the kth folding will have at least W base pairs that are at least a distance W from any of the base pairs in the first $k-1$ foldings. As W increases, the number of predicted foldings decreases. If W is not specified, *mfold* selects a value by default based on sequence length, as displayed in Table 3.

MAXBP : A base pair $i.j$ will not be allowed to form (in linear RNA) if $j - i > MAXBP$. For circular RNA, a base pair $i.j$ cannot form if $\min\{j-i, n+i-j\} > MAXBP$. Thus small values of MAXBP ensure that only short range base pairs will be predicted. By default, $MAXBP = +\infty$, indicating no constraint.

MAX : This is the maximum number of foldings that *mfold* will compute (50 by default). It is better to limit the number of foldings by careful selection of the P and W parameters.

ANN : This parameter currently takes on 3 values. 1. 'none' : secondary structures are drawn without any special annotation. Letters or outline are in black, while base pairs are red lines or dots for GC pairs and blue lines or dots for AU and GU pairs. 2. 'p-num' : Colored dots, colored base characters or a combination are used to display in each folding how well-determined each base is according to the P-num values in the 'fold_name.ann' file. 3. 'ss-count' : Colored dots, colored base characters or a combination are used to display in each folding how likely a base is to be single-stranded according to sample statistics stored in the 'fold_name.ss-count' file. Both 2. and 3. were recently described [38].

START : A segment to be folded is selected from the entire sequence. START is the first base, and is 1 by default.

STOP : This is the last base in the folded segment. It is the entire sequence length by default.

REUSE : This parameter is either N (no, the default) or Y (yes). *mfold* creates a large save file, 'fold_name.sav' that contains all the input parameters and the arrays of minimum folding energies for all sub-fragments of the folded sequence. This file, especially for large sequences, is expensive to create. If REUSE is 'Y', then a file named 'fold_name.sav' should exist from a previous run. You must specify the sequence file. Any (new) constraint file will be ignored, since constraints from the initial folding will be used. Similarly, NA, LC, T, MAXBP, START and STOP are determined from the initial run. However, RUN_TYPE, P, W, MAX and ANN may be altered to give different numbers of foldings, different *energy dot plots*, and/or different types of structure annotation.

Additional command line parameters for future development are discussed in the section on "Future plans".

5.2 Output

As described above, a prefix name, called 'file_name', is derived from the name of the input sequence file. All output files begin with this prefix. *mfold* produces 2 kinds of output, the *energy dot plot* and a selection of foldings within a prescribed increment from the minimum folding energy.

5.2.1 The energy dot plot

A nucleic acid secondary structure dot plot is a triangular plot that depicts base pairs as dots or other symbols. We shall refer to these symbols as dots. A dot in column i and row j of a triangular array, $\{(i,j)|1 \leq i \leq j \leq n\}$ represents the base pair $i.j$. The advantage of a dot plot is that it can display the base pairs in more than 1 folding simultaneously. It can be used to compare a few foldings, or the base pair distribution in many millions of foldings.

Mfold computes a number, $\Delta G(i,j)$ for every possible base pair, $i.j$. This is the minimum free energy of any folding that contains the $i.j$ base pair. As above, we let ΔG be the overall minimum folding free energy, and $\Delta\Delta G$ a user selected free energy increment. Clearly

$$\Delta G = \min_{1\leq i<j\leq n} \Delta G(i,j).$$

The energy increment is derived from ΔG and P. That is, $\Delta\Delta G = P \times \Delta G/100$. The current convention is to lower $\Delta\Delta G$ to 12 kcal/mole when it would otherwise be greater, and to raise it to 1 kcal/mole when it would otherwise be smaller. Then the ***energy dot plot*** is defined to be the collection of all base pairs $i.j$ satisfying:

$$\Delta G(i,j) \leq \Delta G + \Delta\Delta G.$$

This dot plot contains the **superposition of all possible foldings** whose folding energy is within $\Delta\Delta G$ of the minimum folding energy. Typically, $|\Delta\Delta G|$ is small compared to $|\Delta G|$, or P is a small percentage. In this case, the *energy dot plot* contains the superposition of all close to optimal foldings.

The *energy dot plot* gives an overall visual impression of how "well-defined" the folding is. A cluttered plot, or cluttered regions, indicate either structural plasticity (the lack of well-defined structure) or else the inability of the algorithm to predict a structure with confidence. A couple of crude measures of "well-definedness" have been introduced in *mfold* . The first is "P-num". $P\!-\!num(i)$ is a measure of the level of promiscuity of r_i in its pairing with other bases in foldings within $\Delta\Delta G$ of ΔG. It is the number of different base pairs, $i.j$, or $k.i$ that can form in this set of foldings, and is simply the number of dots in the ith row and i^{th} column of the *energy dot plot* . If $\delta(expression)$ is defined to be 1 when "expression" is true, and 0 otherwise, then P-num may be defined as:

$$P\!-\!num(i) = \sum_{k<i} \delta(\Delta G(k,i) \leq \Delta G + \Delta\Delta G) + \sum_{i<j} \delta(\Delta G(i,j) \leq \Delta G + \Delta\Delta G).$$

P-num pertains to individual bases. H-num is "well-definedness" measure for a base pair $i.j$. It is the average value of the two P-num quantities, adjusted by removing the "desirable" $i.j$ base pair. That is:

$$H\!-\!num(i,j) = (P\!-\!num(i) + P\!-\!num(j) - 1)/2.$$

A helix, already defined as a collection of two or more consecutive base pairs, may be described as a triple i,j,k, where k is the number of base pairs, and the actual base pairs are $i.j, i+1.j-1, \ldots, i+k-1.j-k+1$. When $k = 1$, the helix becomes a single base pair. With some abuse of notation, we may also write $H\!-\!num(i,j,k)$ to be the H-num value of the helix, i,j,k. This is the average value of H-num over all the base pairs in the helix.

There are 5 files associated with the *energy dot plot* .

'FILE_NAME.PLOT' : This is a text file that contains all the base pairs on the *energy dot plot* , organized into helices for which $\Delta G(i,j)$ is constant. The first record is a header, and each subsequent record describes a single helix. The

```
level  length istart jstart energy
  1      8     206    242   -972
  1      7     319    434   -972
  1      7     108    141   -972
  1      7      53    185   -972
  1      6     334    412   -972
  1      6     308    444   -972
  1      6     288    472   -972
  1      6     247    279   -972
...
  2      4       8     23   -971
  2      2      69     78   -971
  2      4       1     24   -970
  2      2      10     17   -970
  2      3     345    400   -967
  2      2     297    462   -967
...
```

Figure 9: Selected records from a plot file. "level" refers to a free energy range that is to be plotted in the same color, where 1 is always optimal. The "level" parameter is obsolete in the newer plotting programs of ***mfold*** 3.0. "istart", "jstart" and "length" define a helix and refer to i, j, k, respectively. The "energy" is free energy expressed as an integer in 10^{th}s of a kcal/mole. Note that this is not the free energy of the helix, but the mimimum free energy of any folding that contains the helix.

records are usually sorted by $\Delta G(i, j)$, and are often filtered so that short helices or isolated base pairs (helices of length 1) in suboptimal foldings are removed. Figure 9 shows a sample plot file.

'FILE_NAME.ANN' : This file contains P-num information for a particular $\Delta\Delta G$. The ith record contains i and $P-num(i)$. This file is used for annotating plotted structures.

'FILE_NAME.H-NUM' : This file is the same as 'file_name.plot', except that the "energy" column is replaced by an "h-num" column. These files are usually sorted by h-num; lowest to highest, or best determined to worst determined. Often, only helices in optimal foldings are retained. Figure 10 shows part of a sorted and filtered h-num file corresponding to the plot file in Figure 9.

'FILE_NAME.PS' : This is a PostScript file of the *energy dot plot* .

'FILE_NAME.GIF' : This is an image of the *energy dot plot* in "gif" format, suitable for display on web pages.

5.2.2 Optimal and suboptimal foldings

Mfold predicts a number of optimal and suboptimal foldings. They are automatically predicted in order of increasing free energy, although this order may change when the more exact *efn2* program is used to re-evaluate free energies. The num-

level	length	istart	jstart	h-num
1	4	38	194	6.8
1	4	215	232	7.3
1	5	31	201	8.4
1	7	53	185	8.4
1	2	47	189	11.0
1	8	206	242	11.9
1	6	61	176	13.7
1	4	89	163	13.8
1	3	255	271	14.0
1	3	104	145	15.0
1	1	68	79	16.0
1	4	121	131	17.0
1	6	288	472	17.3
...				
1	2	353	389	35.0
1	3	364	377	38.7
1	3	297	459	39.0

Figure 10: The beginning and end of an h-num file sorted by h-num and filtered to include only helices in optimal foldings. As with P-num, H-num values are relative to a particular sequence and free energy increment.

ber of computed foldings is limited directly by the MAX parameter, and in more subtle ways by the P and W parameters. It should be stated clearly here that while the *energy dot plot* rigorously displays all possible base pairs that can take part in all possible foldings within $\Delta\Delta G$ of ΔG, the computation of foldings is arbitrary. They do not represent a statistical sample of likely foldings, but rather a collection of foldings that show the variation that is possible within optimal and suboptimal foldings.

The collection of triples, $i, j, \Delta G(i, j)$, for all possible base pairs is sorted in order of increasing $\Delta G(i, j)$. The algorithm to construct foldings proceeds as follows:

1. The base pair at the top of the list is selected, and an optimal folding *containing the selected base pair* is computed.
2. All base pairs in the computed folding, as well as all those within a distance of W of base pairs in the computed folding, are crossed off the list.
3. The computed folding is retained if it contains at least W base pairs that were not found in previous foldings.

The first structure is always retained, even if it contains fewer than W base pairs. Steps 1 to 3 are repeated until either MAX structures have been computed and retained, or until there are no more base pairs on the list.

Mfold creates a number of files associated with predicted structures. The files marked with an optional "html" are created only when RUN_TYPE is html. Files

```
FOLDING BASES     1 TO    76 OF tRNA
Initial ENERGY  =       -22.3

                   10
----          UUA       AGU
     GCGGAU      GCUC    U
     CGCUUA      CGAG    G
ACCA        --A       AGG
     70               20

                   30
                G       CUG
                 CCAGA
                 GGUCU     A
                -        AGA
                   40

                      50
                AGGUC  |  UUC
                      CUGUG
                      GACAC    G
                -----  ^  CUA
                          60
```

(a) Text

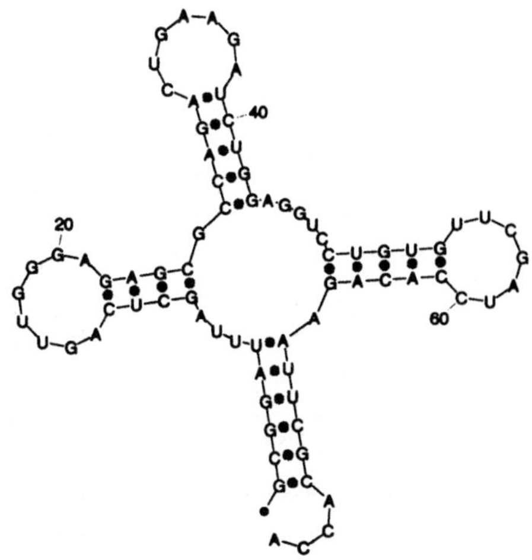

(b) Plot

Figure 11: The second and final folding of *S. cerevisiae* Phe-tRNA at 37°, with P=5% and W=3 (default values). (a) The selected base pair is G^{51}-C^{63}. The base numbers are placed so that the least significant digit, always a 0, is above or below the enumerated base. (b) The usual plotted representation. The *efn2* program has adjusted ΔG from -22.3 to -22.7 kcal/mole

that contain an underscore, '_', in their names enumerate the individual foldings, so that 'file_name_i.ct' refers to the ct file for the ith predicted structure.

'FILE_NAME.OUT(.HTML)' : This is a text file (html file) containing a plain text form of output for each of the predicted foldings. It is useful because it can always be displayed and is intelligible for foldings on short sequences. The selected base pairs for computing each structure are specially marked with a '|' above and a '^' below. A sample output is shown in Figure 11.

'FILE_NAME_i.CT' : The "ct" file (connect table) contains the sequence and base pair information, and is meant to be an input file for a structure drawing program. In addition to containing base pair information, it also lists the 5′ and 3′ neighbor of each base, allowing for the representation of circular RNA or multiple molecules. The ct file also lists the historical base numbering in the original sequence, as bases and base pairs are numbered according from 1 to the size of the folded segment. A portion of a ct file is displayed in Figure 12.

'FILE_NAME.DET(.HTML)' : This is a text file (html file) containing the detailed breakdown of each folding into loops, and the corresponding decomposition of the overall free energy, ΔG, into the free energy contributions, $\delta\delta G$, for each loop. A sample output is shown in Table 4.

'FILE_NAME.SS-COUNT' : If l foldings are predicted, then $ss-count(i)$ is the number of times that r_i is single stranded in these foldings. Thus $\frac{ss-count(i)}{l}$

TABLE 4: Free energy details for the second and final folding of *S. cerevisiae* Phe-tRNA at 37°, with default folding parameters. This layout mimics the html output.

Loop Free-Energy Decomposition

Structure 3

tRNA.seq Initial Free energy = -22.3

Structural element	$\delta\delta G$	**Information**
External loop:	-1.7	4 ss bases & 1 closing helices.
Stack:	-3.4	External closing pair is G^{1}-C^{72}
Stack:	-2.4	External closing pair is C^{2}-G^{71}
Stack:	-1.5	External closing pair is G^{3}-C^{70}
Stack:	-1.3	External closing pair is G^{4}-U^{69}
Stack:	-1.1	External closing pair is A^{5}-U^{68}
Helix	-9.7	6 base pairs.
Multi-loop:	1.0	External closing pair is U^{6}-A^{67} 10 ss bases & 4 closing helices.
Stack:	-2.1	External closing pair is C^{49}-G^{65}
Stack:	-2.1	External closing pair is U^{50}-A^{64}
Stack:	-2.2	External closing pair is G^{51}-C^{63}
Stack:	-2.1	External closing pair is U^{52}-A^{62}
Helix	-8.5	5 base pairs.
Hairpin loop:	4.8	Closing pair is G^{53}-C^{61}
Stack:	-3.3	External closing pair is C^{27}-G^{43}
Stack:	-2.1	External closing pair is C^{28}-G^{42}
Stack:	-2.1	External closing pair is A^{29}-U^{41}
Stack:	-2.4	External closing pair is G^{30}-C^{40}
Helix	-9.9	5 base pairs.
Hairpin loop:	5.7	Closing pair is A^{31}-U^{39}
Stack:	-3.4	External closing pair is G^{10}-C^{25}
Stack:	-2.1	External closing pair is C^{11}-G^{24}
Stack:	-2.4	External closing pair is U^{12}-A^{23}
Helix	-7.9	4 base pairs.
Hairpin loop:	3.9	Closing pair is C^{13}-G^{22}

```
76   ENERGY = -22.7  [initially  -22.3] yeast tRNA Phe
 1 G        0    2   72    1
 2 C        1    3   71    2
 3 G        2    4   70    3
 4 G        3    5   69    4
 5 A        4    6   68    5
 6 U        5    7    0    6
 7 U        6    8    0    7
 8 U        7    9    0    8
...
67 A       66   68    0   67
68 U       67   69    5   68
69 U       68   70    4   69
70 C       69   71    3   70
71 G       70   72    2   71
72 C       71   73    1   72
73 A       72   74    0   73
74 C       73   75    0   74
75 C       74   76    0   75
76 A       75    0    0   76
```

Figure 12: The ct file for the second and final folding of *S. cerevisiae* Phe-tRNA at 37°, with default parameters. The first record displays the fragment size (76), ΔG and sequence name. The ith subsequent record contains, in order, i, r_i, the index of the 5′-connecting base, the index of the 3′-connecting base, the index of the paired base and the historical numbering of the ith base in the original sequence. The 5′, 3′ and base pair indices are 0 when there is no connection or base pair.

is a ***sample based probability*** for single strandedness. The ss-count file contains the number of computed foldings in the first record. The ith subsequent record contains i and $ss-count(i)$. This file may be used to predict which regions of an RNA are likely to be single stranded, and values of ss-count, averaged over a window of perhaps 5 to 25 base pairs, are often plotted. This file is also used for annotating plotted structures.

'FILE_NAME_*i*.PLT2' : This is an intermediate, device independent plot file. It is the output of ***mfold***'s adaptation of the ***naview*** program for plotting secondary structures. This file is used as input to the ***plt22ps*** and ***plt22gif*** programs. It was originally intended to be used as input to the ***plt2*** plotting package [39], but this software is now old and not maintained.

'FILE_NAME_*i*.PS' : This is a PostScript file of a secondary structure. It is the output of the ***plt22ps*** program.

'FILE_NAME_*i*.GIF' : This is an image file (gif) of a secondary structure. It is the output of the ***plt22gif*** program.

The progression from ct file to images of secondary structures is:
'file_name_i.ct' → ***naview*** → 'file_name_i.plt2' → ***plt22ps*** → 'file_name_i.ps'

or
'file_name_i.ct' → naview → 'file_name_i.plt2' → *plt22gif* → 'file_name_i.gif'

'FILE_NAME.HTML' : This is a simple html file that links together some of the output files. It is an early version of a format originally used by the *mfold* web server.

'FILE_NAME.LOG' : This is a log file containing the standard output and standard error of the various programs and scripts that make up *mfold* . It can be useful for debugging.

'FILE_NAME.PNT' : This is a human readable file containing the entire input sequence. Every 10th base is labeled. In addition, auxiliary information is incorporated, if there is any. Bases that are forced to be double stranded have the letter 'F' underneath. Those that are forced to be single stranded have the letter 'P' underneath. Pairs of rounded brackets '(' and ')' underline forced base pairs, and pairs of curly brackets '{' and '}' underline prohibited base pairs. If 2 disjoint segments are prohibited from pairing with one another, then these segments are highlighted by underlining the residues of the first with a common lowercase letter, and the residues of the second with the same letter in uppercase. Different letters are used for different prohibited pairs. 'F' and 'P' are not used in this case.

5.3 AUXILIARY AND INDIVIDUAL PROGRAMS

The *mfold* package contains a script, also named *mfold* , that performs a folding according to information entered on the command line. This script is itself composed of scripts and (Fortran or C) executable programs, many of which can be run separately. Some of these programs are now described.

1. *auxgen*: This program creates the 'FILE_NAME.PNT' file.

2. *boxplot97_ng*: This program creates an *energy dot plot* in PostScript or gif form from 'FILE_NAME.PLOT'. The file BOXPLOT97_NG.DOC contains instructions.

3. *ct_boxplot*: This program creates a dot plot containing only those base pairs found in a collection of structures (in "ct format") that are specified on the command line. These must all be foldings of the same sequence fragment. At present, the *mfold* script does not use this program.

4. *ct_compare*: This program compares 2 "ct files". The first contains a single reference structure. The second contains 1 or more foldings of the same sequence. The number of bases and helices from the first structure that are conserved in the other structures are computed and displayed. For the purposes of this program, a "helix" may contain bulge or interior loops of size 1 or 2 and must have at least 3 base pairs [17, 40].

5. *efn*: This program computes ΔG for all the foldings in a "ct file". The energy rules correspond exactly to what is used in *mfold* .

6. *efn2*: This program computes ΔG for all the foldings in a "ct file" using a more precise free energy computation that takes into account coaxial stacking and Jacobson-Stockmeyer theory for multi-branched loops. (See equation 7 and Figure 8, respectively.)

7. *h-num*: This program computes an "h-num" file from a "plot" file.

8. *nafold*: This is the principle folding program of the *mfold* package, and corresponds to the program "lrna" and "crna" in earlier versions of *mfold* . It has 2 command line arguments. The first takes on the values 'l' or 'c', for linear (default) or circular folding, respectively. The second has values "text" (default) or "html" for plain text output, and for some html output, as described above. It is run twice by the *mfold* script. It may be run alone, as it prompts the user for input on an interactive basis. In this mode, the user may fold all the sequences contained in a single file, an option not available with the *mfold* script. The instructions for running this program have been described in [19] and also on the WWW at `http://www.ibc.wustl.edu/~zuker/mfold-2.0`. The interactive *energy dot plot* is not functional in version 3.0.

9. *naview*: This is a modified version of the *naview* program described in [41]. Actually, in *mfold* , *naview* is a script that runs a binary called *naview.exe*. Both *naview* and *naview.exe* may be run alone in interactive form. The *mfold* script uses the files "bases.nav" and "lines.nav", stored in MFOLDLIB to direct "naview". The first produces an output that displays individual residues as letters, while the second gives a structure outline only.

10. *newtemp*: The *newtemp* program creates free energy files for folding RNA or DNA at different temperatures. The latest RNA parameters consist only of free energies measured at 37°. Since there are no corresponding enthalpy files, it is not possible to fold RNA at temperatures other than 37°. If the user wishes to fold RNA at different temperatures, then the "dg" and "dh" files from *mfold* version 2.3 should be copied into MFOLDLIB. These older free energy parameters will be supplied with *mfold* version 3.0. DNA folding may be done at arbitrary temperatures between 0 and 100 degrees. However, it should be remembered that the DNA loop parameters are all estimated from values published in the literature or by comparison with RNA.

11. *plt22ps*: This program takes a "plt2" file from *naview* and creates a PostScript file of a plotted structure. It can use an "ann" file or an "ss-count" file to annotate with P-num or ss-count values, respectively.

12. *plt22gif*: This program is the same as *plt22ps*, except that the structure output file is in gif format.

13. *sav2p-num*: This interactive program uses an existing 'fold_name.sav' file to create a P-num file.

14. *sav2plot*: This interactive program uses an existing 'fold_name.sav' file to create a "plot" file. The resulting "plot" file may be sorted and filtered using the command: FILTER-SORT NAME.PLOT N, where name.plot is a "plot" file, and N is the minimum helix size that is desired. Helices shorter than N and not removed in optimal foldings.

15. *scorer*: This program is similar to *ct_compare*. It compares foldings helix by helix, displaying a more detailed output.

16. *split_ct.awk*: This is a Unix awk script that splits a ct file into multiple files, containing 1 folding in each. These individual files may be processed by *naview*, *plt22ps* and *plt22gif*.

17. *ss-count*: This program computes single stranded sample statistics from a collection of foldings stored in a single ct file.

6 Sample foldings

This section illustrates the appearance of *mfold* PostScript output files and is meant also to give the reader some insight into the use of of these programs.

6.1 Example 1

The *energy dot plot* is an integral part of the folding prediction. Consider the folding of a short RNA sequence:
`ACCCCCUCCU UCCUUGGAUC AAGGGGCUCA A,`
using default parameters. $\Delta G = -9.8$ kcal/mole at 37°, so $\Delta\Delta G = 1.0$ rather than 5% of ΔG. A single, optimal folding is computed. A glance of the *energy dot plot* , shown in Figure 13, reveals the optimal folding in black dots (symbols), but another set of yellow dots, indicating base pairs in at least 1 other suboptimal folding. The default value of 'W' (2, from Table 3) is too large for this other folding to be predicted, but a glance at the dot plot shows that something else is there. When the sequence is refolded with 'W'=0, a second, totally different folding is predicted. Figure 14 displays these foldings with individual bases drawn.

6.2 Example 2

Important alternative foldings might not appear in the *energy dot plot* if $\Delta\Delta G$ is too small. This is especially true in the folding of short sequences. When the short sequence:
`AAGGGGUUGG UCGCCUCGAC UAAGCGGCUU GGAAUUCC,`
is folded, also with default parameters, a single optimal folding is computed. However, the *energy dot plot* contains only the optimal, black dots from Figure 15. Changing the window size would not reveal anything new. When the value of P is increased to 25 (25%), the *energy dot plot* now reveals a very distinct alternate folding as shown in Figure 15. The *mfold* program now computes 2 foldings, plotted in Figure 16, using the default value of W.

6.3 Example 3

Here we present some results from the folding of an RNA that is related to a Human adenovirus pre-terminal protein (U52533). This RNA exhibits both "well-

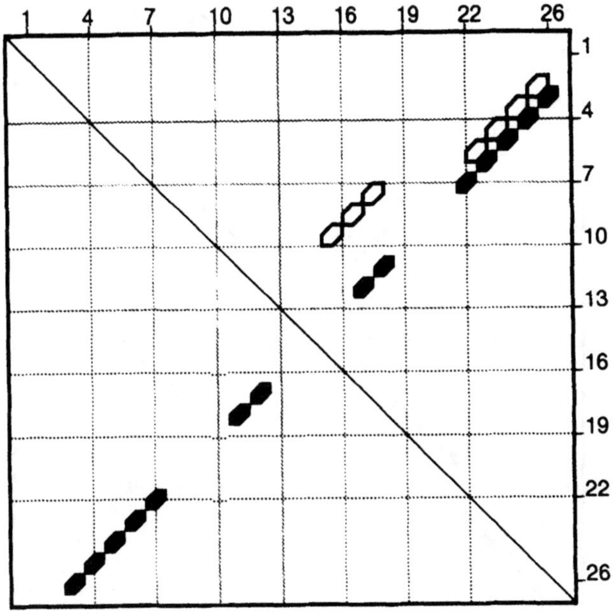

Figure 13: The ***energy dot plot*** for the "Example 1" sequence. Surrounding annotation, which would not be legible at this scale, has been removed. The yellow dots indicate base pairs in foldings within 0.3 kcal/mole of the optimal folding free energy of -9.8 kcal/mole.

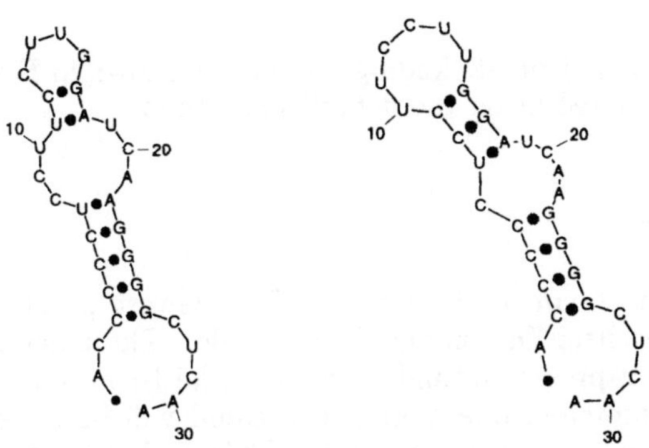

(a) (b)

Figure 14: The 2 predicted foldings for the "Example 1" sequence. (a) The optimal folding with $\Delta G = -9.8$ kcal/mole. (b) The suboptimal fold ($\Delta G = -9.5$ kcal/mole) found after refolding with 'W'=0.

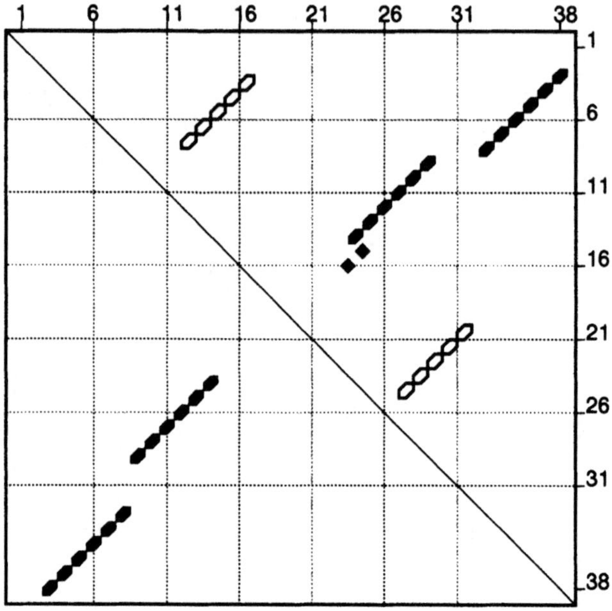

Figure 15: The ***energy dot plot*** for "Example 2" sequence with $\Delta\Delta G$ increased to 25% of 10.1, or 2.5 kcal/mole. The value of $\Delta\Delta G$ in the plot may be less than this maximum value, since there may be no base pairs in foldings that are ***exactly*** $\Delta\Delta G$ from the minimum free energy. The 2 green dots represent base pairs that can be in foldings with ΔG between -9.4 and -8.6 kcal/mole. These numbers are -8.6 and -7.9 for the yellow dots. In this case, the black dots comprise the optimal folding, and the yellow dots comprise the single suboptimal folding that is computed. The green dots would only be found in a folding if the value of W were lowered sufficiently.

defined" and "poorly-defined" folding regions, as shown in Figure 17. A total of 7 foldings were computed using the default parameters.

7 Future plans

The Unix version of *mfold* will remain. The newest programs, such as *plt22ps* and *plt22gif* are written in command line mode. The older programs have shell scripts or Perl "wrappers" around them to make them appear as single binaries that operate in command line mode. The trend will be to replace older code as necessary with non-interactive programs. This makes it easier to piece together different programs to create new forms of output.

When *mfold* was first created, the limitations of personal computers did not make a PC version practical. This has changed radically in the past 10 years, and an Intel/Windows PC is now a fine environment for running *mfold*. The basic Fortran and C programs have already been ported and will run under Widows, but a Unix-like shell is necessary. The *RNAstructure* program is now a faithful recreation of *mfold* in Windows, with a convenient user interface. The major problem with the

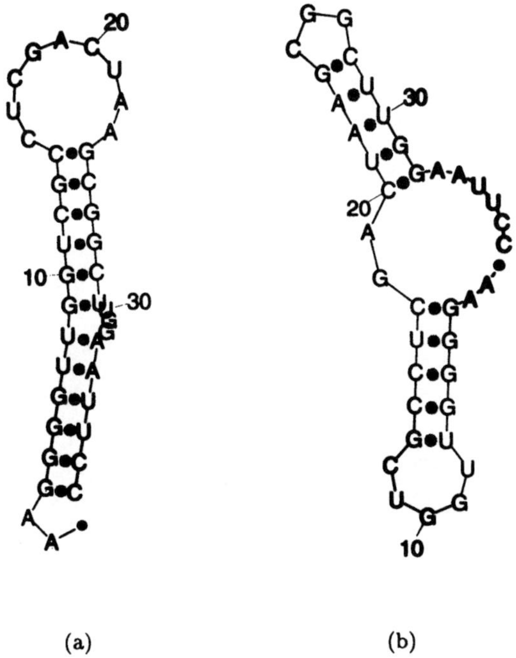

Figure 16: The 2 predicted foldings for the "Example 2" sequence. (a) The optimal folding with $\Delta G = -10.1$ kcal/mole. (b) The suboptimal fold ($\Delta G = -7.9$ kcal/mole) found after refolding with 'P'=25.

existing setup is that the Unix and Windows versions are totally different and will have to be updated in parallel to keep them equivalent.

The *mfold* programs running on SGI/Irix, SPARC/Solaris and Intel/Solaris have been incorporated into a world wide web (WWW) server that allows users from around the world to submit sequences for folding. This server has some extra features not available in the *mfold* package, and goes well beyond the very simple HTML output of the current *mfold* software. However, this software offers nothing new in terms of predictions, and it will be described elsewhere. Rapid developments in web browsers and languages such as Java-script and Java may make an HTML (or similar) interface to *mfold* better than others. As things stand now, the *mfold* server can run on a (local) Unix computer and be accessed by web browsers running on personal computers.

Additional parameters will be added to the command line version of *mfold* in the future. These will be described when the *mfold* command is given without parameters and the documentation will be altered accordingly. In the near future, a base labeling frequency will be added so that the user can specify the frequency of base enumeration. Other controls on secondary structure, such as zooming on images about specified coordinates, could be added, but these are already available

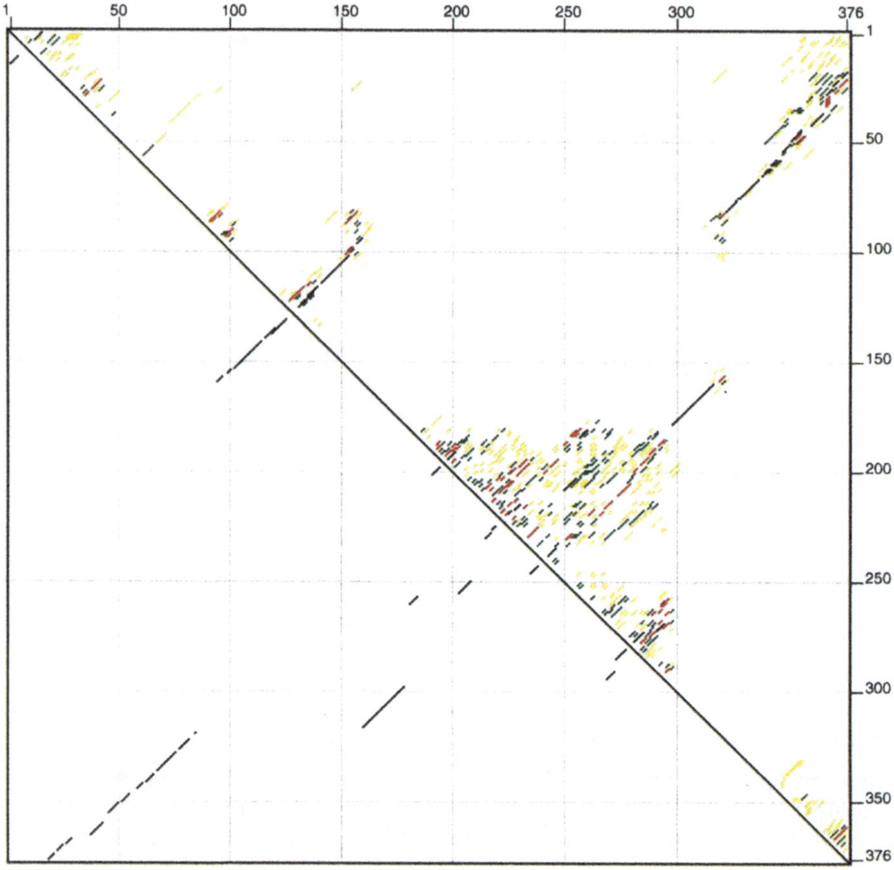

Figure 17: The ***energy dot plot*** for the "Example 3" sequence, with ΔG = -196.3 kcal/mole and $\Delta\Delta G$ = 9.8 kcal.mole. The region from bases 180 to 295 shows a great deal of uncertainty in its folding. This may be interpreted as a large ensemble of different foldings, or simply an "unstructured" region. In contrast, the long stem pairing bases 160-178 with 316-298, respectively, is extremely well determined. Also well-determined is a stem loop region that stretches from 37-81 and 322-365.

TABLE 5: Additional command line variables that could be added to *mfold* .

Parameter	Description	Default value
ddSTACK	extra $\delta\delta G$ per stack	0
ddBULGE	extra $\delta\delta G$ per bulge loop	0
ddILOOP	extra $\delta\delta G$ per interior loop	0
ddHLOOP	extra $\delta\delta G$ per hairpin loop	0
MAXILOOP	maximum size of an internal loop	30
MAXLOP	maximum asymmetry of an internal loop	30

through the use of the *plt22ps* and *plt22gif* programs.

The "energy" parameters from the older, interactive versions of *mfold* could easily be introduced in command line form. This would make it unnecessary to run the *nafold* program directly to alter them. Table 5 lists these parameters that are not defined in the MISCLOOP file.

Current plans call for the addition of coaxial stacking to the folding algorithm and possibly the creation of a special version that uses Jacobson-Stockmeyer theory to assign more realistic free energies to multi-branch loops, as in equation 7. In addition, a practical approach to pseudoknots will be attempted, where pairs of mutually exclusive helices that create pseudoknots are identified in the *energy dot plot* . In these case, the bases involved in 1 or perhaps 2 of these pseudoknots can be constrained to be single stranded, and foldings predicted to fill in the rest of the secondary structure.

Another future development will be the introduction of a 2 molecule folding system. This immediately complicates the problem, since concentration now becomes important. In addition, the folding of certain very simple bi-molecular systems is at least as hard as predicting pseudoknots in the folding of a single sequence.

Acknowledgment

We thank Darrin Stewart for creating the new *energy dot plot* and structure display programs. The plt22gif and boxplot97 programs use the *tgd* program of Bradley K. Sherman, and this uses the *gd 1.2* graphics library developed by Thomas Boutell, 1994, 1995, Quest Protein Database Center, Cold Spring Harbor Laboratory. We thank Ann Jacobson for her suggested improvements in the software. MZ is supported by NIGMS grant GM54250. DHM and DHT are supported, in part, by NIGMS grant GM22939.

References

1 P.M. MacDonald. (1990) *Bicoid* mRNA localization **signal**, phylogenetic conservation of function and RNA secondary structure. *Development*, **110**, 161–171.

2 M.H. de Smit and J. van Duin. (1990) Control of prokaryotic translation initiation by mRNA secondary structure. *Progress in Nucleic Acid Research in Molecular Biology*, **38**, 1–35.

3 D.R. Mills, C. Priano, P.A. Merz, and B.D. Binderow. (1990) Qβ RNA **bacteriophage**, mapping cis-acting elements within an RNA genome. *J. Virol.*, **64**, 3872–3881.

4 C.I. Brannan, E.C. Dees, R.S. Ingram, and S.M. Tilghman. (1990) The product of the h19 gene may function as an RNA. *Mol. Cell. Biol.*, **10**, 28–36.

5 C.J. Brown, A. Ballabio, J.L. Rupert, R.G. Lafreniere, M. Grompe, R. Tonlorenzi, and H.F. Willard. (1991) A gene from the region of the human X inactivation centre is expressed exclusively from the inactive X chromosome. *Nature*, **349**, 38–44.

6 T.R. Cech and B.L. Bass. (1986) Biological catalysis by RNA. *Ann. Rev. Biochem.*, **55**, 599–629.

7 T.R. Cech. (1990) Self-splicing of group I introns. *Ann. Rev. Biochem.*, **59**, 543–568.

8 S.C. Darr, J.W. Brown, and N.R. Pace. (1992) The varieties of Ribonuclease P. *Trends Biochem. Sci.*, **17**, 178–182.

9 S.H. Kim, F.L. Suddath, G.J. Quigley, A. McPherson, and J.L. Sussman. (1974) Three dimensional tertiary structure of yeast phenylalanine transfer RNA. *Science*, **185**, 435–440.

10 J.D. Robertus, J.E. Ladner, J.T. Finch, D. Rhodes, and R.S. Brown. (1974) Structure of yeast phenylalanine tRNA at 3 Å resolution. *Nature*, **250**, 546–551.

11 H.W. Pley, K.M Flaherty, and D.B. McKay. (1994) Three-dimensional structure of a hammerhead ribozyme. *Nature*, **372**, 68–74.

12 R.R. Gutell, (1995) personal communication.

13 F. Michel and E. Westhof. (1990) Modelling of the three-dimensional architecture of group I catalytic introns based on comparative sequence analysis. *J. Mol. Biol.*, **216**, 585–610.

14 F. Major, M. Turcotte, D. Gautheret, G. Lapalme, E. Fillion, and R.J. Cedergren. (1991) The combination of symbolic and numerical computation for three- dimensional modeling of RNA. *Science*, **253**, 1255–1260.

15 F. Major, D. Gautheret, and R. Cedergren. (1993) Reproducing the three-dimensional structure of a tRNA molecule from structural constraints. *Proc. Natl. Acad. Sci. USA*, **90**, 9408–9412.

16 M. Zuker. (1989) On finding all suboptimal foldings of an RNA molecule. *Science*, **244**, 48–52.

17 J.A. Jaeger, D.H. Turner, and M. Zuker. (1989) Improved predictions of secondary structures for RNA. *Proc. Natl. Acad. Sci. USA.*, **86**, 7706–7710.

18 J.A. Jaeger, D.H. Turner, and M. Zuker. (1990) Predicting optimal and suboptimal secondary structure for RNA. *Meth. Enzymol.*, **183**, 281–306.

19 M. Zuker. (1994) *Prediction of RNA Secondary Strcture by Energy Minimization.*, volume 25 of *Computer Analysis of Sequence Data, Part II, A.M. Griffin & H.G Griffin, Eds.*, chapter 23, pages 267–294. CRC Press, Inc., Totowa, NJ.

20 D.H. Mathews, T.C. Andre, J. Kim, D.H. Turner, and M. Zuker. (1998) *An Updated Recursive Algorithm for RNA Secondary Structure Prediction with Improved Free Energy Parameters.*, chapter 15, pages 246–257. American Chemical Society Symposium Series 682. American Chemical Society, Washington, DC.

21 D. Sankoff, J.B. Kruskal, S. Mainville, and R.J. Cedergren. (1983) *Fast algorithms to determine RNA secondary structures containing multiple loops.*, chapter 3, pages 93–120. Time warps, string edits, and macromolecules: the theory and practice of sequence comparison, Sankoff D., Kruskal J.B., Eds. Addison-Wesley, Reading, MA.

22 M. Zuker and D. Sankoff. (1984) RNA secondary structures and their prediction. *Bull. Math. Biol.*, **46**, 591–621.

23 M. Zuker. (1986) RNA folding prediction: The continued need for interaction between biologists and mathematicians. *Lectures on Mathematics in the Life Sciences*, **17**, 86–123.

24 C.W. Pleij and L. Bosch. (1989) RNA pseudoknots: structure, detection, and prediction. *Meth. Enzymol.*, **180**, 289–303.

25 J.P.Abrahams, M. van den Berg, E. van Batenburg, and C.W. Pleij. (1990) Prediction of RNA secondary structure, including pseudoknotting, by computer simulation. *Nucleic Acids Res.*, **18**, 3035–3044.

26 R.R. Gutell and C.R. Woese. (1990) Higher order structural elements in ribosomal RNAs: Pseudo-knots and the use of noncanonical pairs. *Proc. Natl. Acad. Sci. USA*, **87**, 663–667.

27 E. Dam, K. Pleij, and D. Draper. (1992) Structural and functional aspects of RNA pseudoknots. *Biochemistry*, **31**, 11665–11676.

28 C.W. Pleij. (1994) RNA pseudoknots. *Curr. Opin. Struct. Biol.*, **4**, 337–344.

29 Z. Du, D.P. Giedroc, and D.W. Hoffman. (1996) Structure of the autoregulatory pseudoknot within the gene 32 messenger RNA of bacteriophages T2 and T6: A model for a possible family of structurally related RNA pseudoknots. *Biochemistry*, 35 **(13)**, 4187–4198.

30 H. Jacobson and W.H. Stockmayer. (1950) Intramolecular reaction in polycondensations. I. The theory of linear systems. *J. Chem. Phys.*, **18**, 1600–1606.

31 S.M. Freier, R. Kierzek, J.A. Jaeger, N. Sugimoto, M.H. Caruthers, T. Neilson, and D.H. Turner. (1986) Improved free-energy parameters for predictions of RNA duplex stability. *Proc. Natl. Acad. Sci. USA*, **83**, 9373–9377.

32 D.H. Turner, N. Sugimoto, J.A. Jaeger, C.E. Longfellow, S.M. Freier, and R. Kierzek. (1987) Improved parameters for prediction of RNA structure. *Cold Spring Harb. Symp. Quant. Biol.*, **52**, 123–133.

33 D.H. Turner. N. Sugimoto, and S.M. Freier. (1988) RNA structure prediction. *Annu. Rev. Biophys. Biophys. Chem.*, **17**, 167–192.

34 M. Wu, J.A. McDowell, and D.H. Turner. (1995) A periodic table of symmetric tandem mismatches in RNA. *Biochemistry*, **34**, 3204–3211.

35 A.E. Walter, D.H. Turner, J. Kim, M.H. Lyttle, P. Muller, D.H. Mathews, and M. Zuker. (1994) Coaxial stacking of helixes enhances binding of oligoribonucleotides and improves predictions of RNA folding. *Proc Natl Acad Sci USA*, **91**, 9218–9222.

36 J.Jr. SantaLucia. (1998) A unified view of polymer, dumbbell, and oligonucleotide DNA nearest-neighbor thermodynamics. *Proc. Natl. Acad. Sci. USA*, **95**, 1460–1465.

37 N. Sugimoto, S. Nakano, M. Katoh, A. Matsumura, H. Nakamuta, T. Ohmichi, M. Yoneyama, and M. Sasaki. (1995) Thermodynamic parameters to predict stability of RNA/DNA hybrid duplexes. *Biochemistry*, **34**, 11211–11216.

38 M. Zuker and A.B. Jacobson. (1998) Using Reliability Information to Annotate RNA Secondary Structures. *RNA*, **4**, 669–679.

39 R.C. Beach. (1981) *The Unified Graphics System for Fortran 77 Programming Manual.* Stanford Linear Accelerator Center Computational Research Group, Stanford, CA, Technical Memo 203.

40 M. Zuker, J.A. Jaeger, and D.H. Turner. (1991) A comparison of optimal and suboptimal RNA secondary structures predicted by free energy minimization with structures determined by phylogenetic comparison. *Nucleic Acids Res.*, **19**, 2707–2714.

41 R.E. Bruccoleri and G. Heinrich. (1988) An improved algorithm for nucleic acid secondary structure display. *Comput. Appl. Biosci.*, **4**, 167–173.

RECURRENT RNA MOTIFS

Analysis at the basepair level

N.B. LEONTIS
Chemistry Department, Bowling Green State University
Bowling Green, OH 43403, USA

E. WESTHOF
Institut de Biologie Moléculaire et Cellulaire du CNRS
15 rue René Descartes, 67084 Strasbourg Cedex, FRANCE

Abstract. Single-stranded regions of structured RNAs often participate in complex interactions involving non-canonical basepairs. A dictionary of isosteric pairings that substitute for each other while preserving the 3D structure of conserved motifs can be inferred from phylogenetic analysis of conserved «internal loops» for which high-resolution structures also exist, such as loop E of 5S rRNA and the sarcin/ricin loop of 23S rRNA. Such «dictionaries» of isosteric, context-specific pairings allow one to identify occurrences of the same motif in other molecules.

1. Introduction

Single-stranded, structured RNA molecules fold on themselves to form short double-helices consisting of classical Watson-Crick basepairs. These regular A-type helices comprise the secondary structure. Phylogenetically based covariation analysis is the most reliable means for determining secondary structure in homologous RNA molecules [1, 2]. The regular helices are interrupted by stretches of nominally single-stranded nucleotides comprising so-called internal loops, hairpin loops, and multi-helix junction loops, as first proposed nearly 40 years ago [3]. These regions of the primary sequence are frequently observed to comprise evolutionarily conserved bases that correlate with each other in highly specific ways. The fact that many of these regions are highly structured was first clearly indicated by chemical probing experiments of 16S ribosomal RNA [4]. Nucleotides in «single-stranded» regions that participate in tertiary interactions can be identified with confidence using covariation analysis when they vary in a concerted fashion so as to maintain Watson-Crick pairing, and with less confidence when they exchange in a fashion suggesting non-Watson-Crick association [5, 6]. Our goal is to determine patterns of sequence variation that can identify specific non-canonical pairing geometries thus

J. Barciszewski and B.F.C. Clark (eds.), RNA Biochemistry and Biotechnology, 45–61.

allowing one to confidently extend covariation analysis to «single-stranded» nucleotides that in fact participate in non-canonical base-pairing. Non-canonical pairs often occur in blocks or tracts which present unique three-dimensional structures, recur in multiple contexts, and show specific patterns of co-variation. This suggests a modular approach to sequence analysis of RNA. Rather than relying purely on covariation analysis, the modular approach proceeds on the assumption that recurrent, conserved elements of primary and secondary structure share common 3D structures and, therefore, constitute building blocks for organizing the tertiary structure of RNA [7]. Each new high resolution RNA structure of a conserved «loop» region provides, therefore, welcome data to guide efforts to understand and predict RNA 3-dimensional structure.

An RNA motif may be defined as a set of RNA sequences that fold into 3D structures sufficiently close to each other to be considered essentially identical. The criteria for evaluating structural similarity include (1) the path followed by the sugar-phosphate backbone of the RNA in 3D space, (2) the hydrogen-bonding patterns between bases, backbone, and solvent atoms, and (3) the geometry of the stacking, as well as other hydrophobic interactions, especially between the bases. The local and idiosyncratic interactions generating the motif maintain the positioning of chemical groups on the surface of the structure, where they are disposed to interact either with distant regions of the same molecule or with other molecules (proteins, substrates, other RNAs). The modular view of RNA structure anticipates that certain motifs will recur to mediate crucial interactions in multiple contexts.

Our strategy, therefore, is to examine sequence variations in structural motifs for which high-resolution structural information has been obtained for at least one representative molecule, with an eye toward defining the available sequence space sampled by each particular motif. For comparing complex systems, it is not appropriate to use consensus sequences solely; the identity, or «molecular personality» of motifs is best established by comparing the full range of variations at each position in the motif among the available sequences. The approach assumes that during evolution the sequences compatible with a given 3D fold are adequately sampled.

Once all the sequence substitutions compatible with a particular 3D motif have been identified, one can investigate whether that motif occurs in other RNA molecules for which a collection of homologous sequences exists. A second powerful tool to aid in that endeavor is the available database of chemical probing data. Indeed, each motif, if recurrent in an autonomous way, displays definite chemical and enzymatic signatures.

In this chapter we will review the application of the approach outlined above to two related structures. The first is the symmetrical internal «loop E» of bacterial 5S ribosomal RNA (rRNA) and the second is the bulge-containing sarcin/ricin loop of 23S rRNA (Fig. 1).

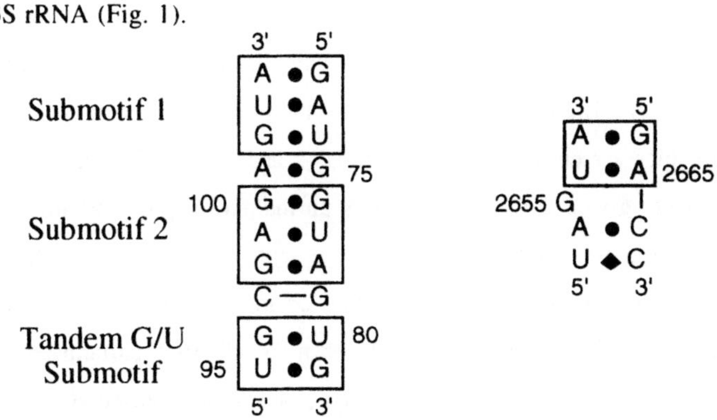

Figure 1: Basepairing in loop E from 5S rRNA (left) and the sarcin/ricin loop from 23S rRNA (right). Filled circles indicate non-canonical basepairs, lines, W.C. basepairs, and diamonds, non-identified pairs. Sequences and numbering correspond to *E. coli* molecules. White boxes identify isosteric submotifs in loop E related by 180° rotational symmetry Shaded box in loop E indicates tandem G/U motif with cross-strand, G upon G stacking. Shaded box in sarcin loop indicates A/G sheared and U/A trans Hoogsteen pairs exhibiting cross-strand, A upon A stacking.

The two structures are closely related. In fact, the sarcin loop is identical to loop E of eucaryal 5S rRNA [8]. However, to avoid confusion with the bacterial 5S loop E, we refer to this motif as the sarcin loop. The structure of the symmetrical internal «loop E» of bacterial 5S rRNA, recently solved to high-resolution by x-ray crystallography [9], reveals that it is not a «loop» at all. All bases actually pair. Moreover, only one canonical Watson-Crick pairing (between positions 79 and 97) occurs within a stretch of 10 basepairs (see Figure 1). Likewise, NMR analyses of the sarcin/ricin loop of 23S rRNA indicate that all bases pair with the exception of one (the «bulged» G2655) [9, 10]. An x-ray structure will soon be available for the sarcin loop [11]. The two motifs are compared in Figure 1, using the numbering for the *E. coli* 5S and 23S rRNA sequences.

2. Description of Structures

Fortuitously, the *E. coli* sequence corresponds closely to the sequence consensus for bacterial loop E, the one minor exception being that the canonical Watson-Crick pairing C97/G79 in *E. coli* 5S rRNA is reversed in the consensus, G97/C79. In previous work [12], we identified two nearly identical submotifs in the structure, related by rotational symmetry (submotif 1: G72-U74/G102-A104; submotif 2: G76-A78/G98-G100). Each submotif comprises three non-canonical basepairs. The corresponding basepairs in the two motifs are the sheared purine/purine pairs A104/G72 and A78/G98, the trans-Hoogsteen pairs U103/A73 and U77/A99, and the bifurcated pairs G102/U74 and G76/G100. Manipulation of the structure on the computer graphics screen shows that G102/U74 and G76/G100 are isosteric, a surprising result, given that one pair is a purine/purine pair and the other a purine/pyrimidine pair. The two submotifs are separated by a self-isosteric A/G basepair that we refer to as «open» A/G [12], or more accurately as «cis water-inserted» A/G, to draw attention to the water molecule that participates in the basepairing. Submotifs 1 and 2 exhibit cross-strand purine-purine stacking with A104 stacking on A73 and A78 stacking on A99. In addition, loop E contains tandem G/U pairs that also exhibit cross-strand stacking with G96/U80 stacking on U97/G81 [9].

The sarcin loop also contains adjacent sheared purine/purine (A2657/G2664) and trans-Hoogsteen (U2656/A2665) pairs with cross-strand stacking of A2657 and A2665. G2655 is bulged and forms a base triple with U2656/A2665, while the backbone is locally reversed at A2654, which pairs with C2666 via the Hoogsteen faces of each base. In eucaryal 5S loop Es, this pair is A/A.

3. Sequence Analysis

A great number of 5S rRNA sequences are available (888 total species, 332 bacterial species) and they are distributed rather uniformly between the various phylogenetic groups [13]. The sequence variations observed were classified as «conservative» or «concerted» [12]. We define a "conservative change" as the substitution of a single base or basepair in the consensus sequence by a potentially isosteric pairing without changes in immediately adjacent pairs in a particular sequence of the database. For example, the substitution of G by A in a sheared A/G pair is conservative, while the reversed base pair G/A is not. Concerted changes, on the other hand, refer to instances in which a change of one or more bases in a pair is accompanied by changes in one or both flanking pairs.

3.1 SHEARED A/G PAIRINGS

Two sheared purine-purine pairings are found in the Loop E structure, A104/G72 and A78/G98. Both are expected to exchange with A/A, which can adopt the same geometry, but not with G/A or G/G, which cannot. In fact, A/A is the most commonly observed, but not the only, substitution at both positions [12]. The following conservative substitutions were observed for the 104/72 sheared pairing: A104/A72, A104/C72, A104/U72, all of which preserve A104, and C104/A72, C104/C72, C104/G72, C104/U72, all of which have C104. Of these, the most frequent pairings found in the database are A/G (consensus), A/A, C/C and C/U, stressing the characteristic H-bond of the sheared base pair between the amino group (N6 of A or N4 of C) and the N3 of a purine or the O2 of a pyrimidine. All of these pairings can be accommodated isosterically, with the exception of the single C/G that would require considerable readjustment to accommodate the two amino groups. The consensus A78/G98 covaries primarily with A/A. As also found for A104/G72, a small number of sequences in which G98 is replaced by C or U are observed. Isosteric structures were generated with the FRAGMENT program [14] using the crystal structure as a template. Interestingly, all alternative pairings, which would necessarily involve U or G at position 104, are not feasible and are not observed. Thus, the database appears to display all geometrically possible variants for this pairing.

Sheared A/A pairs have been observed in the crystal structures of the P4-6 domain of Group I introns (URX053.PDB) in the J4/5 «internal loop» and at positions 13/22 in the D-stem of glutaminyl tRNA complexed to its synthetase (PTE003.PDB). Comparing sheared A/A with sheared G/A pairs, a shorter distance is observed between C2 of the A replacing G and N7 of the conserved A. This may be due to interaction between the polarized AH2 atom and AN7 in the A/A pair, emphasizing that small adjustments occur to accommodate base substitutions while still retaining the overall geometry of specific motifs.

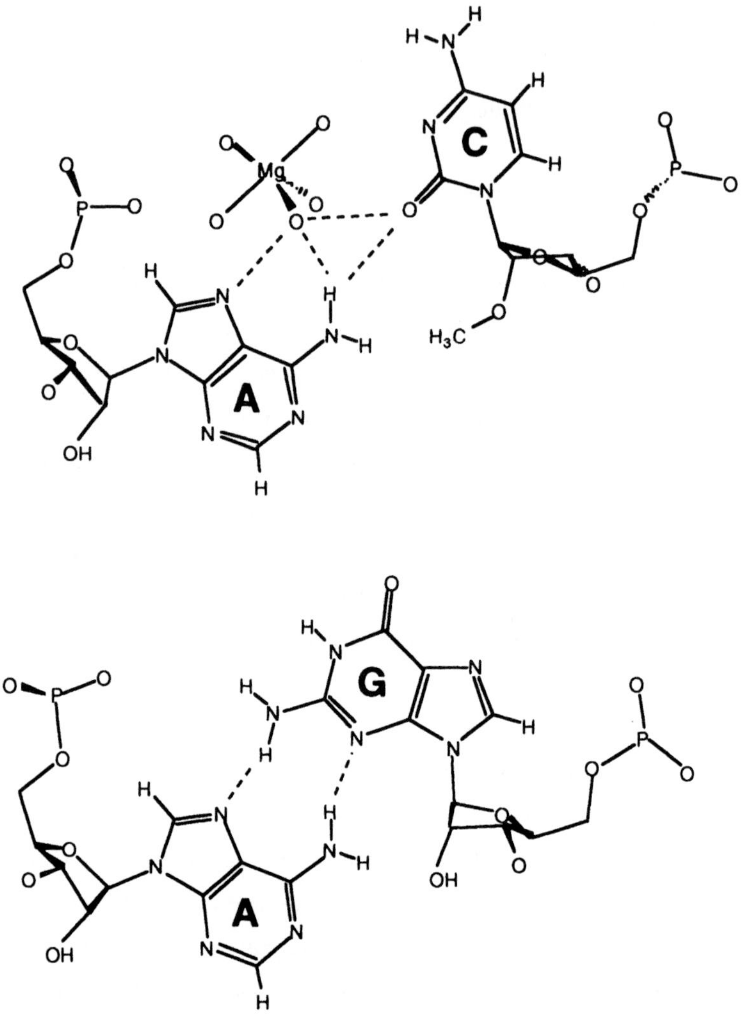

Figure 2: H-bonding geometry of sheared A/G basepair from bacterial 5S loop E compared with isosteric A38/O-methyl C32 at the base of the anti-codon loop of tRNA(Phe) from yeast.

The O-methyl C32/A38 pairing at the base of the anti-codon loop of tRNA Phe from yeast (see for example, TRNA09.pdb) provides an example of a C/A pairing isosteric with sheared G/A, as predicted from modeling based on sequence analysis (see above). H-bonding interactions between C32 and A38 are mediated by a water molecule coordinated to a magnesium ion specifically bound in the deep groove of the loop, illustrating the role that water molecules often play to complete H-bonding interactions, even within basepairs. The sheared A/G from loop E is compared to the A/C pair from tRNA in Figure 2.

3.2 TRANS-HOOGSTEEN U/A PAIRS

The U103/A73 and U77/A99 pairings in 5S loop E are trans-Hoogsteen paired with H-bonding between AN6 and UO2 and between AN7 and UN3. U/A is almost universally conserved at both positions, as expected for this special type of pairing. Only two conservative substitutions of C103/A73 pairings are observed. The C/A pairing is isosteric with the U/A that it replaces, but lacks one H-bond (should CN3 remain unprotonated), which could explain its low incidence. However, a slight lateral movement of the cytosine residue would lead to the formation of two H-bonds (CN3 to AN6 and CN4 to AN7).

3.3 BIFURCATED PAIRS (G102/U74 and G76/G100)

The two pairings G102/U74 and G76/G100 are discussed together because they share a common structure. Both pairs exhibit bifurcated H-bonds from a carbonyl group (UO4 or GO6) to the Watson-Crick positions (N1 and N2) of a guanosine (G102 or G76). Note that as for wobble G/U pairs, bifurcated G/U pairs are not isosteric with their reversed pairs. The conservative substitutions observed for the G102/U74 pairing are A/A, A/C, A/U, G/A, G/C and G/G, of which A/A and A/C are statistically favored. The G76/G100 pairing covaries almost exclusively with A/A with a strong statistical bias against A/G pairs (giving a signature opposite to that of the A101/G75 pair, see below). In addition, conservative A/C and G/A substitutions are also observed, but, as for the 102/74 pair, no C/A or A/G. The conservative G/G substitutions for G102/U74 need no further discussion, as they demonstrate explicitly that bifurcated G/U and G/G are interchangeable. As already mentioned, the G76/G100 pair may be superimposed almost exactly upon the G102/U74 in 3D space: the sugar phosphate backbones superimpose precisely while substitution of G for U74 merely pushes the guanosine base approximately 1.3Å out into the shallow groove side. Thus, submotifs 1 and 2 of loop E, which differ only in the third pairing (G102/U74 in submotif 1 and G76/G100 in submotif 2) are seen to be essentially identical geometrically.

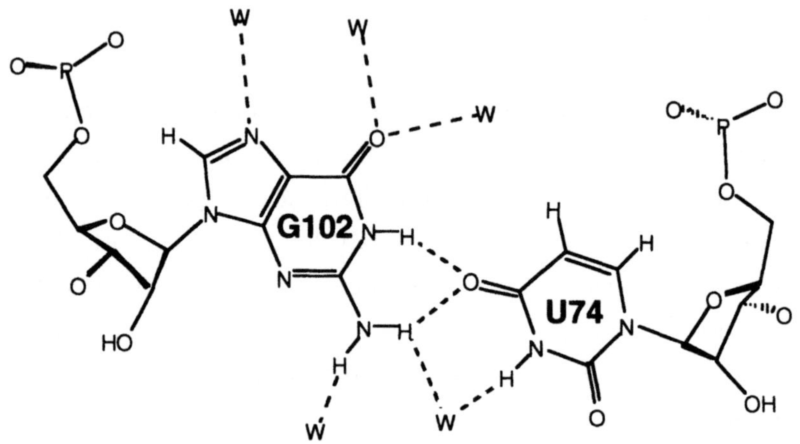

Figure 3: H-bonding geometry of bifurcated G/U basepair (isosteric to bifurcated G/G) as found in 5S rRNA loop. Note the water molecule bridging between the UN3 imino and GN2 amino protons.

The A/A substitutions for the 102/74 and 76/100 pairings can be accommodated by A74 (or equivalently A100) H-bonding via its N6 amino group to the N1 of A102 (equivalently A76). A slight reorientation leads to the formation of an additional H-bond between N7 of A74 and N6 of A102. A/C is equally easily accommodated at these positions by using AN1 as the acceptor and CN4 as the H-bond donor in place of the GN1-UO4 H-bond. A water molecule can potentially bridge from CN3 (or AN1 in the A/A pairing) to the polarized base proton H2 of A74 (or equivalently A102).

The G/A substitutions for these pairings require geometrical adaptation, as the amino group of the A would otherwise be directed toward the imino of G. For example, the amino of A could pair with the carbonyl of G, as in the A101/G75 basepair. The A102/U74 pairing is geometrically possible as an isosteric replacement for G102/U74, but is expected to be less stable than G/U unless AN1 is protonated.

Therefore, a correlation is observed between conservative substitutions and isosteric pairings for the bifurcated pairings. Moreover, one sees an overlapping set of substitutions for the 102/74 and 76/100 pairings, which taken together, clearly define the sequence variation signature that can identify these pairings in other contexts.

3.4 «OPEN» A101/G75 PAIRING

This pairing varies almost exclusively with G/A, which symmetry considerations alone indicate must be isosteric. Although three instances of A/A pairing are also observed in the database of bacterial 5S sequences, there is a strong statistical bias against homopurine pairings (no cases of G/G pairing are observed). A shift in geometry would be required to prevent clash of the amino groups of the two adenosines, perhaps allowing N6 of one base to pair with N1 of the other, so this is not an isosteric pairing. The isosteric G101/A75 pairing may be generated by simply rotating the A101/G75 basepair around the (pseudo-symmetric) axis passing between the bases, perpendicular to the axis of the double helix. It is found on the computer screen that the sugar-phosphate backbones of the original and the rotated pair superimpose exactly.

4. Identification of Loop E Submotifs in Other Contexts

Using the observed basepair substitutions discussed above, we have identified the loop E submotif in other contexts. An example is the symmetrical internal loop at positions 581-583/758-760 in 16S rRNA. Based on the observed sequence variations, we concluded that three non-canonical pairs form with the same geometries as those observed in the loop E submotifs: Pair 583/758 is a sheared purine/purine, 582/759 a trans-Hoogsteen pyrimidine/adenine, and 581/760 a bifurcated pairing showing covariation between G/G and G/U. The location of this motif in Helix 20 of 16S rRNA is shown in Figure 4. Also shown in this figure are the locations in 16S rRNA of two motifs that we recently proposed are structurally homologous to the sarcin/ricin loop of 23S based on analysis of sequence and chemical probing data [15].

5. Comparison to tRNA Structure

It is interesting that almost all of the non-canonical pairings observed in the loop E and sarcin loop motifs were first observed in tRNA although in different contexts. In tRNA these interactions occur in contexts generally classified as

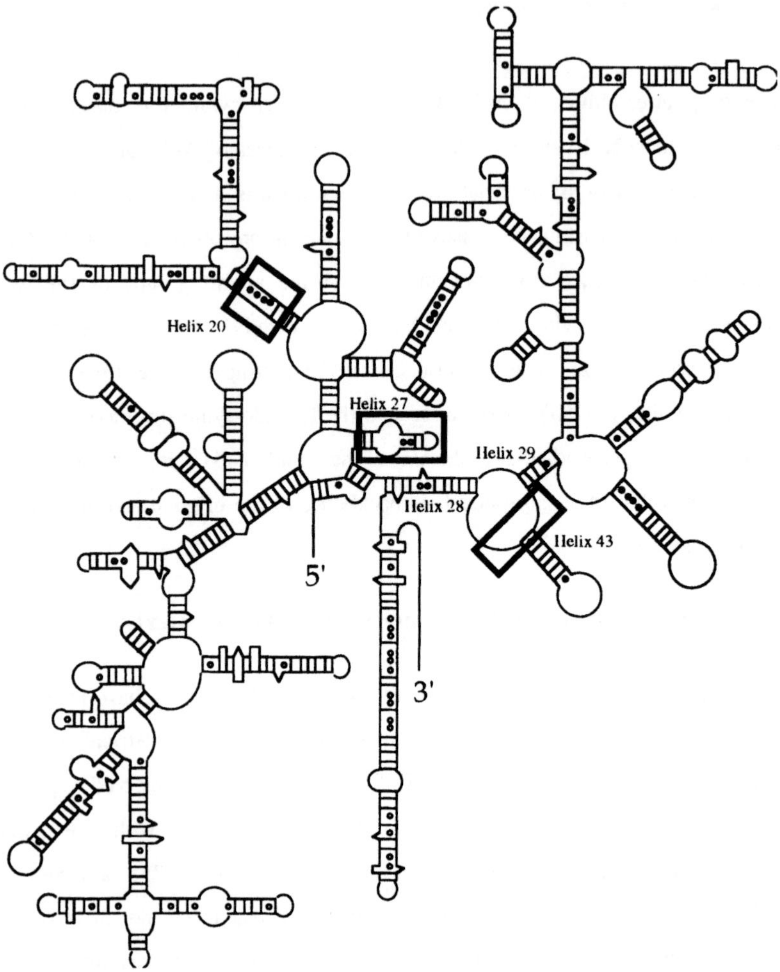

Figure 4: Schematic of 16S rRNA (E. coli version) showing location of loop E motif and sarcin/ricin loop motifs discussed in text.

tertiary interactions, since canonical Watson-Crick pairs define the secondary. The trans-Hoogsteen U/A pairing occurs at the nearly universally conserved U8/A14 as well as T54/A58. Interestingly, a small number of C/A pairings are observed for the 8/14 pairing (in archael tRNAs). As noted above, a small number of C/A pairs are the only covariations observed for trans-Hoogsteen U/A pairs in 5S rRNA.

The last pairing of the D-stem is usually a Watson-Crick or wobble pairing, but in a significant number of sequences (belonging largely to Class II tRNAs) G13/A22 and A13/A22 occur. In fact, G13/A22 occurs in $tRNA^{SER}$ from

T. thermophilusis, for which a crystal structure exists, and it is seen that this is a purine/purine sheared pairing.

The A9/A23 tertiary interaction in the base triple A9/A23/U12 in yeast Phe tRNA (TRNA09.PDB) is identical to the A/A pair adjacent to the bulged G in the sarcin loop motif. Many substitutions are observed for the A/A in the sarcin loop and its sister motifs in other locations in the ribosomal RNAs. These include A/N (and N/A as expected from symmetry, where N is any nucleotide), C/U, U/U, U/C, and G/C, but never G/G [15]. Interestingly, G/G is the one juxtaposition that can definitely be excluded based on H-bonding considerations (Gs only offer H-bond acceptors on their Hoogsteen faces).

In the tRNAs, sequence analysis is complicated by the fact that in Class II tRNA structures, the base at position 9 interacts with pair 22/13 rather than 23/12. Nonetheless, correlations are seen in the crystal structures available. In the crystal structure of initiator tRNA (TRNA012.PDB), 1mG9 interacts with C23 in a pairing isosteric to the A9/A23 interaction, whereas in glutaminyl tRNA (PTE003.PDB) this interaction is reversed (C9/G23), but with retention of the pairing geometry, as anticipated from the symmetry of the interaction. Again, small adjustments are observed when comparing the C/G, G/C, and A/A pairings. These serve to optimize local interactions. Nonetheless the overall geometry of the interactions is preserved.

Bifurcated pairings involving Gs also occur in tRNA, with the first example being the conserved bifurcated pairing G18 and Ψ55. The O4 of Ψ55 is positioned to form a bifurcated H-bond with N1 and N2 of G18 while exposing the N3 of Ψ55. A phosphate oxygen (A58), positioned at a site equivalent to the water molecules in the two loop E pairings, H-bonds to ΨN3. The observation that the phosphodiester backbones are locally parallel in the G18/Ψ55 base pair suggests that "bifurcated" G/U pairs could occur with a locally parallel orientation of the strands (with O2 of U interacting with N1 and N2 of G) and that "bifurcated" G/Ψ could also occur with a locally antiparallel orientation. The second example is the tertiary interaction between G45 and U25 in yeast tRNA(Asp). U25 is wobble-paired to G10. The O4 carbonyl forms bifurcated H-bonds to the N1 and N2 positions of G45. This suggests that the U in the bifurcated pairing of loop E could interact with another base using its Watson-Crick face.

6. Phylogenetic and geometric analysis of basepairs substituting for G/U in loop E.

The *E. coli* and also the consensus bacterial 5S rRNA loop E contain a pair of tandem G/U basepairs (U80/G96 and G81/U95), shown highlighted in Figure 1. We analyzed the sequence variations at these positions with the aim of determining possible isosteric basepairs that preserve the 3D structure of this motif. U80/G96 is highly conserved in bacterial 5S rRNAs, whereas G81/U95 shows a complex pattern of variation, partly due to the variations in the length of the stem that joins loop E to the closing hairpin loop D. Therefore, we grouped the bacterial 5S rRNAs according to the length of the stem between the tandem G/U motif of loop E and loop D. Only those molecules having at least three Watson-Crick basepairs between the hairpin loop and the motif were included in the analysis. Of the total of 297 sequences that meet these criteria, 288 have U80/G96. The remaining nine sequences all have tandem Y80/R96 - R81/Y95 Watson-Crick pairings and were therefore excluded from analysis. Table 1 shows that G81/U95 is the most common pairing. Besides Watson-Crick R81/Y95 pairings, all possible Y/Y pairings occur, with a statistically significant preference observed for C81/C95 and U81/C95.

Table 1. Covariations between positions 81 and 95 in loop E of bacterial 5S rRNAs. The number of sequences having the indicated pair substituting for the consensus G81/U95 is shown in each cell. The database is the subset of 288 bacterial 5S RNA sequences described in the text. The number of statistically expected occurrences of each pairing is shown in parentheses.

G81 / U95	A	C	G	U
A	**0** (0.2)	**0** (0.1)	**0** (1.2)	**1** (0.5)
C	**0** (5.4)	**15** (3.6)	**21** (41)	**30** (15)
G	**2** (0.2)	**0** (0.2)	**0** (1.8)	**0** (0.7)
U	**21** (18)	**4** (12)	**159** (134)	**35** (50)

The C/C and U/U pairings were modeled with FRAGMENT using a wobble U/G basepair as the template. The results suggest H-bonding between the substituent at position 4 of the pyrimidine substituting for G81 with N3 of the pyrimidine occupying position 95. This is shown for the U/U pairing in Figure 5 (upper panel). An equivalent C/C pairing can be generated due to the interchange of the H-bond donors and acceptors in U vs. C at the N3 and O4/N4 positions. As anticipated, the H-bonding distances are greater than ideal in the pair so generated, but the results are tantalizing since small readjustments should suffice to shorten the distance. Moreover, water molecules may also participate in the H-bonding between the bases as in fact occurs for U/C pairs discussed below.

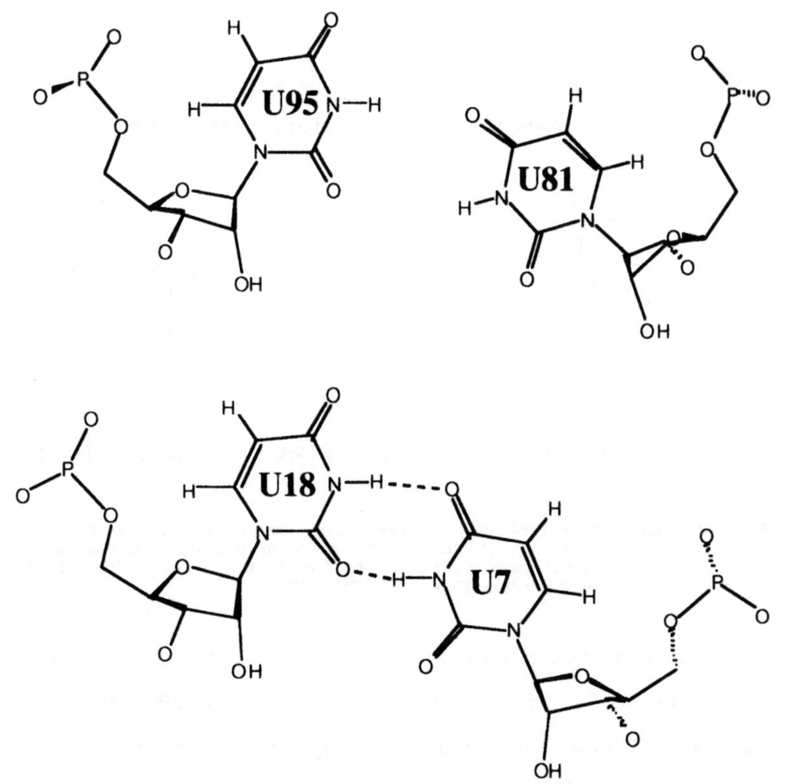

Figure 5: (Upper) Modeling of the U95/U81 pair in 5S loop E using the wobble U95/G81 pair observed in the loop E crystal structure as a template. (Lower) The corresponding U18/U7 pair from the tandem U/U motif of the oligonucleotide crystal structure URL050.PDB.

U/U basepairs occur in several recent x-ray crystal structures of RNA oligonucleotides. Relevant for the present discussion is URL050.PDB, which has tandem U/U pairs flanked on both sides by Watson-Crick pairs (Fig. 6), as is also the case for the tandem G/U motif of loop E. Also relevant for this discussion is the structure AR0001.PDB which features tandem A(+)/C basepairs flanked by

W.C. pairs (Fig. 6). The tandem A(+)/C pairs exhibit wobble geometry isosteric to that of the tandem G/U pairs in loop E. Both motifs exhibit identical cross-strand, purine-purine stacking. This is shown in stereo for the tandem A(+)/C in Figure 7. Both U/U pairs in URL050.PDB also have wobble geometry. Since the two uridines of a U/U wobble pair are not equivalent, two orientations are possible. For example, in the U7/U18 pair of URL050 shown in Figure 5, the H-bond interactions could be reversed, with N3 and O2 of U7 H-bonding to O4 and N3 of U18. Interestingly, in the tandem U/U structure of URL050.PDB, the U/U wobble pairs show the same orientation as the wobble pairs of the tandem G/U and A(+)/C motifs. This geometry results in cross-strand stacking between U7 and U19, exactly as observed for the tandem wobble pairs containing purines (see Fig. 7).

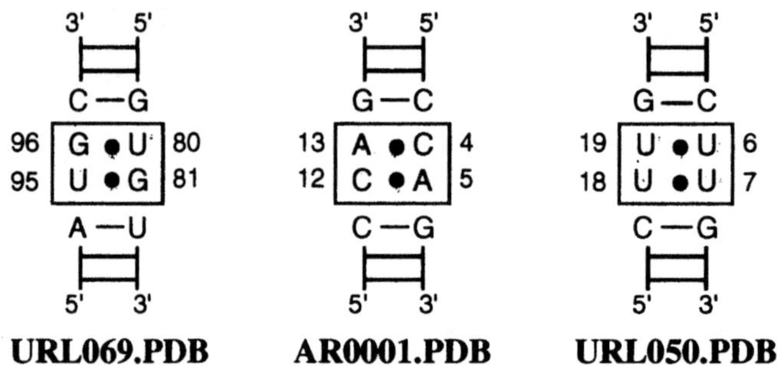

Figure 6: Sequences of RNA oligonucleotides having tandem wobble pairing geometries and exhibiting cross-strand stacking. (Left) The tandem G/U motif in the consensus 5S rRNA loop (URL069.PDB). (Center) Oligonucleotide containing Tandem A(+)/C motif (AR0001.PDB). (Right) Oligonucleotide containing tandem U/U motif (URL050.PDB).

Based on these comparisons, we propose that the mixed tandem motifs observed in the 5S database, for which the conserved G96/U80 is flanked by U95/U81 or C95/C81, have the same geometry as that of the tandem G/U, A(+)/C, and U/U motifs of Figure 6. Thus we are led to anticipate wobble pairing between Y81 and Y95 as shown in Figure 5 and cross-strand stacking of Y81 on G96 as in Figure 7.

Regarding the C/U and U/C pairings, modeling based on a U/G wobble pair as a template results in structures requiring readjustment, due to juxtaposition of two H-bond acceptors (UO4 opposite CN3) or two donors (CN4 opposite UN3). A small readjustment places UO4 opposite CN4. Such pairing has in fact been observed in two existing crystal structures of RNA oligonucleotides which have U/C pairs adjacent to U/G just as in one variant of 5S loop E (ARL037.pdb and

AR0005.pdb). In the crystal structure U/C basepairs are «open» pairings (cis, water-inserted) with H-bonding between UO4 and CN4 and a water molecule bridging between the N3 positions of the pyrimidines to complete the H-bonding. This pairing is isosteric with C/U, in agreement with the observation of both orientations in the 5S sequences at position 81/95.

Thus we can make sense, in terms of isostericity, of the conservative substitutions observed in the tandem G/U motif of loop E, just as we could with the rest of loop E. This suggests further that Y/Y pairs may be interchangeable with G/U pairs in other contexts.

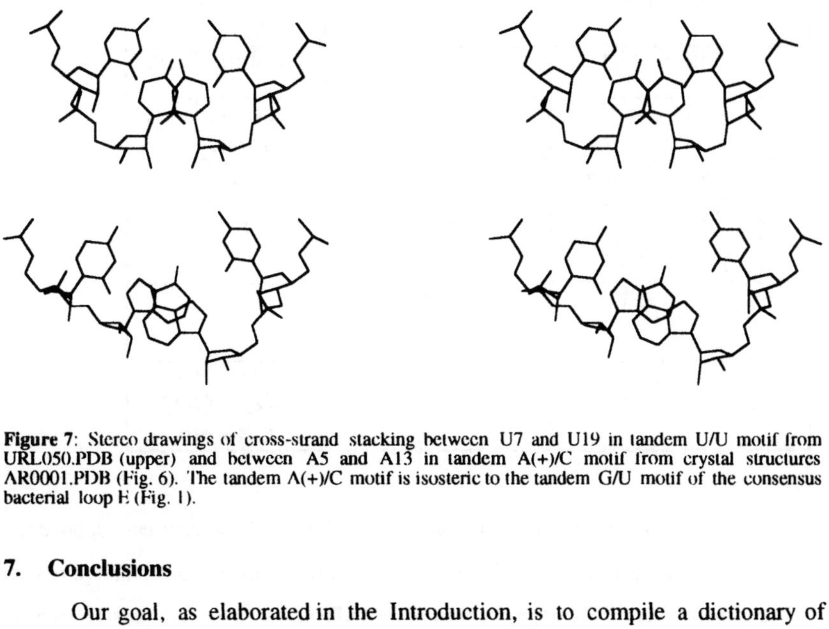

Figure 7: Stereo drawings of cross-strand stacking between U7 and U19 in tandem U/U motif from URL050.PDB (upper) and between A5 and A13 in tandem A(+)/C motif from crystal structures AR0001.PDB (Fig. 6). The tandem A(+)/C motif is isosteric to the tandem G/U motif of the consensus bacterial loop E (Fig. 1).

7. Conclusions

Our goal, as elaborated in the Introduction, is to compile a dictionary of isosteric pairings that substitute for each other while preserving the 3D structures of conserved motifs. We expect that such a dictionary will be useful in identifying occurrences of the same motif in other molecules and in extending covariation analysis with more confidence to «single-stranded» nucleotides in RNA that in fact participate in non-canonical base-pairing.

Table 2. Isosteric relationships between basepairing geometries observed in RNA.

Basepair Type	Isosteric Upon Reversal	Isosteric Pair
Cis Watson-Crick **G/C or A/U**	**Yes**	Any cis W.C. Basepair
Trans Watson-Crick **G/C or A/U**	**Yes**	-
Cis Watson-Crick **A/G**	**Yes**	**G/A**
Sheared **A/G** Sheared **C/A**	**No**	Sheared **A/N** Sheared **C/Y**
Wobble **G/U**	**No**	Wobble **A(+)/C**
"Wobble" **U/U**	**No**	"Wobble" **C/C**
Bifurcated **G/U** Bifurcated **G/G**	**No**	Bifurcated **A/C** Bifurcated **A/A**
Cis water-inserted ("Open") **G/A**	**Yes**	**A/G**
Cis water-inserted ("Open") **C/U**	**Yes**	**U/C**
Trans-Hoogsteen **A/U**	**No**	-
Trans-Hoogsteen-Hoogsteen **A/A**	**Yes**	Trans-H.H. **A/C, G/C, C/G** etc.

We present our present compilation in Table 2. Each row corresponds to a particular pairing geometry. In the first column, we provide a representative pair for that geometry and in the second column we indicate whether that pair is isosteric with its reverse. In the third column we indicate other (isosteric) pairings compatible with each geometry. This tabulation shows that basepairing in RNA is highly context dependent. For example, an A/G pair may exhibit sheared, trans-Hoogsteen, cis-Watson Crick, or even «open» geometries with an inserted water molecule. All the factors that determine which geometry is adopted in a particular context remain to be elucidated. In the meantime, however, differences in the sequence variations compatible with a given geometry allow one in many cases to exclude certain possibilities in favor of others.

Acknowledgements. This work was supported in part by NIH grant 1R15-GM/OD55898-02 (NBL). E.W. acknowledges support from the Institut universitaire de France.

References

1. Woese, C.R. and N.R. Pace (1993) Probing RNA Structure, Function and History by Comparative Analysis, in R.F. Gesteland and J.F. Atkins, Editors, *The RNA World*, Cold Spring Harbor Laboratory Press, Cold Spring Harbor, NY, .

2. Michel, F., A. Jaquier, and B. Dujon (1982) Comparison of fungal mitochondrial introns reveals extensive homologies in RNA secondary structure, *Biochimie* **64**, 867-881.

3. Fresco, J.R., B.M. Alberts, and P. Doty (1960) Some Molecular Details of Secondary Structure of Ribonucleic Acid, *Nature* **188**, 98-101.

4. Moazed, D., S. Stern, and H.F. Noller (1986) Rapid chemical probing of conformation in 16S ribosomal RNA and 30S ribosomal subunits using primer extension, *J. Mol. Biol.* **187**, 399-416.

5. Woese, C.R. and R.R. Gutell (1989) Evidence for several higher order structural elements in ribosomal RNA, *Proc. Natl. Acad. Sci. USA* **86**, 3119-3122.

6. Larsen, N. (1992) Higher order structure in 23S rRNA, *Proc. Natl. Acad. Sci. USA* **89**, 5044-8.

7. Michel, F. and E. Westhof (1990) Modelling of the three-dimensionnal architecture of group-I catalytic introns based on comparative sequence analysis, *J. Mol. Biol.* **216**, 585-610.

8. Wimberly, B., G. Varani, and I. Tinoco, Jr (1993) The conformation of loop E of eukaryotic 5S ribosomal RNA, *Biochem.* **32**, 1078-1087.

9. Correll, C.C., *et al.* (1997) Metals, Motifs, and Recognition in the Crystal Structure of a 5S rRNA Domain, *Cell* **91**, 705-712.

10. Szewczak, A.A. and P.B. Moore (1995) The sarcin/ricin loop, a modular RNA, *J. Mol. Biol.* **247**, 81-98.

11. Seggerson, K. and P.B. Moore (1998) Structure and stability of variants of the sarcin-ricin loop of 28S rRNA: NMR studies of the prokaryotic SRL and a functional mutant, *RNA* **4**, 1203-1215.

12. Leontis, N.B. and E. Westhof (1998) The 5S rRNA loop E: Chemical probing and phylogenetic data versus crystal structure, *RNA* **4**, 1134-1153.

13. Szymanski, M., *et al.* (1998) 5S rRNA Data Bank, *Nucleic Acids Res.* **26**, 156-159.

14. Westhof, E. (1993) Modelling the three-dimensional structure of ribonucleic acids, *J. Mol. Struct.* **286**, 203-210.

15. Leontis, N.B. and E. Westhof (1998) A common motif organizes the structure of multi-helix loops in 16S and 23S ribosomal RNAs, *J. Mol. Biol.* **283**, 571-583.

TOWARDS THE 3D STRUCTURE OF 5S rRNA

M. PERBANDT[1], S. LORENZ[1], M.VALLAZZA[1],
V.A. ERDMANN[1] & C. BETZEL[2]

[1] *Freie Universität Berlin, Thielallee 63, D-14195 Berlin*
[2] *Universitätskrankenhaus Eppendorf, Inst. f. Physiol. Chemie, AG Makro. Strukturanalyse, c/o DESY, D-22603 Hamburg*

1. Abstract

The ribosomal 5S RNA is approximately 120 nucleotides long and is an integral part of the large ribosomal subunit. Several parts of the 5S rRNA interact specifically with several ribosomal proteins [1-3]. Nevertheless its precise role in protein synthesis remains unclear. It is clear that reconstituted 50S ribosomal subunits, lacking the 5S rRNA, are inactive in proteinbiosynthesis [4]. Structural studies will support a more detailed understanding of the 5S rRNA function. Our extensive attempts to crystallise ribosomal 5S RNA have led to crystals, which diffract to about 7.5Å [5]. For the investigation of some special features like GU basepairs and hairpin-loops in the ribosomal 5S RNA from Thermus flavus we have divided the 5S rRNA into five domains [6] (A to E) and have solved three crystal structures (Figure 1): Helix I of domain A has been determined at 2.4Å [7]. Helix V of domain E without the terminal hairpin-loop has been determined at 1.6Å and Helix V of domain E including the terminal hairpin-loop has been determined at 3.0Å [8]. Furthermore it was possible to observe a general mode of intermolecular interaction, which occur only in RNA structures and it was possible to demonstrate the structural importance of water molecules for the stability of RNA-molecules.

J. Barciszewski and B.F.C. Clark (eds.), RNA Biochemistry and Biotechnology, 63–71.

2. Introduction

Large RNA molecules such as those in the ribosome show secondary structures that have only stretches of Watson-Crick duplex connected frequently by internal and terminal loops and bulges whose secondary structures cannot be accurately predicted. One assumption for a prediction of non-Watson-Crick structures is a pool of structural data to understand the language of special RNA motifs. The repeated occurrence of those RNA motifs like tandem GU wobble basepairs [9] and GNRA tetraloops [10] in large RNA molecules shows the importance of those motifs. The difficulties in crystallisation and structure solution to high resolution of large RNA-molecules has led to the synthesis of shorter RNA fragments, which contain certain motifs.

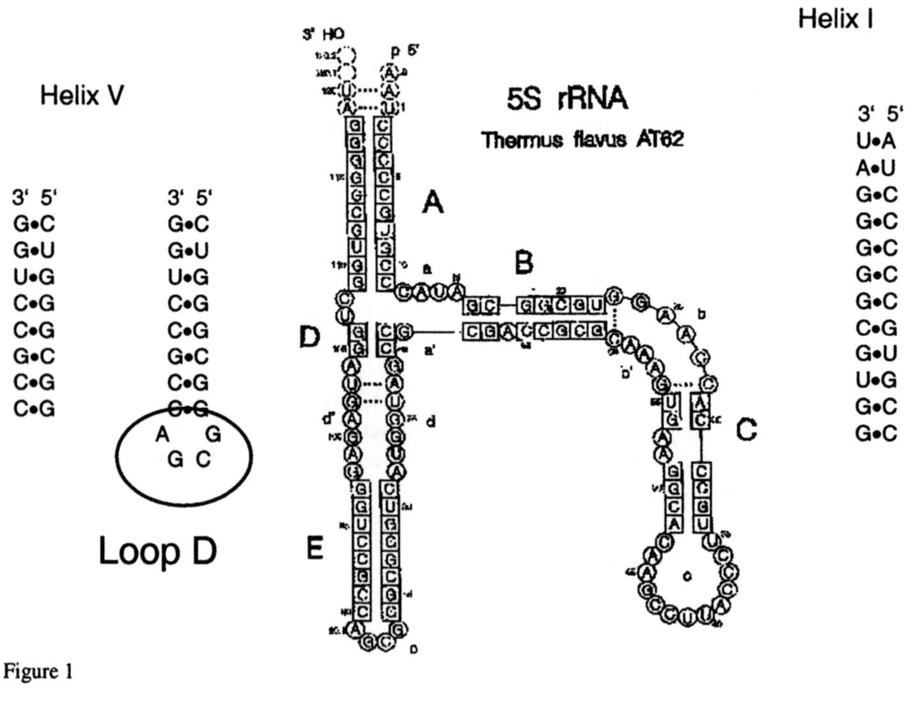

Figure 1

The condition to use RNA fragments as a model-system is, that these motifs have the same conformation in different structures and it is important to compare them with motifs in natural molecules. The tandem wobble GU basepair formation is the most frequent "mismatch formation" in the ribosomal 5S RNA [9] and are supposed to be possible protein binding sites [11]. The hairpin-loop in domain E is near the peptidyltransferase centre [12] and its residues can be crosslinked to the 23S rRNA. The loop belongs to the highly conserved class of highly stable 5'-GNRA tetraloops occurring in the ribosomal RNAs [10]. Recently the crystal structure of the 62 nt fragment of *E.coli* containing was presented [11]. This molecules is of special interest because it was yielded by mild nuclease digestion of the entire 5S rRNA and represents in this way a natural molecule. In domain E the GU bp in tandem formation are conservative for the sequences of *E.coli* and *Thermus flavus*. In contrast to that, the sequences differ in the region of the terminal hairpin-loop D. In *E.coli* we have an 5'UCU pyrimidine terminal loop instead of an 5'GCGA tetraloop in *Thermus flavus*. Fortunately we are able to compare both loops, because the solution structure of the terminal loop D from *E.coli* was recently presented [13].

3. Material and Methods

For the synthesis of the ribosomal 5S rRNA an 3200 bp fragment of the genomic DNA from *Thermus flavus* AT 62 was ligated into the vector pT7T3 18U (2890 bp) and transformed into cells of *E.coli* strain X90. PCR's with the isolated plasmids and different 3'- and 5'-end primers have been done. The PCR products were ligated into the pUC 18 vector and transformed into cells of the *E.coli* strain X90. The isolated plasmids were tested of their correct inserts by DNA sequencing. After linerisation of the plasmids we sythesized the 5S rRNA variants during run off transcriptions. The 5S rRNA products were purified by a Sephadex G150 gel chromatography and by a hydrophobic interaction chromatography. The crystals were grown with the hanging drop vapour diffusion method in Linbro 24-well plates. Different precipitation agents, additives and buffers were tested. The best crystals we obtained show the following solutions: 40-50% saturated ammoniumsulfate, 300mM cesiumchloride, 5mM cobaltchloride, 15mM calciumchloride, 20mM spermine, 50mM sodiumcacodylate. The hanging drops contained 2μl 0.1mM 5S rRNA and 2μl of the described crystallization solutions, the reservoir 300μl of the crystallization solution. The crystals grow in few days up to a size of about $0.5x0.2x0.2mm^3$.

All synthetic fragments of 5S rRNA of *Thermus flavus* were prepared by solid phase chemical synthesis and further purified by reversed phase HPLC. Crystals of the domains suitable for X-ray analysis were obtained by vapour diffusion as reported before [14-16].

The data sets were processed using the program DENZO [17,18]. The structures of the domains were solved by molecular replacement applying the program AMoRe [19,20] distributed with CCP4 [21] and refined with the program X-PLOR [22] as reported before. The parameters of the data collection and the refinement statistics are summarised in Table 1.

Table 1: Data-collection and Refinement statistics

	5S rRNA	Helix V	Helix V/Loop D	Helix I	
Cellconstants a=b	110.3Å	41.9Å	42.7Å	30.1Å	
Cellcontant c	387.6Å	127.1Å	161.2Å	86.8Å	
Solvent V_M	3.3Å^3/Da	2.4Å^3/Da	3.7Å^3/Da	2.6Å^3/Da	
x-ray source	ELETTRA	ELETTRA	DESY/EMBL	Sealed tube	DESY/EMBL
Wavelength	1.0Å	1.0Å	0.92Å	0.71Å	0.92Å
Resolution	20-7.5Å	20-1.6Å	30-2.8Å	20-3.0Å	20-2.3Å
All reflexes	34178	43171	10753	5388	8797
Unique reflexes	1051	5994	4702	1477	2170
R_{symm}	7.9%	6.6%	7.2%	9.0%	
R_{symm} last shell	28.6	29.2	33.7	32.4	
Completeness	99.0%	96.1%	93.0%	81.0%	
R-value/Rfree	-	22.4/27.1%	24.1/31.8%	18.3/26.3%	
Nucleicacid-atome	-	338	856	508	
Water-molecules	-	75	221	156	
Distances [Å]	-				
Bonds(1-2)	-	0.023	0.012	0.030	
Angles(1-3)	-	0.046	0.047	0.070	
Planar groups [Å]	-	0.027	0.040	0.014	
Chiral volumes [Å^3]	-	0.039	0.125	0.036	
Angles [°]	-				
Bondangles	-	1.4	1.7	3.8	
Torsion angles	-	4.6	6.7	5.9	
Dihydral angles	-	11.4	25.1	13.3	

4. Results and discussion

4.1 WOBBLE GU bp IN TANDEM FORMATION

The crystal structures of Helix I and Helix V reveal the conformation of the tandem wobble GU basepairs in domain A and E. In contrast to the three direct hydrogen bonds in GC Watson-Crick-basepairs those GU wobble basepairs have only two direct hydrogen bonds: GUA-N1•URI-O2 and GUA-O6•URI-N3. In some regions of the ribosomal 5S RNA, which are suggested as a protein-binding region, those GU-basepairs occur [11]. The conformation of the GU-basepairs in that region give us some hints of possible protein-RNA-interaction.

The two adjacent GU-basepairs in tandem from a so called "cross strand G stack", because the G of the one GU basepair stacks from the G of the other GU

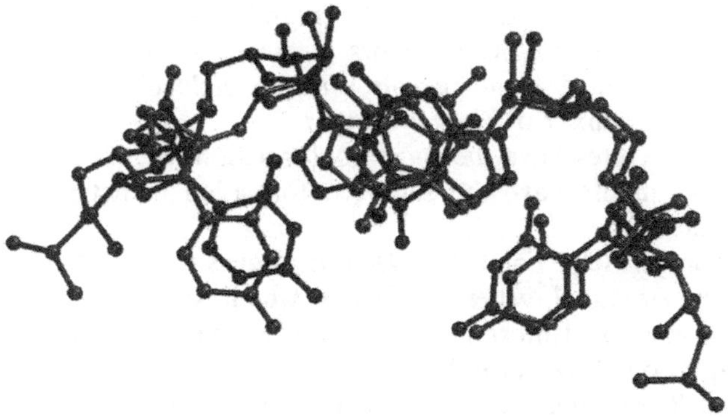

Figure 2: A view along the helical axis of the superimposed GU-tandem basepairs of the synthetic Helix V of *Thermus flavus* and the natural molecule of *E.coli.* They have r.m.s. deviation of 0.54Å. The cross strand purin stack is clearly visible.

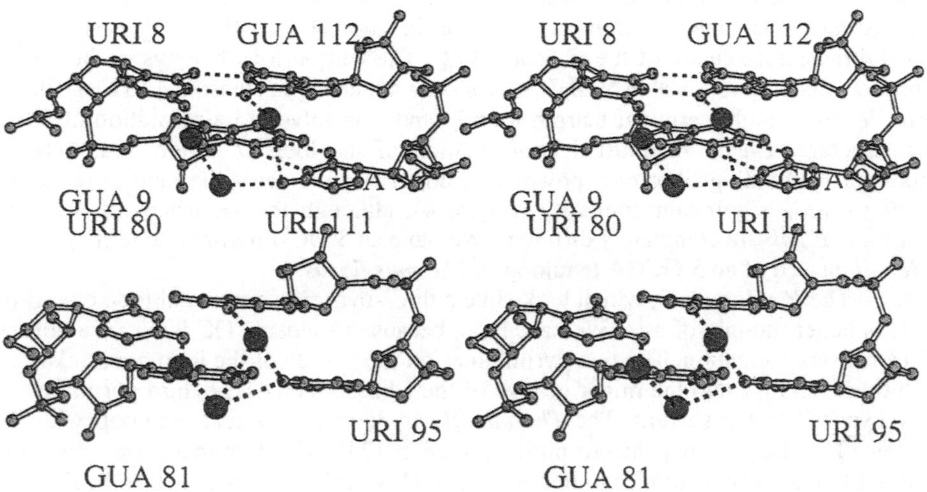

Figure 3: Conformation of the tandem GU-basepairs in Helix I and V including the structural waters.

basepair, which comes from the opposite strand. This was also observed by Corell et. al [11] for the same GU basepairs in the 62 nt fragment of the *E.coli* 5S rRNA. The 62nt fragment of *E.coli* reveals a native molecules in comparison to the chemically synthesised Helix V of *Thermus flavus*.
Nevertheless both structures have the same conformation (Figure 2). This is a strong evidence, that those motifs have the same conformation in natural and chemically synthesised molecules.

Furthermore in Helix I and V of *Thermus flavus* this GU tandem formation is stabilised by three structural water molecules which show an identical hydrogen bonding scheme (Figure 3). Unfortunately this was not visible for the crystal structure Helix V including Loop D, because that part at the open 5'-3' end of the structure is thermally distorted. The sequences of the domains are shown in Figure 1.
We can now say, that this is a general mode of RNA-water-interaction to compensate missing hydrogen bonds in those structural motifs in RNA. We call those water molecules "structural waters", because they reappear in different structures and interact with the same atoms.

4.2 CONFORMATION OF THE 5'GNRA TETRALOOP IN HELIX V

Domain E is of major interest because crosslinks between 5S rRNA and 23S rRNA demonstrated that the location of the hairpin loop D in Helix V is near the peptidyltransferase centre of the ribosome [12]. The hairpin loop belongs to the highly conserved class of very stable 5'GNRA tetraloops occurring in ribosomal RNA [10].
Helix V contains the terminal hairpin loop D and was solved to a resolution of 3.0Å. The structure reveals an unusual conformation of the loop D region, that may be important for binding ribosomal proteins. A comparison towards the loop D region of *E.coli* shows a significant similarity (Figure 4), although the sequences of *Thermus flavus* and *E.coli* are completely different. We have an 5'UCU pyrimidine terminal loop in *E.coli* instead of an 5'GCGA tetraloop in *Thermus flavus*.

The *E.coli* loop D, which looks like a three-pyrimidine terminal loop closed by a GC is better thought of as a five-base loop, because its closing GC is not in a normal Watson-Crick formation. The two pyrimidines on the 5'-side of the loop are stacked on each other, and tilt into the minor groove of the adjacent helix. The third pyrimidine is fully exposed to the solvent. The *Thermus flavus* loop D is a real tetraloop with the closing GA basepair in a hetero-purin-formation [23]. The terminal bases are fully exposed to the solvent, similar to that in *E.coli*. Obviously the position U90 in *E.coli* is occupied by G90 in *Thermus flavus* and U87 by C88. Furthermore both loop-regions are quite flexible. This is expressed by two different conformations for the two molecules in the asymmetric unit in the crystal structure of the *Thermus flavus* loop D.

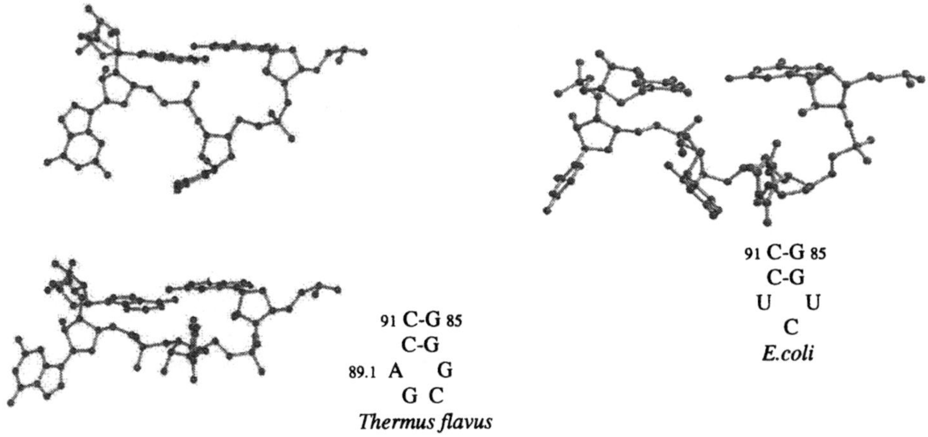

Figure 4: Structural homologies between the loop-regions of *Thermus flavus* and *E.coli*.

4.3 RNA-RNA INTERACTIONS

The crystal structure of Helix I and also the crystal structure of Helix V revealed an unusual GC basepair, which is not in the normal Watson-Crick-formation, although both structures form an almost perfect double helix. The three established hydrogen bonds for a GC basepair do not exist for the basepair G118•C2 and G7•C113 in Helix I and G92•C84 in Helix V (see Figure 5). In both structures have only two instead of three hydrogen bonds. The new hydrogen bonds GUA-O6•CYT-N3 and GUA-N1•CYT-O2 between guanine and cytosine are unusual but according to the electrondensity (Figure 6) the atoms show typical distances for hydrogen bonds. Furthermore the hydrogen bond GUA-O6•CYT-N3 is not possible in the normal amino-form of the cytosine. We have two possibilities: Firstly, the cytosine is the imino-form, which is quite unlikely [23]. Secondly, the cytosine has an additional proton at N3, which is more plausible. Unfortunately we were not able to locate the protons at a resolution of 1.6Å. For this goal we need atomic resolution.

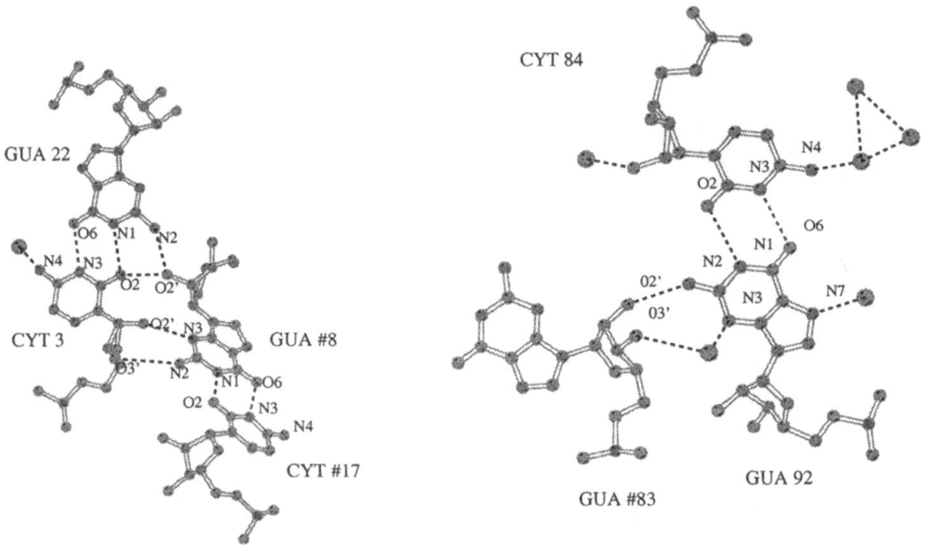

Figure 5: The crystal structures of Helix I and V revea an unusual GC basepairs in non-Watson-Crick formation. They are stabilised by only two direct but several intermolecular interactions and water molecules.

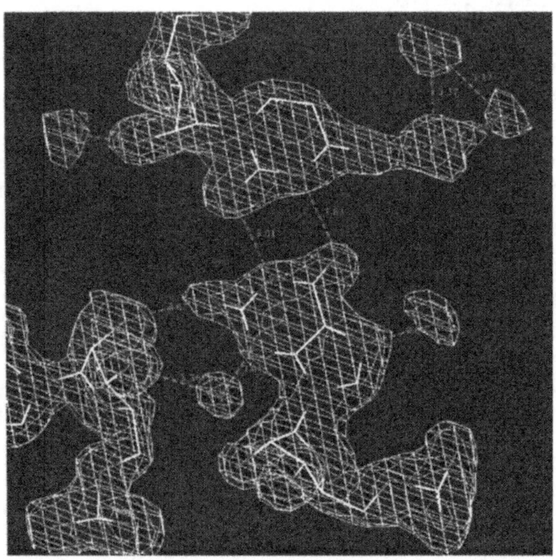

Fıgure 6: A representatıve 2FoFc Electron-densıty around the regıon of the non-Watson-Cnck GC basepaır ın Helix V of domain E.

Acknowledgements: The project has been supported by the Deutsche Forschungsgemeinschaft (SFB 344-D6), the DASA and the Fonds der Chemischen Industrie e.V. The EMBL-Outsation Hamburg and the Synchrotron ELETTRA in Trieste is also acknowledged for the allocated beamtime.

5. References

1. Erdmann, V.A., Fahnestock, K.H. and Nomura, M. (1971). Role of 5S RNA in the function of 50S ribosomal subunits. Proc. Nat. Acad. Sci. **68**, 2932-2936.
2. Hartmann, R.K., Vogel, D.W., Walker, R.T. and Erdmann, V.A. (1988). In vitro incorporation of eubacterial and eucaryotic 5S RNA into large ribosomal subunits of Bacillus stearothermophilus. Nucl. Acid. Res. **16**, 3511-3524.
3. Horne, J. and Erdmann, V.A. (1972). Isolation and Characterization of 5S RNA-Protein Complexes from Bacillus stearothermophilus and Escherichia coli Ribosomes. Molec. gen. Genet. **119**, 337-344.
4. Hartmann, R.K., Vogel, D.W., Walker, R.T. and Erdmann, V.A. (1988). In vitro incorporation of eubacterial and eukaroyotic 5S RNA into large ribosomal subunits of Bacilus stearothermophilus. Nucl. Acid Res. **18** (Suppl.), 2215-2230.
5. Lorenz, S., Betzel, C., Raderschall, E., Dauter, Z., Wilson, K.S. and Erdmann, V.A. (1991). Crystallisation and preliminary X-ray diffraction studies of 5S RNA from the thermophiliv bacterium Thermus flavus. J.Mol.Biol. **219**, 399-402.
6. Specht, T., Wolters, J. and Erdmann, V.A. (1990). Compalition of 5S rRNA gene sequences. Nucleic Acids Res. **18** (Suppl.), 2215-2230.
7. Betzel, Ch., Lorenz, S., Fuerste, J.P., Bald, R., Zhang, M., Schneider, Th.R., Wilson, K.S. and Erdmann, V.A. (1994). Crystal structure of domain A of of Thermus flavus 5S rRNA. FEBS Lett. **351,** 159-164.
8. Perbandt M., Nolte, A., Lorenz, S., Bald, R., Betzel, Ch. and Erdmann, V.A. (1998). Crystal structure of domain E of *Thermus flavus* 5S rRNA: a helical RNA-structure incuding a hairpin loop. FEBS Lett. **429/2,** 211-215
9. Wu, M., McDowell, J.A. and Turner D.H. (1995). A periodic table of symmetric tandem mismatches in RNA. Biochemistry **32**, 3204-3211.
10. Woese, C.R., Winkler, S. and Gutell, R.R. (1990). Architecture of ribosomal RNA: constains on the sequence of tetra-loops. Proc. Natl. Acad. Sci. USA **87**, 8467-8471.
11. Correll, C.C., Freeborn, B., Moore, P.B. and Steitz, T. (1997). Metals, Motifs and Recognition in the Crystal Structure of a 5S rRNA Domain. Cell **91**, 705-712.
12. Dontsova, O., Tishkov, V., Dokudovskaya, S., Bogdanov, A., Döring, T., Rinke-Appel, J., Thamm, S., Greuer, B. and Brimcombe, R. (1994). Stem-loop IV of 5S rRNA lies close to the peptidyltransferase center. Proc. Natl. Acad. Sci. USA **91**, 4125-4129.
13. Dallas, A. and Moore, P.B. (1997). The loopE-loopD region of Escherichia coli 5S RNA: the solution structure reveals an unusual loop that may be important for binding ribosomal proteins. Structure **5/12**, 1639-1653.
14. Lorenz, S., Betzel, Ch., Fuerste, J.P., Bald, R., Zhang, M., Raderschall, E., Dauter, Z., Wilson, K.S. and Erdmann, V.A. (1993). Crystallization and preliminary diffraction studies of the chemically synthesized domain A of Thermus flavus 5S rRNA: an RNA dodecamer double helix. Acta Cryst. **D49**, 418-420.
15. Nolte, A., Klußmann, S., Lorenz, S., Bald, R., Betzel, Ch., Dauter, Z., Wilson, K.S., Fuerste, J.P. and Erdmann, V.A. (1995) Crystallization and preliminary diffraction studies of the structural domain E of Thermus flavus 5S rRNA. FEBS Lett. **374**, 292-294.
16. Vallazza, M., Förster, C., Eickmann, A., Lippmann, C., Perbandt, M., Betzel, Ch. and Erdmann, V.A. (1998). Crystallization and 1.6Å X-ray diffraction data of Thermus flavus 5S rRNA E-helix. In prep..
17. Otwinowski, Z. (1991). DENZO: A film processing program for macromolecular crystallography, Yale University.
18. Otwinowski, Z. (1993). SCALEPACK: Software for the scaling together of intgrated intensities measured on a number of seperate diffraction images, Yale University.
19. Navazza, J. (1997). The molecular replacement. Acta Cryst. **A43**, 645-653.
20. Navaza, J. (1994). AmoRe: An automated package for molecular replacement. Acta Cryst. **D50**, 157-163.
21. Collaborative Computational Project Number 4 (1994). The CCP4 Suite: Programs for Protein Crystallography. Acta Cryst. **D50**, 760-763.
22. Brünger, A.T. (1992). X-PLOR.Version 3.1. A System for Crystallography and NMR. Yale University Press, New Haven.
23. Saenger, W. (1987). Principles of Nucleic Acid Strukture, Springer-Verlag, New York.

STRUCTURE AND DYNAMICS OF ADENOSINE LOOPS IN RNA BULGE DUPLEXES. RNA HYDRATION AT THE BULGE SITE

ŁUKASZ BIELECKI, TADEUSZ KULIŃSKI AND RYSZARD W. ADAMIAK*

Institute of Bioorganic Chemistry, Polish Academy of Sciences, Noskowskiego 12/14, 61-704 Poznań, Poland

1. Abstract

An RNA bulge duplex resulted from annealing of oligoribonucleotides: GUCGAAPAGCUG and CAGCCGAC, where 2-aminopurine (AP) was introduced as a fluorescent conformational probe, has been studied by *in aqua* molecular dynamics simulation (AMBER force field, 250 ps at 300 K, cutoff 15 Å; Cray J916) to complement our results of laser spectrofluorimetry measurements of the same duplex. During the entire MD run, the stem regions maintained their conformational stability. The RNA molecule showed a considerable axis bending. The hydrogen bonds and stacking interactions within the molecule were preserved and sugar puckering remained typical for A-RNA. On the other hand, the bulged nucleotide residues showed a tendency to "bulge out" to the exterior which resulted in destabilisation of their closest neighbourhood, which was clearly indicated by the conformational parameters of the stem/bulge contact region. The hydration pattern observed at the bulge site clearly suggests that this part of RNA is specifically stabilised by water hydrogen bonding network. In addition, the 2-aminopurine residue forms C-H···O hydrogen bonds with water molecules. Dynamics of individual water molecules surrounding the bulge loop region of the RNA duplex has been analysed. The map of hydration network strongly indicates that water has to be treated as an integral part of the RNA structure.

2. Introduction

Double helical RNA can form bulge loops when the helix is interrupted by unpaired nucleotides in one strand only. RNA bulge loops are known to be important for the specificity of RNA-protein binding, intron splicing and other processes [1].

NMR studies on the RNA duplexes containing single base bulges resulted in proposing "looped out" and "intercalated in" states for those bases [2,3]. The NMR structure of the HIV-1 TAR RNA hairpin with UCU bulge loop [4] and its complex with arginine mimicking the *Tat* protein was presented [5,6]. Very recently, the crystal structure of a duplex being part of the TAR hairpin has been solved to the resolution of 1.3 Å [7]. Thermodynamical properties of U_n and A_n (n=1-3) bulge

J. Barciszewski and B.F.C. Clark (eds.), RNA Biochemistry and Biotechnology, 73–87.

loops in the RNA duplexes were studied [8] in order to evaluate their structure in solution and to give the experimental basis for RNA secondary structure prediction algorithms. In terms of the global RNA structure, bulged bases can cause bending of the duplex as detected by its anomalous mobility during electrophoresis under non-denaturing conditions [9,10].

To our knowledge, no studies on dynamic phenomena in the RNA bulge loops have been published so far. Time-resolved spectrofluorimetry studies of fluorophore-labelled RNAs aided by simulation of their molecular dynamics were undertaken in our laboratory [11,12]. We started with RNA bulge duplexes containing adenosines in the loop. Adenosines seem to prevail in the bulges of ribosomal RNAs [1]. The 2'-hydroxyl function of a bulged adenosine is the nucleophile in the cleavage process at the 5'-splice site in both mitochondrial group II introns and nuclear pre-mRNA splicing [13]. The adenosine-rich bulge is an important structural element of the P4-P5-P6 domains in the *Tetrahymena thermophila* group I intron. The structure of that bulge has been solved by the NMR method [14] and the adenosines are intercalated into the helix. On the other hand, the crystal structure of a 160-nucleotide domain [15] containing such a bulge shows the bulge nucleotides oriented towards the outside of the helix. In the same structure, interesting adenine-rich structural motifs were identified [16], described as adenosine platforms and assumed to be responsible for stabilisation of the RNA bulges. A further interesting problem was the intriguing difference in the properties of HIV-1 TAR RNA with homopyrimidine and homopurine bulge loops, as only the former are functionally important for recognition by the *Tat* protein [17].

In this paper, we would like to present the preliminary results of our *in aqua* molecular dynamics studies of one of the members of a family of RNA duplexes containing bulge loops of the $(A)_n$ type labelled with 2-aminopurine (AP) - blue fluorescent isomer of adenine:

```
          AAPA
5'-GUCG        GCUG
3'-CAGC________CGAC
```

It should be stressed that analysis of free energy increments for the bulged loops, $DG^{o}_{37(bulge)}$, allowed us to conclude that the structural properties of 2-aminopurine in RNA bulge loops are very similar to those of the isomeric adenine [11]. The fluorescent 2-aminopurine could therefore be used as a "non invasive" conformational probe. The results of molecular dynamics simulations of the duplex studied will be confronted with the spectrofluorimetry data we obtained for 2-aminopurine labelled RNA [18, cf. Chapter 5.1.]. In interpreting our molecular dynamics results, RNA hydration was a problem of special interest. The hydration pattern observed at the bulge site clearly suggests that this part of RNA is specifically stabilised by water hydrogen bonding network. The 2-aminopurine residue also forms C-H···O hydrogen bonds with water molecules. Dynamics of single water molecules surrounding the bulge loop region of the RNA duplex has also been analysed. The map of the hydration network strongly indicates that water has to be treated as an integral part of the RNA structure.

3. Laser spectrofluorimetry methods

The RNA duplex was prepared by annealing of oligoribonucleotides: GUCGAAPAGCUG and CAGCCGAC, both prepared by automated solid-supported synthesis using 2'-O-tBDMSi protection as described in [11].

Time-resolved fluorescence measurements were performed with a time-correlated single-photon-counting laser spectrometer [19]. AP was excited at wavelength of 295 nm, and the emission was measured through a set of two cut-off filters, WG 345 and UG11, creating a spectral window between 350 and 450 nm. The fluorescence intensity decay and the anisotropy decay were measured with magic and perpendicular orientations of the analysing polarisers, respectively [19].

The analysis of the experimental data in terms of the distribution of fluorescence decay and anisotropy rates was done with a numerical analysis with a non-linear parameterisation procedure [20]. The convolution of the model function with the measured excitation pulse was applied with the assumption that the fluorescence decay consisted of one or several exponential components. The final number of exponentials was selected according to the quality of the fit improvement as judged by the reduced χ^2 value. Rotational correlation times were evaluated from the ratio of the fluorescence decay measured with the orientation of the analysing polariser perpendicular to the polarisation of the exciting beam to that measured with the orientation at the magic angle I_{perp}/I_{magic} [21].

The fluorescence decay of the fluorophore changes in the presence of other residues due to the additional radiationless transitions induced by dynamic or static interactions. With the assumption that all conformational substates *i* are related to the presence of different decay rates of the emission characterised by the different decay rates k^i, the fluorescence decay is described with a formula:

$$I(t) = \sum_i a_i \exp\{-(k^i)t\}dr$$

where a_i are the probability factors (amplitudes) for the different macromolecular conformations.

When polarised light is used for the excitation, the emission also shows a high degree of polarisation. For the system with fluorophores showing some orientational freedom, dynamic averaging of the orientations takes place, resulting in the decay of the anisotropy of fluorescence *r(t)* in time from the initial value *r(0)*. The degree of restricted motion of the fluorophore may be obtained by the use of the information available from polarisation measurements. For the wobbling-in-a-cone model with the assumption that the absorption or emission transition moments are oriented along the wobbling axis of the chromophore:

$$r(t)/r_0 = (1 - S^2)\exp[-t(\phi_p^{-1} + \phi_{eff}^{-1})] + S^2 \exp(-t/\phi_p)$$

where ϕ_p, ϕ_{eff}, are rotational correlation times of the whole macromolecule and fluorophore wobbling diffusion, respectively [22,23].

4. Protocol of *in aqua* MD simulations

The RNA duplex was built using the BIOPOLYMER program (InsightII package, MSI) with canonical A-RNA structural parameters. Its initial sequence was:

```
5'-GUCGAAAGCUG
3'-CAGCUUUCGAC
```

Subsequently, three uridine residues were removed from the second strand and thus we obtained three unpaired nucleotides forming a bulge. The amino group was translocated from the C^6 to the C^2 atom in the 6th adenine of the first strand, which gave us the 2-aminopurine (AP) residue. Its atomic partial charges were calculated using the MOPAC program. Thus, the final sequence of the modified RNA duplex was:

```
          AAPA
5'-GUCG        GCUG
3'-CAGC________CGAC
```

Sodium cations were placed at the distance of 5 Å from the phosphorus atoms at the bisectors of the O-P-O angles and the entire molecular system was located in a 5.5 nm x 5 nm x 5 nm box filled with 4392 water molecules. Periodic boundary conditions (PBC) were introduced with the nonbonded interactions cutoff set to 15 Å [24]. The dielectric constant was independent of the distance.

Calculations were performed using the AMBER force field [25] implemented in the DISCOVER 2.9.7 program working on a Cray J-916 machine (Poznań Supercomputing and Networking Center). The following simulation protocol was applied (no conformational restraints were imposed on the model during the dynamics stage):

1. Initial energy minimisation:
   200 steps using the Steepest Descent algorithm and 500 steps using the Conjugate Gradients algorithm (only for the water molecules, in order to achieve their optimal alignment in the box) and subsequently:
   100 steps using the Steepest Descent algorithm and 500 steps using the Conjugate Gradients algorithm (for the entire system, with 100 kcal/mol*$Å^2$ restraints imposed on distances between the phosphorus atoms and sodium cations to keep the distance value of 5Å)
2. Molecular Dynamics
   5 ps initialisation at 300 K
   245 ps simulation at 300 K

Over the molecular dynamics run, the conformation frames were saved every 500 fs, which brought 501 frames of the trajectory in total. Conformational analysis was subsequently performed on the trajectory files obtained. The problem taken into special consideration was the alignment of the bulge towards the paired regions of the duplex and the hydration pattern of the bulge area.

5. Global structure and dynamics of RNA

5.1. SPECTROFLUORIMETRY RESULTS

5.1.1. *Fluorescence decay of AP*

The fluorescent 2-aminopurine has been introduced as a conformational probe into the bulge loop of the RNA duplex. Free 2-aminopurine riboside in aqueous solution has the fluorescence lifetime of 10 ns and has the absorption and emission bands separated from those of canonical bases; excitation maximum at 305 nm, emission maximum at 375 nm [26]. The fluorescence of 2-aminopurine is kept when incorporated into an oligodeoxynucleotide [12].

The analysis of the AP fluorescence decay usually requires a multiexponential function to describe the time dependence of fluorescence intensity, related to different micro environments sensed by the AP residue [19]. The analysis of fluorescence of AP in the RNA bulge duplex revealed the presence of a minimum of four exponential components. The fluorescence lifetimes and corresponding pre-exponential coefficients are given in Table 1.

Temp.	τ_1		τ_2		τ_3		τ_4		$<\tau>$	χ^2
[°C]	[ns]	ampl.	[ns]	ampl.	[ns]	ampl.	[ns]	ampl.	[ns]	
4	9.60	0.18	3.70	0.19	1.10	0.27	0.24	0.35	2.90	1.08
20	6.60	0.21	1.50	0.23	0.43	0.25	0.10	0.30	1.87	1.02
25	6.40	0.18	1.60	0.23	0.41	0.26	0.08	0.33	1.65	1.02
30	5.20	0.15	1.57	0.23	0.40	0.26	0.06	0.35	1.27	1.03

Table 1. Fluorescence intensity decay of the RNA bulge duplex in different temperatures

The interactions of AP with the surrounding nucleic acid regions influence its fluorescence decay, as it was already observed for AP in DNA oligomers [26] and RNA duplexes [12]. The stacking interaction is assumed to be the main cause of fluorescence quenching found for AP incorporated into nucleic acids. The longest fluorescence lifetime was assigned to the least stacked alignment of the AP residue, in agreement with the long single exponential lifetime observed for free AP base in solution, and consequently, the shorter lifetime components were correlated with better stacked 2-aminopurine. It is known that thermal unfolding of the helical structure leads to a drastic decrease of the fluorescence intensity and shortening of the lifetimes observed above the melting temperature. Our bulge duplex melts at 37.6 °C [11]. Examination of the data in Table 1. shows a temperature dependence of the decay lifetimes much earlier i.e. in the premelting range. That dependence is mostly pronounced for the shortest decay component which decreases by 75% when temperature increases from 4° C to 30° C approaching the melting temperature of the duplex. The multiexponential fluorescence decays suggest that the AP base exists in this oligomer in a range of conformational states characterised by different stacking and collisional quenching. The relative amplitudes remain almost unchanged. The quenching of the 2-aminopurine base fluorescence has apparently a more complex mechanism, where stacking, conformational dynamics of the fluorophore itself and dynamic collisional quenching contribute to a different extent.

5.1.2. *Rotational Correlation Times*

The results of the fluorescence anisotropy decay measurements for AP in the RNA duplex are summarised in Table 2. The measurements were made at different temperatures, where different stabilisation of the structure might be expected. Excitation of AP at 295 nm leads to the high initial anisotropy equal to 0.31.

Temp. [°C]	Φ_1 [ns]	r_{01}	Φ_2 [ns]	r_{02}	χ^2
4	3,65	0,23	0,65	0,07	1,02
20	1,45	0,23	0,22	0,10	0,98
25	1,27	0,21	0,21	0,10	1,04
30	0,74	0,23	0,13	0,09	1,02

Table 2. Fluorescence anisotropy decay of AP in the RNA bulge duplex in different temperatures

The anisotropy decay was found to be adequately described by two correlation times. The short component of around 650-150 ps can be ascribed to local AP motions, the longer one to the overall tumbling of the whole RNA molecule. The correlation time ascribed to the overall rotation of the whole molecule is related to its volume and shape, and global conformational transitions should be reflected in the change of the correlation time value. The observed temperature dependence of that component is due mainly to the change of viscosity of the solvent with temperature.

The short correlation times reflect the internal motions of the AP residue in respect to the whole molecule, and may also involve only its nearest neighbouring adenine residues. The temperature dependence of the short correlation time suggests that the fluorescent residue does not show completely free rotations within the restricted site. The extent of mobility reflected by the initial partial anisotropy values r_{02} increases by 30% with the temperature change from 4° C to 30° C, which indicates that AP exhibits a relatively high local mobility within the bulge.

Both the fluorescence decays and anisotropy decays of the AP in the duplex studied point to the conclusion that the degree of mobility of the AP residue in this bulge is relatively high, suggesting that in this position the base is involved in transient, relatively weak stacking or other interactions with the other parts of the structure which exhibit local flexibility. This strongly suggests that the 2-aminopurine riboside residue, most probably together with both adjacent adenosines, has a tendency to be in the "looped out" conformation.

5.2. *IN AQUA* MOLECULAR DYNAMICS SIMULATION RESULTS

An important problem in the analysis of the simulation results was the estimation of preservation of stacking interactions between neighbour nucleobases in the strand. We have proposed a parameter calculated in the following way to monitor that (cf. Fig. 1.):

$$\mathbf{S = D_1 + D_2 + D_3 + 2\,(|D_1 - D_2| + |D_2 - D_3| + |D_1 - D_3|) - 15.3}$$

The parameter values close to zero indicate that the bases are aligned in accordance with the canonical A-RNA conformation. Values higher than 3-5 Å are observed when nucleobases lose their parallel alignment or stand aside which results in the loss of the stacking interactions in the respective region of the strand.

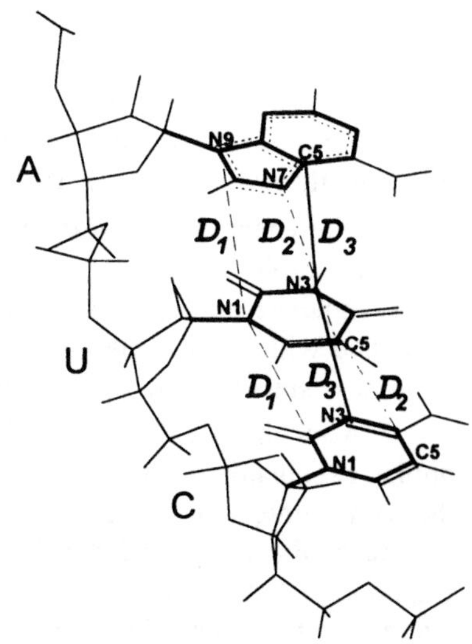

Fig. 1. Definition of the base stacking parameter in RNA strand

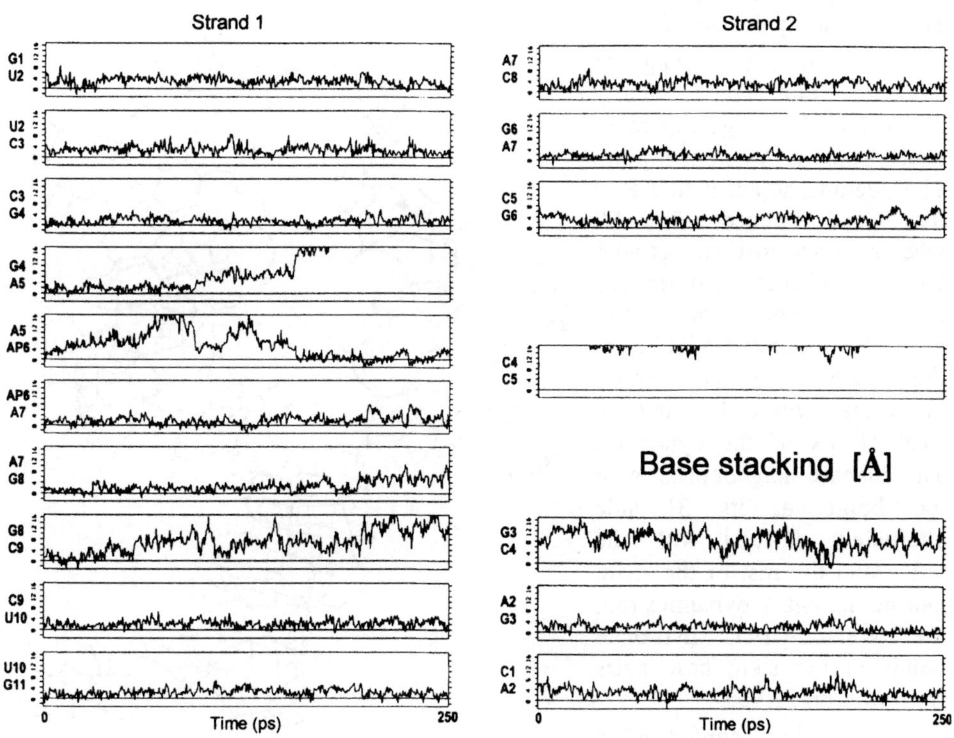

Fig. 2. Base stacking parameter over the molecular dynamics run

The graphs on Fig. 2. present the changes of the parameter of stacking between the neighbouring bases within both strands over the 250 ps molecular dynamics run. One can notice the stability of stacking in the paired (stem) regions of the duplex, as well as within the bulge. In contrast to them, during the dynamics the stem-bulge junctions were very unstable in respect to stacking. The bases from those latter regions showed a tendency to stack alternately with either of their neighbours. That confirms the dynamic character of the bulged region.

The energy minimisation performed before starting the dynamics run resulted in considerable bending of the major duplex axis, which is caused by the presence of three unpaired nucleosides in the middle of the molecule. Still, the unpaired bases remained within the duplex which gave their "bulged in" orientation. There was also good base stacking within the entire molecule.

After the initial 100 ps of dynamics, the structure maintained the tendencies to duplex axis bending and to losing the stacking of AP with its 5'-neighboring adenine. In the same time, one could notice losing the stacking between the stem region at the 3'-side of the bulge and the 3'-guanine adjacent to the bulge. Up to the end of the simulation, that guanine would alternately stack with the adjacent adenine from the bulge (the prevalent situation) or return to the stem region. After the next 50 ps of molecular dynamics, no considerable global conformational changes could be noticed which suggested the achievement of a local optimum. The bending of the major duplex axis has slightly decreased.

At the 200 ps stage of dynamics one could clearly notice that the bulge separated itself from its 5' stem region. The adenine adjacent to AP at its 5'-end stacked with AP whereas it had lost contact with the stem. Also at the other side of the bulge, the latter gradually separated itself from the paired region. Those processes continued during the final 50 ps of the dynamics. The guanine neighbouring with the bulge at its 3' side alternately stacked with the bulge and the rest of the stem. During the entire dynamics run, the Watson-Crick hydrogen bonds in the stem base pairs were well preserved.

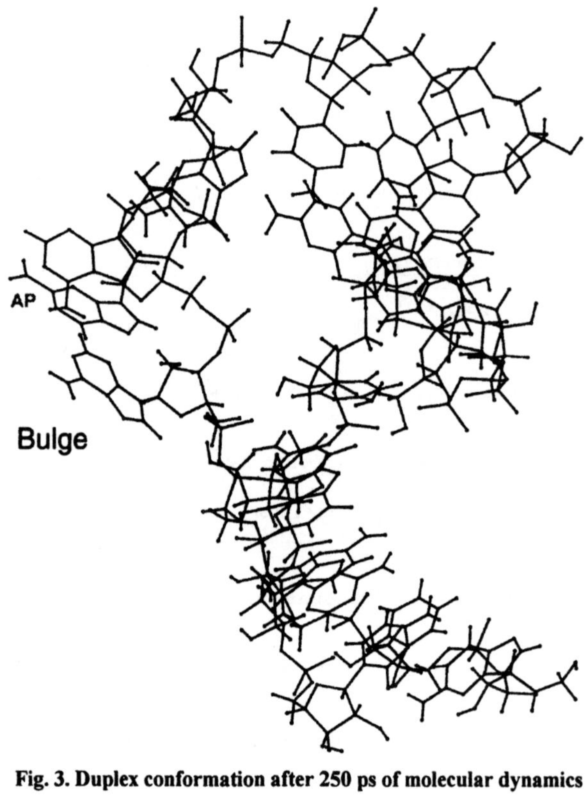

Fig. 3. Duplex conformation after 250 ps of molecular dynamics

The last conformation frame of the trajectory is presented on Fig. 3. (counterions and water molecules not shown). The course of the molecular dynamics simulation experiment in aqueous solution suggests a tendency of the bulged region to "bulge out" towards the solvent, along with destabilisation of adjacent fragments of the stem regions.

The sugar rings conformation of the ribonucleoside residues has also been analysed. Sugar puckering was measured by the pseudorotation phase angle [27]. For the majority of nucleosides, the ribose conformation was stable within the C3'-endo range which is typical for A-RNA. Only the terminal residues showed a considerable instability. During longer periods of the dynamics run, their sugar puckering was often converted to the C2'-endo state which is normally found in B-DNA. Some degree of conformational instability could also be identified in the sugar conformation of the nucleoside residues from the bulge region, which can result from its relatively fast conformational changes. Those results are in good correlation with the stacking measurements and prove the stability of the stem regions in contrast to the dynamic behaviour of the bulge, particularly evident in the stem-bulge transition sites. The graphs on Fig. 4. depict the changes of the pseudorotation angle in all the duplex ribonucleoside residues during the molecular dynamics experiment.

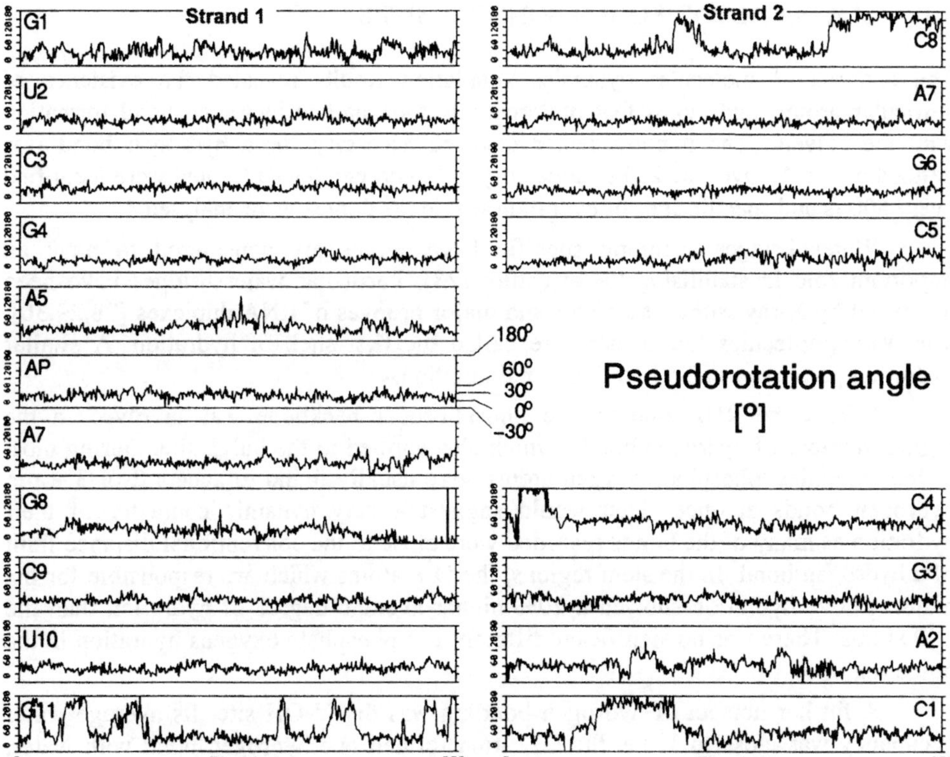

Fig. 4. Sugar puckering parameter over the molecular dynamics run

5.3. SPECTROFLUORIMETRY *VERSUS* MD SIMULATION

The degree of stacking and the relatively high local mobility of 2-aminopurine as the fluorescent probe was directly manifested by the characteristics of fluorescence and anisotropy decay (Table 1. and 2.). These data strongly suggest that the 2-aminopurine riboside residue, most probably together with both adjacent adenosines, has a tendency to be in a "looped out" conformation. Also, our preliminary *in aqua* MD simulation experiment suggests a relatively high conformational variability in the region of the bulge, as observed by the fluctuations of stacking within the bulge and between the neighbouring bases in both strands of the duplex. The stacking between G4/A5 and A5/AP6 changes in the time scale of tens to few hundreds of picoseconds as observed in the 250 ps run. At this stage of our study, both approaches: spectrofluorimetry and MD simulation in aqueous solution have led to a coherent "looped out" picture of the RNA bulge duplex conformation.

6. The RNA hydration pattern

6.1. CLASSICAL HYDROGEN BONDING SITES

The analysis of molecular dynamics simulation results revealed the existence of particular groups and atoms that preferentially took part in hydrogen bond formation with the solvent. The donor-acceptor distance necessary for a hydrogen bond was defined as 2.5 Å. Over the 250 ps trajectory, all hydrogen bonds formed were identified every 500 fs and specific sites of hydration within the duplex were analysed.

Water bridges spanning specific RNA atoms are considered to have an important role in stabilising the structure [28]. Particular water bridges have been identified by X-ray within the minor and major grooves of RNA duplexes [28,29,30]. The water molecules found there are called the first shell of hydration. A similar hydration pattern was also found in our simulation.

Within the RNA duplex, the phosphosugar backbone was involved in the highest number of hydrogen bonds, which also applied to the bulge site. During most of the dynamics, phosphate oxygen atoms were usually found to accept 2 or 3 water hydrogen bonds at once. That would suggest a very dynamic character of their hydration as many of the bonds recorded were close to the conventional distance limit of a hydrogen bond. In the stem regions, the O^1P atoms which are responsible for the major groove hydration showed a 20% higher average degree of hydration than the O^2P atoms. There was no significant difference in phosphate oxygens hydration in the bulge region.

A further hotspot of hydrogen bonding was the 2'-OH site. Its average water occupancy was close to 1, i.e. the O2' atom formed one hydrogen bond with water, which is in a good accordance with the minor groove hydration model [28,29].

The following atoms of the nucleobases were also involved in a high number of hydrogen bonds (water occupancy ranged between 0.5 and 1):

Base	Atoms	
A	N^3 N^7, N^6H_2 N^1	minor groove hydration major groove hydration (in bulge adenines only)
C	O^2 N^4H_2	minor groove hydration major groove hydration
G	N^2H_2, N^3 O^6, N^7	minor groove hydration major groove hydration
U	O^2 O^4	minor groove hydration major groove hydration
AP	N^2H_2, N^3, N^7	bulge

The hydration of the bulge nucleotide residues was in good correlation with its conformation: the number of hydrogen bonds identified was considerably higher during the later stages of the simulation when the bulge adopted the "bulged out" conformation which resulted in a better contact with the solvent. N^1 atoms of the bulge adenines which are normally not accessible to the water molecules also participated in the hydration of the bulge region. The Fig. 5. shows the surrounding of the 2-aminopurine residue and the AP-solvent hydrogen bonds present in example frames from the earliest (left) and last (right) stage of dynamics.

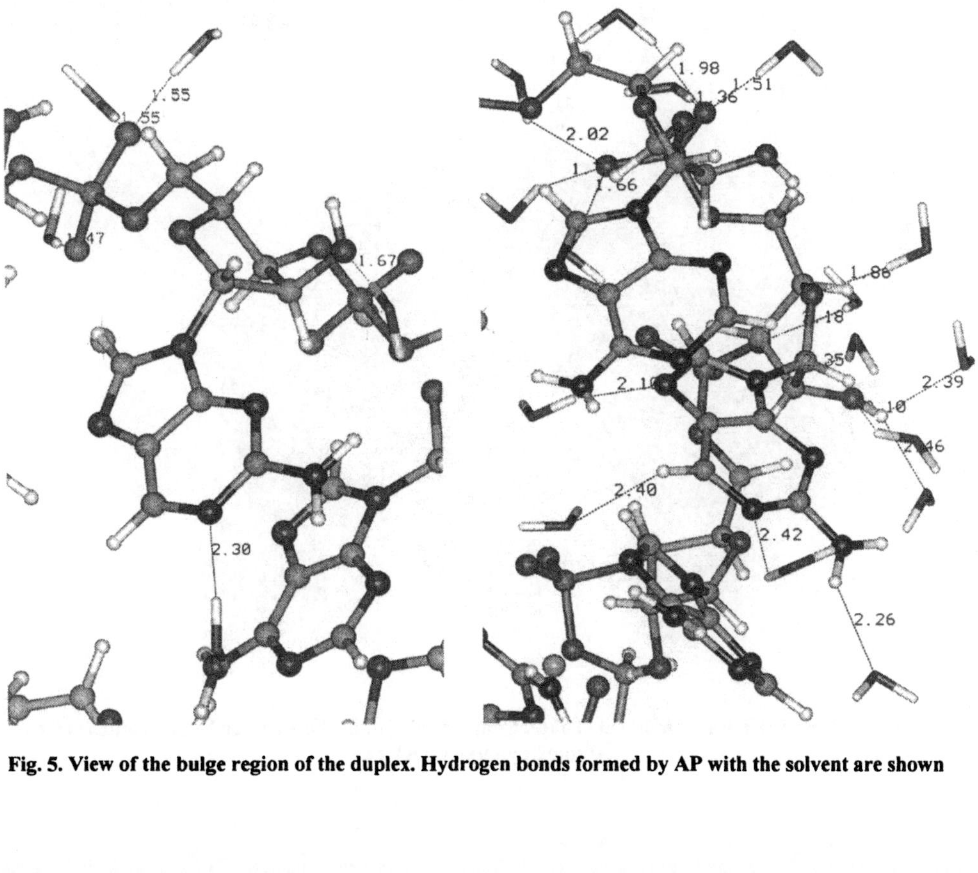

Fig. 5. View of the bulge region of the duplex. Hydrogen bonds formed by AP with the solvent are shown

6.2. C-H···O HYDROGEN BONDING OF 2-AMINOPURINE

The role of the C-H···O hydrogen bonds in tRNA was recently discussed [31]. We tried to evaluate their importance in the hydration network of our duplex. During about 10% of the MD run, there was a hydrogen bond formed by the 2-aminopurine C^6-H with water (also present on Fig. 5, to be identified on the right image with value 2.40 Å). Nevertheless, the parameters of this interaction, as well as those of the other possible C-H···O hydrogen bonds formed by the duplex with water were close to the conventional limit of the H-bond and the occupancies of the sites involved in those interactions were lower than 0.2. This would suggest a relatively minor though traceable role of the C-H···O hydrogen bonds in the hydration of the bulge RNA duplex in our experiment.

6.3. SINGLE WATER MOLECULES DYNAMICS

Many water molecules were identified to remain close to a particular group or atom of the RNA bulge duplex. They were involved in forming hydrogen bonds which existed during several picoseconds (up to 20 ps). A simplified sketch of the route of a water molecule in the neighbourhood of the duplex during the molecular dynamics experiment is presented on Fig. 6.

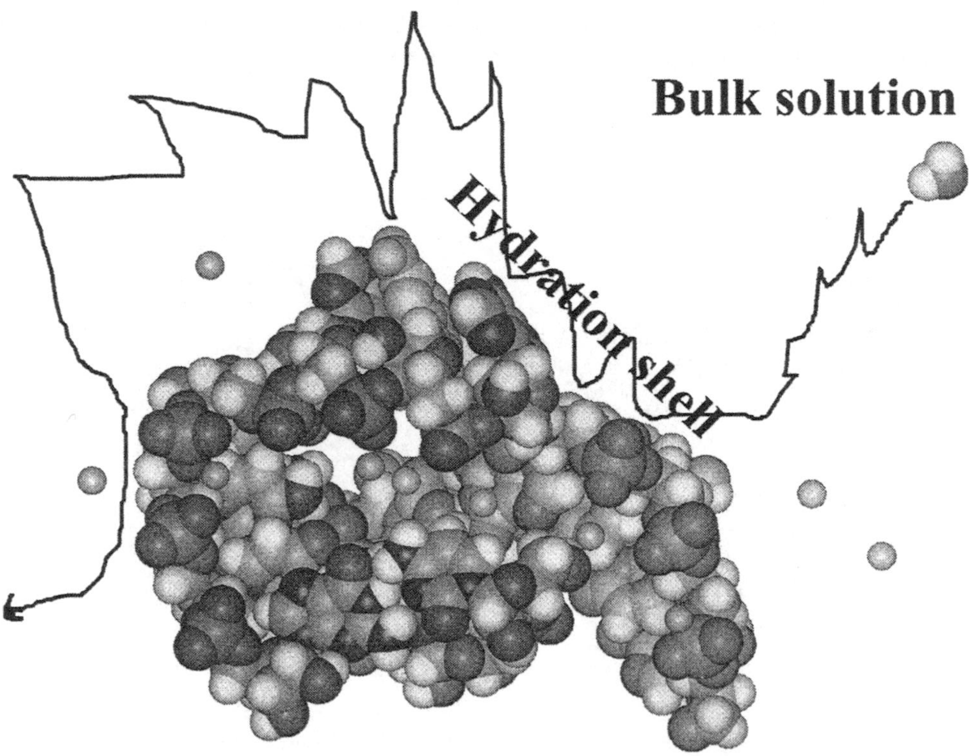

Fig. 6. Scheme of a water molecule route. Counterions out of the Van der Waals radius scale. Duplex dynamics neglected

Longer residencies of individual water molecules were particularly evident in the case of the phosphate groups interacting with the solvent. Another interesting phenomenon was the observed "sliding" along the duplex recorded for the individual water molecules which formed hydrogen bonds to subsequent duplex atoms on their route. This was observed for example for the O5'-, O3'- and phosphate oxygen atoms. Also the hydrogen bonds formed by the solvent molecules with the ribose ring O2' atoms and "bulged out" 2-aminopurine nitrogen atoms were often maintained during longer periods (up to 10 ps). Unlike those, the interactions of other atoms were usually recorded only for short periods and there was a frequent exchange of the water molecules forming the H-bonds with them. Water bridges stabilising the structure by linking two duplex atoms with hydrogen bonding were frequently found. Two graphs on Fig. 7. present the individual history of interactions of two selected water molecules during the molecular dynamics experiment.

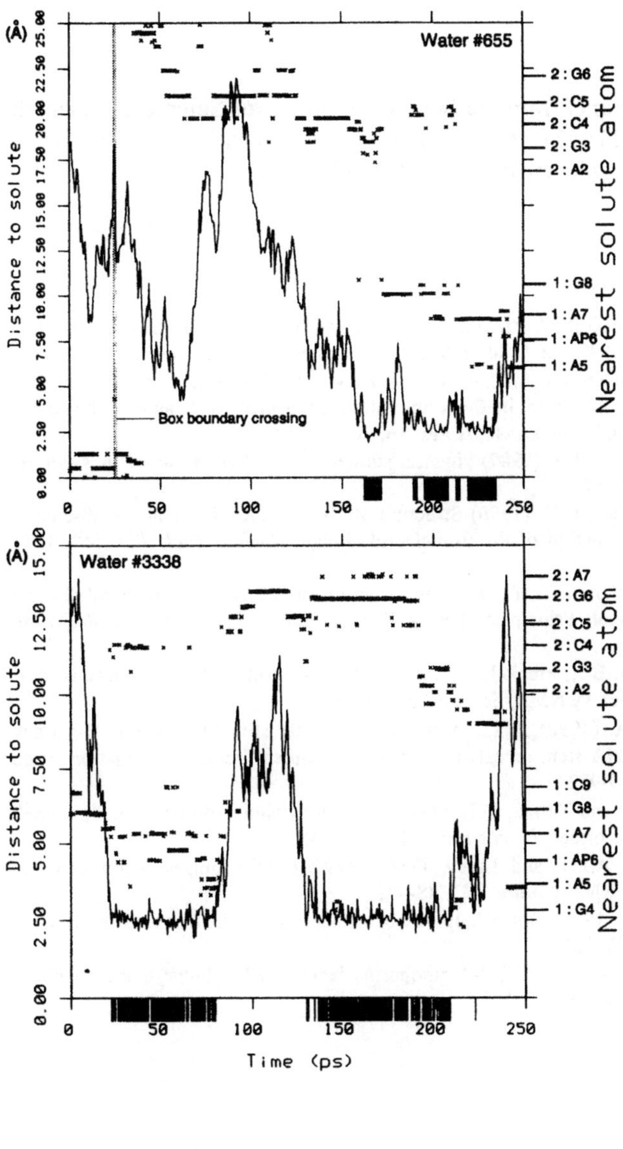

Fig. 7. Water molecules-RNA contacts during the simulation experiment. The main graph line represents the distance between the geometrical centre of the water molecule and the nearest RNA atom at a given moment. Small crosses point to the nearest atom of the duplex which is represented along the right side of each graph (only residue description provided). Periods of hydrogen bond formation marked black along the time axis.

Both examples presented suggest that the molecules move far faster within the bulk solvent than in the hydration shell where they take part in hydrogen bonding with the duplex. Individual water molecules often remain at a particular site of the RNA duplex, usually slowly changing its position in respect to the RNA residues.

Water molecules residing close to the duplex showed a tendency to form more than one hydrogen bond at once, thus taking part in the network of non-covalent interactions that stabilises the conformation of the RNA duplex. The evident importance of considering the non-covalent interactions within the model of an irregular RNA structure in solution prompted us to prepare employing other methods of calculating them, especially the Particle Mesh Ewald (PME) method, to confront the results presented here with alternative approaches.

7. Acknowledgements

This work was supported by grants from the State Committee for Scientific Research, Republic of Poland (3T09A11412). The access to the resources of the Poznań Supercomputer and Networking Center is acknowledged.

8. References

1. Wyatt, J.R., Tinoco Jr., I. (1993) RNA structural elements and RNA function, in R.F. Gesteland and J.F. Atkins (eds.), *The RNA World*, Cold Spring Harbor Laboratory Press, p.465.
2. Van den Hoogen, Y.T., van Beuzekon, A.A., de Vroom, E., van der Marel, G.A., van Boom, J.H., Altona, C. (1988) Bulge-out structures in the single stranded trimer AUA and in the duplex (CUGGUGCGG):(CCGCCCAG), *Nucleic Acids Res.* **16**, 5013-5030.
3. Zhang, P., Popieniek, P., Moore, P.B. (1989) Physical studies of 5S RNA variants at position 66, *Nucleic Acids Res.* **17**, 8645-8657.
4. Aboul-ela, F., Karn, J. and Varani, G. (1996) Structure of HIV -1 TAR RNA in the absence of ligands reveals a novel conformation of the trinucleotide bulge, *Nucleic Acids Res.*, **24**, 3974-3981.
5. Aboul-ela, F., Karn, J. and Varani, G. (1995) The Structure of the Human Immunodeficiency Virus Type-1 TAR RNA Reveals Principles of RNA Recognition by Tat Protein, *J. Mol. Biol.*, **253**, 313-332.
6. Puglisi, J.D., Tan, R., Calnan, B.J., Frankel, A.D. and Williamson, J.R. (1992) Conformation of the TAR RNA-arginine complex by NMR, *Science* **257**, 76-80.
7. Ippolito, J.A. and Steitz, T.A. (1998) A 1.3 Å resolution crystal structure f the HIV-1 trans-activation response region. RNA stem reveals a metal-ion dependent bulge conformation, *Proc. Natl. Acad. Sci.* USA, **95**, 9819-9824.
8. Longfellow, C.E., Kierzek, R. and Turner, D.H. (1990) Thermodynamic and spectroscopic study of bulge loops in oligoribonucleotides, *Biochemistry*, **29**, 278-285.
9. Bhattacharyya, A., Murchie, A.I.H. and Lilley, D.M.J. (1990) RNA bulges and the helical periodicity of double stranded RNA, *Nature*, **343**, 484-487.
10. Tang, C.K. and Draper, D.E. (1990) Bulge loops used to measure the helical twist of RNA in solution, *Biochemistry* **29**, 5232-5237.
11. Zagórowska, I. and Adamiak, R.W. (1996) 2-Aminopurine labelled RNA bulge loops. Synthesis and thermodynamics, *Biochimie*, **78**, 123-130.

12. Kuliński, T., Bielecki, Ł., Zagórowska, I. and Adamiak, R.W. (Special Issue 1996) Introductory data on dynamics of RNA bulge duplexes. 2-Aminopurine labelled adenosine loops, *Collect. Czech. Chem. Commun.*, **61**, 265-267.
13. Sharp, P.A. (1987) Splicing of messenger RNA precursors, *Science* **235**, 766-771.
14. Luebke, K. J., Landry, S.M. and Tinoco Jr, I. (1995) Solution Conformation of a Five-Nucleotide RNA Bulge Loop from a Group I Intron, *Biochemistry*, **36**, 10246-10255.
15. Cate, J.H., Gooding, A.R., Podell, E., Zhou, K., Golden, B.L., Kundrot, C.E., Cech, T.R. and Doudna, J.A. (1996) Crystal structure of a group I ribozyme domain: principles of RNA packing, *Science*, **273**, 1678-1685.
16. Cate, J.H., Gooding, A.R., Podell, E., Zhou, K., Golden, B.L., Szewczak, A.A., Kundrot, C.E., Cech, T.R. and Doudna, J.A. (1996) RNA tertiary structure mediation by adenosine platforms, *Science*, **273**, 1696-1699.
17. Frankel, A.D. (1992) Activation of HIV transcription by Tat, *Curr. Opin. Genet. Dev.*, **2**, 293-298.
18. Kuliński, T., Bielecki, Ł., Zagórowska, I. and Adamiak, R.W. Dynamics of RNA bulge duplexes: 2-aminopurine labelled adenosine loops, in preparation to *Biophysical J.*
19. Rigler, R., Claesens, F. and Lomakka, G. (1984) Picosecond Single photon Fluorescence Spectroscopy of Nucleic Acids, in D.H. Auston and K.B. Eisenthal (eds.), *Ultrafast Phenomena IV*, Springer Verlag, Berlin-Heidelberg, pp. 472-467.
20. Marquardt, D.W. (1963) An algorithm for least-squares estimation of nonlinear parameters, *J. Soc. Indust. Appl. Math.*, **11**, 431-441.
21. Ehrenberg, M. and Rigler, R. (1976) Fluorescence correlation spectroscopy applied to rotational diffusion of macromolecules, *Q. Rev. Biophys.*, **9**, 69-81.
22. Lipari, G., and Szabo, A. (1980) Effect of librational motion on fluorescence depolarization and nuclear magnetic resonance relaxation in macromolecules and membranes, *Biophys. J.*, **30**, 489-506.
23. Kinosita Jr, K., Kawato, S. and Ikegami, A. (1977) A theory of fluorescence polarization decay in membranes, *Biophys. J.*, **20**, 289-305.
24. Auffinger, P., Louise-May, S. and Westhof, E. (1996) Molecular Dynamics Simulations of the Anticodon Hairpin Of $tRNA^{Asp}$ - Structuring Effects of C-H···O Hydrogen Bonds and of Long-Range Hydration Forces, *J. Am. Chem. Soc.*, **118**, 1181-1189.
25. Weiner, S.J., Kollman, P.A., Nguyen, D.T. and Case, D.A. (1986) An All Atom Force Field for Simulations of Proteins and Nucleic Acids, *J. Comp. Chem.*, **7**, 230-252.
26. Nordlund, T.M., Andersson, S., Nilsson, L., Rigler, R., Graslund, A. and McLaughlin, L.W. (1989) Structure and dynamics of a fluorescent DNA oligomer containing the EcoRI recognition sequence: fluorescence, molecular dynamics, and NMR studies, *Biochemistry*, **28(23)**, 9095-10003.
27. De Leeuw, H.P.M., Haasnoot, C.A.G. and Altona, C. (1980) Empirical correlations between conformational parameters in β-D-furanoside fragments derived from a statitical survey of crystal structures of nucleic acid constituents. Full description of nucleoside molecular geometries in terms of four parameters, *Isr. J. Chem.*, **20**, 108-126.
28. Westhof, E. (1993) Structural Water Bridges in Nucleic Acids, in E. Westhof (ed.) *Water & Biological Macromolecules*, The Macmillan Press Ltd, pp. 226-243.
29. Egli, M., Portmann, S. and Usman, N. (1996) RNA Hydration: A Detailed Look, *Biochemistry*, **35**, 8489-8494.
30. Adamiak, D.A., Milecki, J., Popenda, M., Adamiak, R.W., Dauter, Z. and Rypniewski, W. (1997) Crystal Structure of 2'-O-Me$(CGCGCG)_2$, an RNA duplex at 1.30 Å resolution. Hydration pattern of 2'-O-methylated RNA, *Nucleic Acids Res.*, **25**, 4599-4607.
31. Auffinger, P., Louise-May, S. and Westhof, E (1996) Hydration of C-H groups in tRNA, *Faraday Discuss.*, **103**, 151-173.

THE STRUCTURE AND FUNCTION OF THE RIBOZYME RNASE P RNA IS DICTATED BY MAGNESIUM(II) IONS

LEIF A. KIRSEBOM
Department of Microbiology
BMC, Box 581
Uppsala University
S-751 23 Uppsala
Sweden

1. Introduction

The genes encoding tRNA are transcribed as precursors and these precursors have to be processed to generate functional tRNA molecules. The ubiquitous endoribonuclease RNase P is responsible for generating tRNA molecules with matured 5' termini and almost all matured tRNA molecules carry a 5' monophosphate as a result of RNase P cleavage (see for example 1 and references therein). There are, however, examples of tRNA genes among the Archaea and the Eukarya where the first nucleotide in the coding region coincides with the start of transcription (2, 3). In Bacteria, the catalytic activity of RNase P is associated with its RNA subunit, RNase P RNA (4). The RNA alone is able to correctly cleave a large number of different tRNA precursors and other substrates such as RNA stem-loops *in vitro* (Fig 1). Thus, RNase P RNA is a ribozyme. In fact, it is the only naturally isolated transacting ribozyme identified so far. It is generally believed that the catalytic activity of RNase P derived from other organisms than bacteria is also conferred by the RNA, however, this still remains to be demonstrated. The protein subunit of RNase P is essential for activity *in vivo* and reduces its dependence on divalent metal ions for cleavage *in vitro* (see below). Moreover, *in vitro* experiments suggest that its function is to: i) facilitate release of the products, ii) reduce rebinding of matured tRNA to the enzyme, iii) stabilize the conformation of the RNA subunit and iv) facilitate precursor tRNA recognition. In addition, it has been observed that the addition of the protein influences cleavage site recognition on certain tRNA precursors (5-11). In contrast to the bacterial RNase P system, where there is a 1:1 RNA/protein ratio, several protein subunits have been identified in eukaryotic RNase P (12, 13). Li and Williams (14) also concluded that higher vertebrates express multiple

J. Barciszewski and B.F.C. Clark (eds.), RNA Biochemistry and Biotechnology, 89–109.

isoforms of RNase P RNA. These findings clearly suggest that eukaryotic RNase P is considerably more complex than its bacterial counterpart.

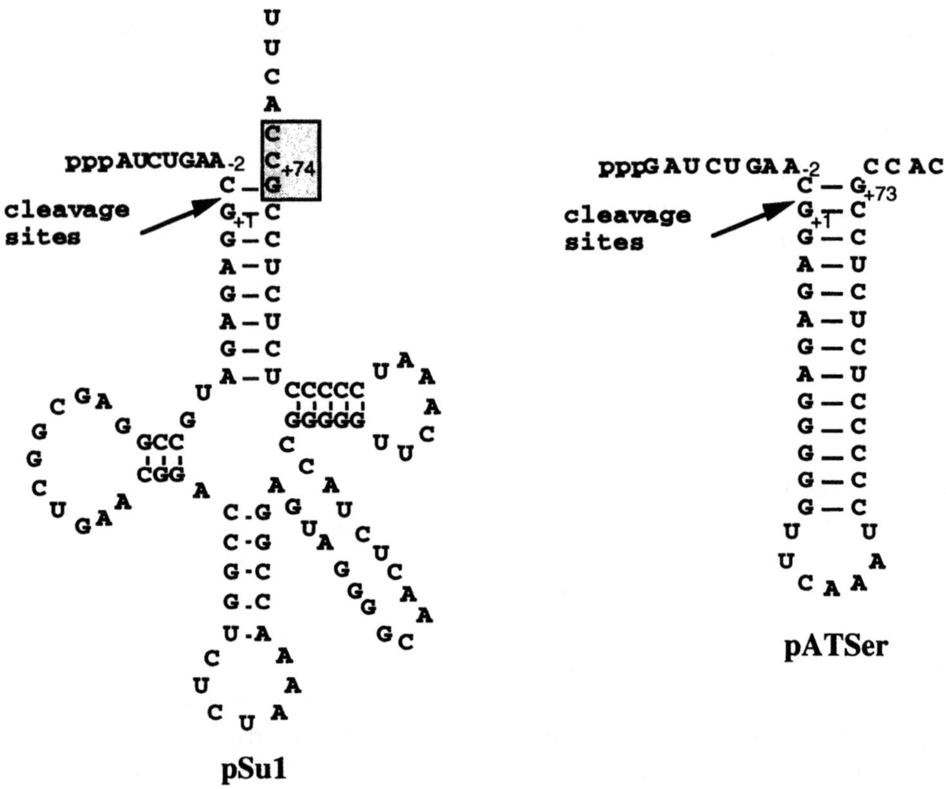

Figure 1 The secondary structures of two examples of RNase P substrates, pSu1, a $tRNA^{Ser}$ precursor (67), and pATSer a stem-loop substrate (71). The RNase P cleavage sites, between positions -1 and +1, are indicated by arrows and the residues which are involved in the "RCCA-RNase P RNA" interaction are boxed (see Fig 5a). The "G_{+73}CC" motif at the 3' end of pATSer represents the corresponding residues that base pairs with the GGU-motif in RNase P RNA and the numbering follows the numbering of a full-size $tRNA^{Ser}$ precursor since the structure of pATSer is based on the acceptor-stem, T-stem and T-loop of pSu1. Thus, pATSer lacks the variable loop, the D-loop, the D-stem, the anticodon stem and the anticodon loop (see Fig 2).

The reaction catalyzed by RNase P requires divalent metal ions (*e.g.* Mg^{2+}, Mn^{2+}, Ca^{2+}), and among these, Mg^{2+} promotes cleavage most efficiently. Comparison of cleavage in the absence and in the presence of the protein subunit reveals that in the

RNA alone reaction a higher concentration of Mg^{2+} is required, typically 100 mM. However, RNase P RNA also cleaves its substrate at lower Mg^{2+}-concentrations (≥10 mM), but at these Mg^{2+}-concentrations a higher concentration of monovalent cation or the presence of spermidine is needed (15-17). In general, Mg^{2+} ions stabilize the conformation of RNA by binding to the highly negatively charged phosphate backbone (see for example 18). Since cleavage of RNA by RNA requires intimate contact between two highly negatively charged polymers the presence of divalent metal ions would clearly influence this process. Furthermore, the chemical step of RNase P cleavage is dependent on divalent metal ions and in fact it has been suggested that RNase P can be considered as a metalloenzyme (19). In this review I will discuss the current status of what is known about the function of Mg^{2+} ions in the RNase P RNA catalyzed reaction. I will focus on bacterial RNase P RNA, in particular M1 RNA, which is the RNase P RNA derived from *Escherichia coli*, and is the most extensively studied RNase P RNA. When applicable I will refer to other systems.

2. A short briefing of Mg^{2+}

For a detailed summary of the chemistry of Mg^{2+} and its biological role I refer to Pan *et al.* (18) and Cowan (20). Here I just want to mention a few characteristics which are relevant to the discussion followed below (see also Table 1). The concentration of magnesium inside cells is in general ≈10^{-3} M (21). The radius for Mg^{2+} is 0.6 Å and due to its high charge density it binds to high negative charge density centres, such as the phosphate backbone of RNA. The preferred co-ordination number for Mg^{2+} is 6 and the bond distance Mg-O in various salts is 2.05±0.05 Å. The ionic radius for two other physiologically important divalent metal ions, Ca^{2+} and Mn^{2+}, are larger, 0.99 Å and 0.78 Å, respectively. The preferred co-ordination numbers for Ca^{2+} is 6 and 8, while it is 6 for Mn^{2+} (21). Furthermore, the exchange rate of H_2O from around Mg^{2+} is 10^5 s^{-1}, which is ≈1000 times slower compared to the exchange rate from around Ca^{2+}. The pK_a for Mg^{2+} is 11.4 which should be compared to Ca^{2+} and Mn^{2+} where the pK_a values are 12.9 and 10.6, respectively. Studies of RNA cleavage using the hammerhead ribozyme suggest a correlation between the pH-rate profile and the pK_a values for various divalent metal ion indicating that a metal hydroxide function as the base in cleavage by this ribozyme (22). There are also important differences of the nature of the ligands that different divalent metal ions use. A Mg^{2+} ion, a hard metal ion, binds preferentially to hard ligands such as phosphates, H_2O, NH_3 and OH^- whereas a thiol is a poor ligand for Mg^{2+}. By contrast a soft metal ion, for example Mn^{2+}, can use thiols and phosphorothioates as ligands.

How are divalent metal ions co-ordinated at specific sites in an RNA molecule? From the crystal structure of yeast $tRNA^{Phe}$ it is evident that Mg^{2+} mainly uses nonbridging oxygens as direct ligands while co-ordination to chemical groups on bases are mediated through bridging H_2O molecules (23). Such indirect binding of Mg^{2+} to bases, in particular guanosine residues, have also been observed in the crystal structure of a 5S rRNA domain. Here the O6 and N7 positions are hydrogen bonded to inner sphere Mg^{2+} co-ordinated H_2O molecules (24). However, O6 and N7 of guanosines can also be used as direct ligands as observed in the crystal structure of the P4/P6 domain of the Group I intron (25; see also 26). Manganese(II) ions prefer different ligands for co-ordination compared to Mg^{2+} (see above). This is also evident from the crystal structure of yeast $tRNA^{Phe}$ where the major Mn^{2+} binding site is ≈2Å away from the strong Mg^{2+} binding site in the T-loop/D-loop domain of the tRNA (Fig 2). Here Mn^{2+} use for example the N7 position of a guanosine as a direct ligand (27). Binding of Mn^{2+} displaces Mg^{2+} and the different way of binding of Mn^{2+} as compared to Mg^{2+} is attributed to the difference in the chemical nature of these two ions. The differences between these two ions have to be considered when interpretating data where Mn^{2+} and Mg^{2+} ions (or any other divalent metal ion) have been used to investigate the function of RNase P RNA.

Mg^{2+} promotes RNase P RNA cleavage most efficiently, but cleavage also occurs in the presence of either Mn^{2+} or Ca^{2+}. Ca^{2+} promotes efficient binding of tRNA, or

TABLE 1. Summary of a few characteristics of some divalent metal ions discussed in this paper.

	Divalent metal ion		
	Mg^{2+}	Mn^{2+}	Ca^{2+}
Concentration in cells	$\approx 10^{-3}$ (M)	$\approx 10^{-8}$ (M)	$\approx 10^{-7}$ (M)
Ionic radius	0.6 Å	0.78 Å	0.99 Å
Preferred Co-ordination number	6	6	6, 8
pK_a value	11.4	10.6	12.9
Exchange rate of H_2O	10^5 (s^{-1})	10^7 (s^{-1})	10^8 (s^{-1})
Promotes cleavage by RNase P RNA	++	++	10^4 reduction in k_{chem}#
Cleavage site recognition		induce miscleavage#	induce miscleavage#

#This is compared to cleavage in the presence of to Mg^{2+}.

tRNA precursor, to RNase P RNA. Together with cross-linking studies as well as studies of the interaction between RNase P RNA and the protein subunit suggest that RNase P RNA is properly folded in the presence of Ca^{2+}. However, the rate of cleavage in the presence of Ca^{2+} is reduced several-fold (Table 1). It should also be noted that addition of Ca^{2+} influences cleavage site recognition (10, 19, 29-36). Mn^{2+} on the other hand is almost as efficient in promoting cleavage as Mg^{2+} (28, 37, 38, Kirsebom and Brännvall, unpublished data). Moreover, Mn^{2+} influences cleavage site recognition on some substrates (Kirsebom and Brännvall, unpublished data). Taken together, the RNase P RNA-catalysed reaction is suggested to require Mg^{2+} not only to promote efficient cleavage rates, but also for cleavage at the correct position (33).

3. RNase P substrate interaction

The current model on how RNase P and its RNA moiety recognize its substrate has been reviewed elsewhere (1, 39) and will only be discussed briefly here. i) RNase P recognizes the overall three dimensional structure of the tRNA domain of a tRNA

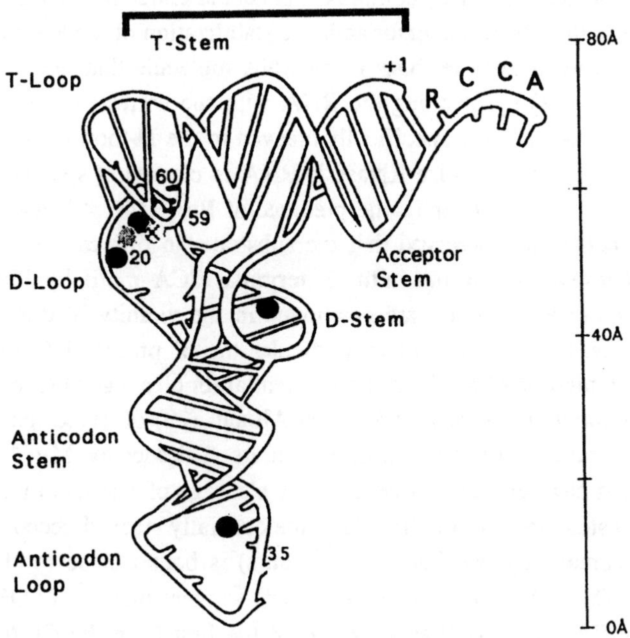

Figure 2 A schematic illustration of the three dimensional structure of yeast $tRNA^{Phe}$ after Brown *et al.* (62 and references therein). Black circles represent the strong Mg^{2+} binding sites, grey circle represent a Mn^{2+} binding site and the squared circle represent binding of a Pb^{2+} ion. The bar indicate the "RNase P ruler" that measures the distance from the T-loop to the RNase P cleavage site.

precursor; ii) the RNA subunit interacts with the T-loop/stem domain of the tRNA; iii) the 3' terminal RCCA sequence (interacting residues underlined) of the substrate, base pairs with a conserved GGU-motif in RNase P RNA (Fig 5a); iv) recent data suggest that nucleotides at specific positions in the 5' leader indeed play a role in cleavage site recognition and subsequent catalysis (40-44); v) cleavage occurs 5' of the residue at the +1 position (Fig 1), which in >80% of the cases is a guanosine. In tRNA precursors the distance between the T-loop and the cleavage site corresponds to 12 base pairs and this distance can be considered as a "ruler" for RNase P cleavage (Fig 2).

4. Mg^{2+} binding to the substrate

The crystal structure of yeast $tRNA^{Phe}$ reveals several Mg^{2+} binding sites (23, 27) and four specific sites are depicted in Figure 2. [RNA in general binds ≈1 Mg^{2+} ion/4 residues (45).] Site 3 is located in the T-/D-loop domain therefore it might be possible that a Mg^{2+} bound at this site influence the interaction with RNase P RNA since this region of a tRNA precursor has been suggested to be in close contact with the ribozyme in the ribozyme-substrate (RS) complex. This could either be through a stabilization of the structure of the tRNA precursor and/or a stabilization of the RS-complex.

It is possible to identify residues in an RNA molecule that are in the vicinity of a divalent metal ion(s) by cleaving the RNA with the corresponding divalent metal ion. For example, yeast tRNA is specifically cleaved in the D-loop in the presence of Pb^{2+} (see for example 18). Likewise RNase P RNA is cleaved at specific positions both in the presence of Mg^{2+} (29) or in the presence of Pb^{2+} (46, 47; and below). Hartmann and coworkers (48) also observed that the substrate, in this case a $tRNA^{Gly}$ precursor, was cleaved at several positions in the 3' terminal CCA motif by Pb^{2+}. This indicates that one or more Pb^{2+} ion(s) are positioned in the vicinity of this sequence element. Recently, it was suggested that in particular the 3' proximal C of the CCA motif participates in binding of Mg^{2+} that are essential for cleavage (49; see also below).

Prior to the latter findings, Kazakov and Altman (29) observed that the $tRNA^{Tyr}$Su3 precursor is cleaved 5' of the -2 position in its 5' leader by Mg^{2+} alone. Moreover, Mg(II)-induced cleavage and RNase P RNA cleavage of variants of a model substrate, comprising a stem and a loop mimicking the coaxially stacked acceptor- and T-stems of a tRNA precursor, suggest that a Mg^{2+} ion(s) is bound close to the RNase P RNA cleavage site (50). More specifically, their data suggest that the 2' OH's at positions -2 and -1 in the 5' leader as well as the 2' OH of the first C in the CCA motif, participate in Mg^{2+} binding (for numbering of the substrate see Fig 1). Indeed, Perreault and Altman (51) proposed that a substrate that does not carry Mg^{2+} at the junction of the double-stranded and single-stranded regions in the vicinity of the CCA motif is not cleaved by RNase P. The importance of the 2' OH at the -1 position is further accentuated by the finding that substitution of this group in the substrate reduced the rate of catalysis dramatically and suggesting that this group might function as a ligand for a functionally important Mg^{2+} (19, 52). Taken together, it is clear that a Mg^{2+} ion(s) bound to the substrate in the vicinity of the cleavage site play an important role with respect to RNase P RNA cleavage.

One way to identify ligands that are involved in the binding of Mg^{2+} ions is to use chemically modified RNA substrates, *e.g.* carrying phosphorothioates at specific positions. The idea here is to substitute an oxygen at the cleavage site with a sulfur. If the oxygen is used as a ligand by a functionally important Mg^{2+}, then this substitution would result in a dramatic decrease in cleavage activity. The reason for this is that sulfur is a poor ligand for Mg^{2+} (53). By contrast a soft metal can use sulfur as a ligand (see above). Consequently if cleavage activity is recovered by the addition of a soft metal such as Mn^{2+} (or Cd^{2+}) one can conclude that that a specific oxygen is directly co-ordinated by Mg^{2+}. This approach has been successfully used to demonstrate that the *pro*-Rp oxygen at the scissile bond is directly coordinated to a Mg^{2+} ion (37, 38; see also Fig 5a). This study also suggested a role for the *pro*-Sp oxygen in the process of cleavage. However, its function is not clear at present (37). From the work of Warnecke *et al.* (37), there is no reason at present to invoke more than two Mg^{2+} ions in the chemistry of cleavage by RNase P RNA. This is in keeping with the two-metal-ion catalysis model proposed by Steitz and Steitz (54). Other Mg^{2+} ions might be needed to stabilize and to position the scissile bond correctly in the active site (see also below).

5. RNase P RNA and Mg^{2+} binding

The work of Beebe *et al.* (45) suggest that ≈100 Mg^{2+} ions bind to one RNase P RNA molecule. One of the major goals is to identify those Mg^{2+} ions that are important for function. This includes Mg^{2+} that are important for: i) folding, ii) substrate interaction and iii) chemistry of cleavage.

The primary structure of bacterial RNase P RNA varies considerably, but, it is possible to fold these different RNase P RNA into conserved secondary structures. Based on secondary structure RNase P RNA can be devided into two types, type A and type B (Fig 3). The *E. coli* RNase P RNA, M1 RNA, is a representative of the more common former type, whereas RNase P RNA derived from the low G+C content Gram-positive bacteria such as, *Bacillus subtilis* and *Mycoplasma hyopneumoniae*, is more unusual (55; Mattsson *et al.*, unpublished data). In spite of this and the fact that differences in the catalytic behaviour comparing these two types have been observed (41 and references therein) RNase P RNA needs to fold into an active conformation in order to carry out its function. As has been demonstrated for other large RNA molecules (56, 57) separate folding domains of RNase P RNA have been identified. The folding of the active conformation of RNase P RNA is dependent on the presence of Mg^{2+} and this is a cooperative process that is completed at a Mg^{2+} concentration of 5-6 mM (58-61). The role of the Mg^{2+} in folding is likely to be co-ordinated to neighbouring as well as to distal none-bridging oxygens and to other chemical groups in the RNA which as an outcome results in structural constraints for its folding. In this context it is noted that the folding transition of RNase P RNA requires the binding of at least 2-3 additional Mg^{2+} ions (59, 60).

Catalytically active RNase P RNA is cleaved at specific positions in the presence of Mg^{2+} (29; Fig 3). Replacement of Mg^{2+} with Pb^{2+} also results in cleavage at approximately the same positions, however, the Pb(II)-induced cleavage is more efficient than the Mg(II)-induced cleavage (46, 47). The Pb(II)-induced cleavage is normally performed in the presence of Mg^{2+}, and an increase in the Mg^{2+}-concentration results in a reduced cleavage by lead. This suggests that Mg^{2+} and Pb^{2+} bind to, if not the same sites, overlapping sites. This is in keeping with earlier findings when lead(II)-induced cleavage of $tRNA^{Phe}$ was studied (62 and references therein).

From our preliminary data it appears that the Pb(II)-induced cleavage at least at site Ia (Fig 3) is dependent on the presence of a low Mg^{2+}-concentration (≥0.5 mM; Brännvall and Kirsebom, unpublished results). This suggests that Mg^{2+} is necessary for the folding of this region of RNase P RNA. Structural changes resulting from substitutions or deletions in this region near site Ia also influence Pb^{2+} cleavage at this site as well as at other sites in some of the cases, in particular cleavage at site IIb (63-64; Fig 3). This is in keeping with the finding that these sites (Ia and IIb) are part of the same folding domain (59, 60). From the work of Zito *et al.* (47) it is also apparent that a deletion of helices P13 and P14 affect lead(II)-induced cleavage at several sites including site Ia indicating that the presence of these helices influence the folding of the region surrounding site Ia and/or binding of Pb^{2+} in the vicinity of this site. Interestingly, an RNase P RNA of the B type is still substantially cleaved at site Ia by Pb^{2+} in spite the fact that this RNase P RNA lacks P13 and P14 (65; Fig 3). This further emphasize the differences between a type A and a type B RNase P RNA (41). In this context it is also noted that it has been suggested that Mg^{2+} ion(s) binds in the vicinity of and influence the structure of the corresponding domain of yeast RNase P RNA (66).

The 3' terminal <u>RCC</u>A motif base pairs with a conserved GGU-motif in RNase P RNA, interacting residues underlined (67-69; see also Fig 5a). In M1 RNA this GGU-motif is part of an internal loop structure, the P15-loop (Figs 3 and 4). The structure of this loop in the context of a 31 residues long model RNA was determined by NMR (70) and biochemical data suggest that the folding of the P15-loop is similar, irrespective of whether it is part of the full-length ribozyme or part of the model RNA molecule (71). This model RNA also represents an autonomous divalent metal ion binding domain of *E. coli* RNase P RNA, since Pb^{2+} (or Mg^{2+})-induced cleavage occurs at the same positions in the loop in the full-size ribozyme and in the model RNA. This model RNA

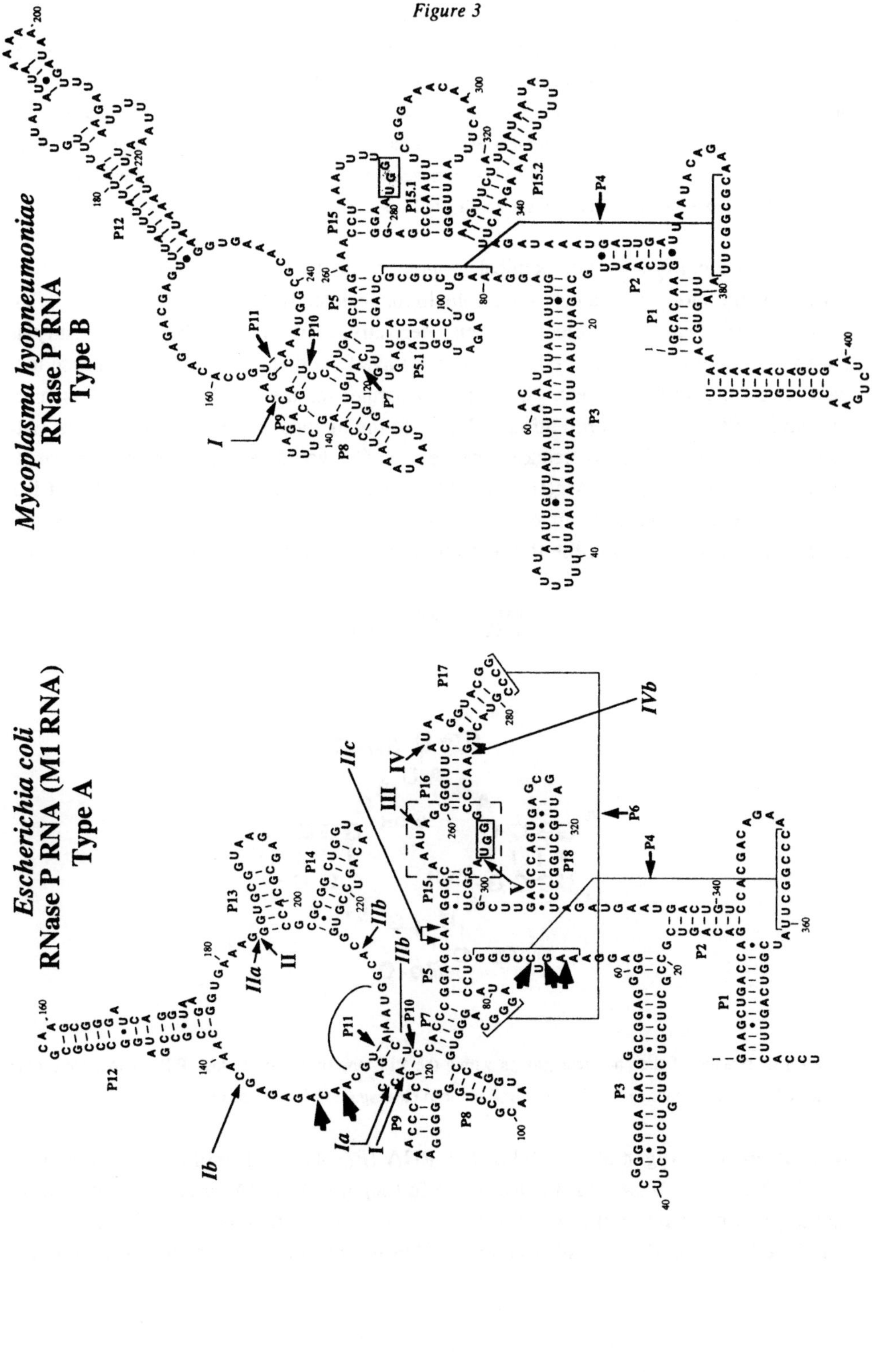
Mycoplasma hyopneumoniae
RNase P RNA
Type B
Escherichia coli
RNase P RNA (M1 RNA)
Type A

Figure 3

Figure 3 Illustration of the secondary structures of *E. coli* RNase P RNA (type A) and *M. hyopneumoniae* RNase P RNA (type B) (55; Mattsson *et al.*, unpublished data). Roman numerals represent Mg^{2+} induced cleavage sites whereas roman numerals in italic represent Pb^{2+} induced cleavage sites. The internal P15-loop is indicated by a box (dashed lines) in M1 RNA while shaded nucleotides in both types represent the GGU-motif that interact with the 3' terminal RCCA sequence element in the substrate. The large arrow heads indicate those *pro*-Rp oxygens that have been suggested to be involved in Mg^{2+} binding (for details see text). The numbering of the helices is according to Haas *et al.* (55).

was used to identify specific chemical groups in the P15-loop that are important for divalent metal induced cleavage within the loop, *i.e.* sites III and V (Figs 3 and 4). The strategy here was to subject various derivatives of the model RNA carrying phosphorothioates at specific positions to cleavage by either Pb^{2+} or Mg^{2+}. These ions were used since sulfur is a poor co-ordination partner for the latter while Pb^{2+} is more thiophilic (53). Thus, co-ordination of Mg^{2+} to the ligands at the sites substituted with sulfur will be significantly reduced, whereas Pb^{2+} would retain or even increase its binding. By following the Mg(II)- and Pb(II)-induced cleavage patterns of different Rp-phosphorothioate-modified model RNA molecules it was then possible to identify specific *pro*-Rp oxygens in the P15-loop important for Mg^{2+} and Pb^{2+} binding. In this

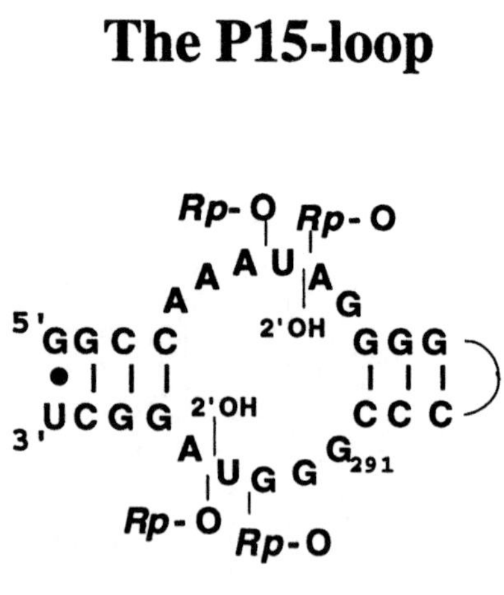

Figure 4 Illustration of the chemical groups within the P15-loop of *E. coli* RNase P RNA that have been identified to be important for Mg^{2+} and Pb^{2+} induced cleavage, for details see text.

way the *pro*-Rp oxygen 3' of G293 in M1 RNA (Fig 4) was identified to be important for Mg^{2+} binding. Based on studies, where full-length M1 RNA carrying a substitution of the *pro*-Rp (or *pro*-Sp) oxygen with a sulfur 3' of G293 was used (Fig 3), it was concluded that the Mg^{2+} positioned at site V in the loop contributes to the function of

RNase P RNA. Its role could either be to stabilize the local structure of the P15-loop such that the GGU-motif is in an optimal configuration for base-pairing with the 3' end of the substrate and/or its function could merely be to stabilize the "RCCA-RNase P RNA" interaction in the RS-complex (71; Fig 5a). This is in keeping with the data that suggest that the structural integrity of this internal loop in M1 RNA is important not only for divalent metal ion binding but also for cleavage site recognition (34) as well as the finding that Mg^{2+} plays a significant role in the folding of this domain (9). These findings strengthen the concept that small RNA building blocks remain essentially unchanged when removed from their structural context. In addition, such small RNA molecules can be used as models for studies of their potential function and structure within native RNA molecules.

6. Mg^{2+} binding in the ribozyme-substrate (RS) complex

Studies of complex-formation between RNase P RNA and tRNA (or tRNA precursor) suggest that at least two Mg^{2+} stabilize this complex (31, 45). The work of Beebe *et al.* (45) also identified a low-affinity Mg^{2+} binding site involved in stabilization of the transition state in cleavage of a tRNA precursor. Moreover, it has been suggested that ≥3 Mg^{2+} are required for optimal RNase P RNA cleavage activity. These findings are consistent with a multiple divalent metal ion mechanism for cleavage by RNase P RNA (19). The question then arises where functionally important Mg^{2+} are positioned in the RS-complex.

Modification-interference approaches have been used in order to identify *pro*-Rp oxygens in RNase P RNA which are used as ligands for functionally important Mg^{2+}-ions in the RS-complex (72, 73). The former group used a gel retardation assay to identify phosphorothioates that interfere with binding of tRNA. This assay relies on the fact that RNase P RNA recognizes the tRNA domain of a tRNA precursor and identifies phosphorothioates that interfere with binding of the tRNA in the ground state in the presence of Mg^{2+}. In this way several Rp-phosphorothioates were identified to interfer with tRNA binding and relevant to the present discussion Rp-phosphorothioates 5' of residues U69, C70, A130 and A132 (Fig 3). A suppression of the phosphorothioate interference as a result of addition of the more thiophilic Mn^{2+} would reveal whether specific *pro*-Rp oxygens are used as direct ligands for metal ion binding (see above). Indeed the result suggested that the *pro*-Rp oxygens 5' of these nucleotides are engaged in Mg^{2+} binding. Interestingly, A130 and A132 are close to the C128/G230 interaction, which has been demonstrated to be important for Pb^{2+}-induced cleavage at in particular site Ia (64; see above). In addition, disruption of this interaction results in a significant decrease in cleavage at sites IIb and IIb', *i.e.* 3' of C226 and A234, respectively (64). Several reports also suggest that this region plays an important role in

RS-complex formation, specifically in the interaction with the T-loop/T-stem of the tRNA (63-65, 74-76). Thus, it is conceivable that the role of the Mg^{2+}-ion(s) located in this region of RNase P RNA is to stabilize its structure and/or the interaction with the substrate.

RNase P RNA is a transacting ribozyme, however, it is possible to obtain cis cleavage by linking the substrate covalently to the ribozyme (77-79). In a modification-interference experiment Harris and Pace (73) used such a cis-cleaving construct with a tRNA sequence attached to residue 331 in *E. coli* RNase P RNA to identify phosphorothioates important for catalysis. Here the authors identified phosophorothioates that interfere with cleavage. Constructs carrying phosphorothioates A67, G68, U69 and A352 clearly interfered with cleavage and the interference at one site, A67, was partially suppressed in the presence of Mn^{2+}. Interference at positions A67, G68, U69 and A352 were also detected by Hardt *et al.* (72) but suppression of the interference by Mn^{2+} was only observed at U69 and C70. A possible reason to the observed differences could be that Harris and Pace did use a cis-cleaving construct. Nevertheless, these data suggest that specific phosphorothioates in this region are important for Mg^{2+} binding and the function of RNase P RNA. Whether this Mg^{2+} ion(s) is part of the active site is at present less certain. A possibililty is that binding of Mg^{2+} in this region is necessary for stabilization of the structure of the RNA in a similar way as has been observed in the crystal structure of the P4/P6 domain of a Group I intron (25). In this context we note that a change of C70 to U70 gives an RNase P RNA that is cleaved more efficiently in the presence of Ca^{2+} as compared to cleavage by wild-type RNase P RNA (80). However, in this report, again a cis-cleaving construct, with a tRNA sequence attached to residue 331 in RNase P RNA, was used. This has to be considered in the interpretation of the data.

As discussed above RNase P RNA is cleaved at specific positions in the presence of divalent metal ions (Fig 3). Addition of a tRNA precursor results in a significant reduction of Pb(II)-induced cleavage at site V and the appearance of Pb^{2+} cleavage 3' of residues U284 and G285 corresponding to site IVb (Fig 3; 46; Svärd and Kirsebom, unpublished data). This suggests that RNase P RNA undergoes a structural change in this region as a result of RS-complex formation. This is not surprising given the fact that the GGU-motif in the P15 internal loop is engaged in base-pairing with the 3' terminal <u>RCC</u>A sequence of the substrate (Fig 5a). In fact, disruption of the "<u>RCC</u>A-RNase P RNA" interaction, due to a change in the substrate (deletion of or a substitution in the 3' terminal <u>RCC</u>A motif), still results in Pb^{2+} cleavage at site V and significantly reduced cleavage at site IVb (Svärd and Kirsebom, unpublished data). This suggests that cleavage at the latter site depends on the establishment of the "<u>RCC</u>A-RNase P RNA" interaction and a likely conformational change in the P15-loop.

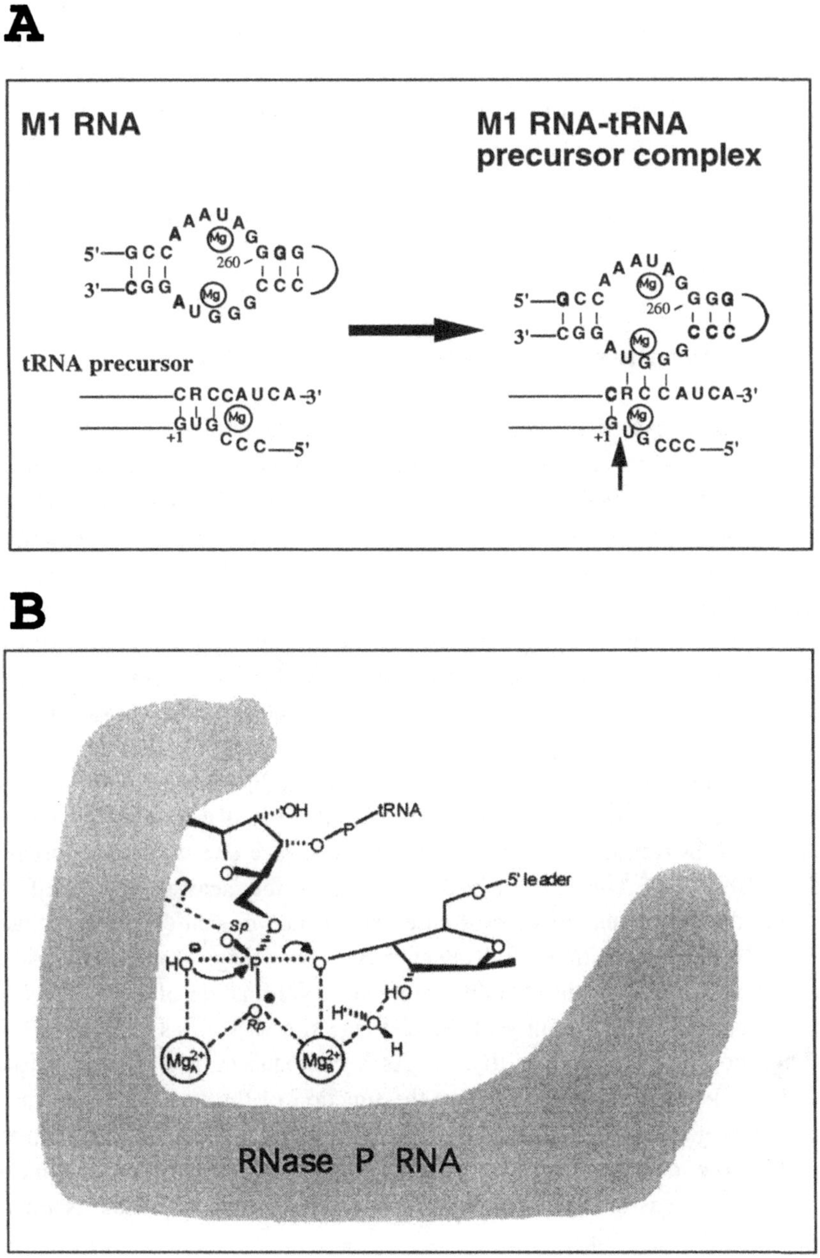
A
M1 RNA
M1 RNA-tRNA precursor complex
tRNA precursor
260
+1
B
?
OH
tRNA
5' leader
Sp
Rp
HO
H
RNase P RNA

Figure 5

Figure 5 **A**. Illustration of the "RCCA-M1 RNA interaction". The R in the tRNA precursor denotes an A or a G resulting either in a A-U Watson-Crick base pair or a G-U wooble base pair in the RS-complex. Three Mg^{2+} ions are depicted in the figure, this does not imply that these are involved in the chemistry of cleavage but they might be, for details see text. The vertical arrow indicate the RNase P cleavage site.
B. Illustration of a proposed transition state in cleavage of a tRNA precursor by RNase P RNA taken from Warnecke *et al.* (37). The question mark denotes that an interaction between RNase P RNA and the *pro*-Sp oxygen has not been demonstrated but it is not excluded that this interaction exist. The small circles indicate negative charges. For details see text.

7. A model for cleavage of RNA by RNase P RNA

The data discussed is summarized in the following model of cleavage of RNA by RNase P RNA. i) One or more Mg^{2+} ion(s) are co-ordinated in the substrate close to the scissile bond. Since divalent metal ions preferentially bind in the junction between single stranded and double stranded regions the binding of Mg^{2+} to the substrate in this region is dependent on whether the residues 5' of the scissile bond are engaged in base-pairing with bases at the 3' end of the substrate, *i.e.* the 3' terminal RCCA sequence, or not. ii) In the RS-complex a Mg^{2+} ion is directly co-ordinated to the *pro*-Rp oxygen at the cleavage site. iii) Several functionally important Mg^{2+} have been localized close to residues U69, A130, G293 in RNase P RNA. iv) A GGU-motif in RNase P RNA base pairs with the 3' terminal RCCA sequence in the RS-complex. If residues -1 and -2 (or just the -1 residue) are base paired to the 3' terminal RCCA motif in the substrate then the "RCCA-RNase P RNA" interaction will result in a disruption of the base paired region in the substrate and as a consequence the cleavage site will be exposed. Thus, a function of the "RCCA-RNase P RNA" is to expose the cleavage site (67; Fig 5a). v) Given that disruption and creation of base pairs result in conformational changes and consequently re-co-ordination of divalent metal ions (Mg^{2+}) the "RCCA-RNase P RNA" interaction will result in a change in the co-ordination of Mg^{2+} ions bound in the substrate both in the vicinity of the cleavage site as well as in RNase P RNA (at site V but possibly also at site III). vi) As a consequence Mg^{2+} will be positioned correctly for cleavage to take place. Here the function of the two Mg^{2+} at the cleavage site are: a) to generate the nucleophile and b) to stabilize the oxyanion of the leaving group. Existing data are consistent with that either a hydroxide or a metal-bound hydroxide function as the nucleophile (19). A proposed transition state is illustrated in Figure 5b.

8. Concluding remark

Many antibiotics have been shown to specifically interact and inhibit the function of RNA (81). Recently it was suggested that the aminoglycoside neomycin inhibits cleavage by the hammerhead ribozyme by interfering with binding of Mg^{2+} at the cleavage site (82 and references therein). Thus, to study the consequences of binding of divalent metal ion such as Mg^{2+} to RNA is not only relevant in our efforts to elucidate the underlying mechanism of cleavage of RNA by RNA but also pertinent in the process to identify novel drugs directed against RNA targets.

9. Acknowledgements

I would like to thank all my coworks and Drs S. Dasgupta, H. Johansson, S. G. Svärd and A. Virtanen for critical reading of the manuscript. The ongoing work in my laboratory is supported by grants from the Swedish Natural Research Council and the Foundation for Strategic Research.

10. References

1. Kirsebom, L.A. (1995) RNase P - a 'Scarlet Pimpernel', *Mol. Microbiol.* **17**, 411-420.
2. Gupta, R. (1984) *Halobacterium volcanii* tRNAs, *J. Biol. Chem.* **259**, 9461-9471.
3. Lee, B.J., De La Pena, P., Tobian, J.A., Zasloff, M. and Hatfield, D. (1987) Unique pathway of expression of an opal suppressor phosphoserine tRNA, *Proc. Natl. Acad. Sci. USA* **84**, 6384-6388.
4. Guerrier-Takada, C., Gardiner, K., Marsh, T., Pace, N. and Altman, S. (1983) The RNA moiety of ribonuclease P is the catalytic subunit of the enzyme, *Cell* **35**, 849-857.
5. Reich, C., Olsen, G.J., Pace, B. and Pace, N.R. (1988) Role of the protein moiety of RNase P, a ribonucleoprotein enzyme, *Science* **239**, 178-181.
6. Tallsjö, A. and Kirsebom, L.A. (1993) Product release is a rate-limiting step during cleavage by the catalytic RNA subunit of *Escherichia coli* RNase P, *Nucl. Acids Res.* **21**, 51-57.
7. Vioque, A., Arnez, J. and Altman, S. (1988) Protein-RNA interactions in the RNase P holoenzyme from *Escherichia coli*, *J. Mol. Biol.* **202**, 835-848.

8. Talbot, S.J. and Altman, S. (1994) Gel retardation analysis of the interaction between C5 protein and M1 RNA in the formation of the ribonuclease P holoenzyme from *Escherichia coli*, *Biochemistry* **33**, 1399-1405.
9. Westhof, E., Wesolowski, D. and Altman, S. (1996) Mapping in three dimensions of regions in a catalytic RNA protected from attack by an Fe(II)-EDTA reagent, *J. Mol. Biol.* **258**, 600-613.
10. Kurz, J.C., Niranjanakumari, S. and Fierke, C.A. (1998) Protein component of *Bacillus subtilis* RNase P specifically enhances the affinity for precursor-$tRNA^{Asp}$, *Biochemistry* **37**, 2393-2400.
11. Svärd, S.G. and Kirsebom, L.A. (1993) Determinants of *Escherichia coli* RNase P cleavage site selection: a detailed *in vitro* and *in vivo* analysis, *Nucl. Acids Res.* **21**, 427-434.
12. Chamberlain, J.R., Lee, Y., Lane, W.S. and Engelke, D.R. (1998) Purification and characterization of the nuclear RNase P holoenzyme complex reveals extensive subunit overlap with RNase MRP, *Genes and Development* **12**, 1678-1690.
13. Jarrous, N., Eder, P.S., Guerrier-Takada, C., Hoog, C. and Altman, S. (1998) Autogenic properties of some protein subunits of catalytically active complexes of human ribonuclease P, *RNA* **4**, 407-417.
14. Li, K. and Williams, S.R. (1995) Cloning and characterization of three new murine genes encoding short homologues of RNase P RNA, *J. Biol. Chem.* **270**, 25281-25285.
15. Guerrier-Takada, C., Haydock, K., Allen, L. and Altman, S. (1986) Metal ion requirements and other aspects of the reaction catalyzed by M1 RNA, the RNA subunit of ribonuclease P from *Escherichia coli*, *Biochemistry* **25**, 1509-1515.
16. Haas, E.S., Brown, J.W., Pitulle, C. and Pace, N.R. (1994) Further perspective on the catalytic core and secondary structure of ribonuclease P RNA, *Proc. Natl. Acad. Sci. USA* **91**, 2527-2531.
17. Mikkelsen, N.E., Brännvall, M., Virtanen, A. and Kirsebom, L.A. (1998), Inhibition of RNase P cleavage by aminoglycosides. *Manuscript in preparation.*
18. Pan, T., Long, D.M. and Uhlenbeck, O.C. (1993) Divalent metal ions in RNA folding and catalysis. In *The RNA world* (Gesteland, R.F. and Atkins, J.F., eds.), pp. 271-302. Cold Spring Harbour Laboratory Press, New York.
19. Smith, D. and Pace, N.R. (1993) Multiple magnesium ions in the ribonuclease P reaction mechanism, *Biochemistry* **32**, 5273-5281.
20. Cowan, J.A. (1995)*The Biological Chemistry of Magnesium* VCH Publishers, Inc. New York, NY 10010.
21. Frausto da Silva, J.J.R. and Williams, R.J.P. (1991) *The Biological Chemistry of the Elements: The inorganic chemistry of life*, Clarendon Press, Oxford.

22. Dahm, S.C., Derrick, W.B. and Uhlenbeck, O.C. (1993) Evidence for the role of solvated metal hydroxide in the hammerhead cleavage mechanism, *Biochemistry* **32**, 13040-13045.
23. Holbrook, S.R., Sussman, J.L., Warrant, R.W., Church, G.M. and Kim, S-H. (1977) RNA-ligand interactions: (I) magnesium binding sites in yeast $tRNA^{Phe}$, *Nucl. Acids Res.* **4**, 2811-2820.
24. Correll, C.C., Freeborn, B., Moore, P.B. and Steitz, T.A. (1997) Metals, motifs, and recognition in the crystal structure of a 5S rRNA domain, *Cell* **91**, 705-712.
25. Cate, J.H., Hanna, D.L. and Doudna, J.A. (1997) A magnesium ion core at the heart of a ribozyme domain, *Nature Struct. Biol.* **4**, 553-558.
26. Cate, J.H. and Doudna, J.A. (1996) Metal-binding sites in the major groove of a large ribozyme domain, *Structure* **4**, 1221-1229.
27. Jack, A., Ladner, J.E., Rhodes, D., Brown, R.S. and Klug, A. (1977) A crystallographic study of metal-binding to yeast phenylalanine transfer RNA, *J. Mol. Biol.* **111**, 315-328.
28. Gardiner, K.J., Marsh, T.L. and Pace, N.R. (1985) Ion dependence of the *Bacillus subtilis* RNase P reaction, *J. Biol. Chem.* **260**, 5415-5419.
29. Kazakov, S. and Altman, S. (1991) Site-specific cleavage by metal ion cofactors and inhibitors of M1 RNA, the catalytic subunit of RNase P from *Escherichia coli*, *Proc. Natl. Acad. Sci. USA* **88**, 9193-9197.
30. Smith, D., Burgin, A.B., Haas, E.S. and Pace, N.R. (1992) Influence of metal ions on the ribonuclease P reaction, *J. Biol. Chem.* **267**, 2429-2436.
31. Hardt, W-D., Schlegl, J., Erdmann, V.A. and Hartmann, R.K. (1993) Gel retardation analysis of *E. coli* M1 RNA-tRNA complexes, *Nucl. Acids Res.* **21**, 3521-3527.
32. Kufel, J. and Kirsebom, L.A. (1994) Cleavage site selection by M1 RNA, the catalytic subunit of *Escherichia coli* RNase P, is influenced by pH, *J. Mol. Biol.* **244**, 511-521.
33. Kufel, J. and Kirsebom, L.A. (1996) Different cleavage sites are aligned differently in the active site of M1 RNA, the catalytic subunit of *Escherichia coli* RNase P, *Proc. Natl. Acad. Sci. USA* **93**, 6085-6090.
34. Kufel, J. and Kirsebom, L.A. (1996) Residues in *Escherichia coli* RNase P RNA important for cleavage site selection and divalent metal ion binding, *J. Mol. Biol.* **263**, 685-698.
35. Talbot, S and Altman, S. (1994) Kinetic and thermodynamic analysis of RNA-protein interactions in the RNase P holoenzyme from *Escherichia coli*, *Biochemistry* **33**, 1406-1411.

36. Harris, M.E., Kazantsev, A.V., Chen, J-L. and Pace, N.R. (1997) Analysis of the tertiary structure of the ribonuclease P ribozyme-substrate complex by site-specific photoaffinity crosslinking, *RNA* **3**, 561-576.
37. Warnecke, J.M., Fürste, J.P., Hardt, W-D., Erdmann, V.E. and Hartmann, R.K. (1996) Ribonuclease P (RNase P) RNA is converted to a Cd^{2+}-ribozyme by a single Rp-phosphorothioate modification in the precursor tRNA at the RNase P cleavage site, *Proc. Natl. Acad. Sci. USA* **93**, 8924-8928.
38. Chen, Y., Li, X. and Gegenheimer, P. (1997) Ribonuclease P catalysis requires Mg^{2+} coordinated to the pro-Rp oxygen of the scissile bond, *Biochemistry* **36**, 2425-2438.
39. Kirsebom, L.A. (1997) RNase P and its substrate, In *The Many Faces of RNA*, Eds. Eggleston, Prescott and Pearson., Academic Press, UK, pp 127-144.
40. Meinnel, T. and Blanquet, S. (1995) Maturation of pre-$tRNA^{fMet}$ by *Escherichia coli* RNase P is specified by a guanosine of the 5'-flanking sequence, *J. Biol. Chem.* **270**, 15908-15914.
41. Brännvall, M., Mattsson, J.G., Svärd, S.G. and Kirsebom, L.A. (1998) RNase P RNA structure and cleavage reflect the primary structure of tRNA genes, *J. Mol. Biol.* in press.
42. Crary, S.M., Niranjanakumari, S. and Fierke, C.A. (1998) The protein component of *Bacillus subtilis* ribonuclease P increases catalytic efficiency by enhancing interactions with the 5' leader sequence of pre-$tRNA^{Asp}$, *Biochemistry* **37**, 9409-9416.
43. Lazard, M. and Meinnel, T. (1998) Role of base G_{-2} of pre-$tRNA^{fMet}$ in cleavage site selection by *Escherichia coli* RNase P *in vitro*, *Biochemistry* **37**, 6041-6049.
44. Loria, A. and Pan, T. (1998) Recognition of the 5' leader and the acceptor stem of a pre-tRNA substrate by the ribozyme from *Bacillus subtilis* RNase P, *Biochemistry* **37**, 10126-10133.
45. Beebe, J.A., Kurz, J.C. and Fierke, C.A. (1996) Magnesium ions are required by *Bacillus subtilis* ribonuclease P RNA for both binding and cleaving precursor $tRNA^{Asp}$, *Biochemistry* **35**, 10493-10505.
46. Ciesiolka, J., Hardt, W-D., Schlegl, J., Erdmann, V.A. and Hartmann, R.K. (1994) Lead-ion-induced cleavage of RNase P RNA, *Eur. J. Biochem.* **219**, 49-56.
47. Zito, K., Hüttenhofer, A. and Pace, N.R. (1993) Lead-catalyzed cleavage of ribonuclease P RNA as a probe for integrity of tertiary structure, *Nucl. Acids Res.* **21**, 5916-5920.
48. Hardt, W-D., Schlegl, J., Erdmann, V.A. and Hartmann, R.K. (1995) Kinetics and thermodynamics of the RNase P RNA cleavage reaction: Analysis of tRNA 3'-end variants, *J. Mol. Biol.* **247**, 161-172.

49. Oh, B-K., Frank, D.N. and Pace, N.R. (1998) Participation of the 3'-CCA of tRNA in the binding of catalytic Mg^{2+} ions by ribonuclease P, *Biochemistry* **37**, 7277-7283.
50. Perreault, J-P. and Altman, S. (1992) Important 2'-hydroxyl groups in model substrates for M1 RNA, the catalytic RNA subunit of RNase P from *Escherichia coli*, *J. Mol. Biol.* **226**, 399-409.
51. Perreault, J-P. and Altman, S. (1993) Pathway of activation by magnesium ions of substrates for the catalytic subunit of RNase P RNA from *Escherichia coli*, *J. Mol. Biol.* **230**, 750-756.
52. Forster, A.C. and Altman, S. (1990) External guide sequences for an RNA enzyme, *Science* **249**, 783-786.
53. Jaffe, E.K. and Cohn, M. (1979) Diastereomers of the nucleoside phosphorothioates as probes of the structure of the metal nucleotide substrates and of the nucleotide binding site of yeast hexokinase, *J. Biol. Chem.* **254**, 10839-10845.
54. Steitz, T.A. and Steitz, J.A. (1993) A general two-metal-ion mechanism for catalytic RNA, *Proc. Natl. Acad. Sci. USA* **90**, 6498-6502.
55. Haas, E.S., Banta, A.B., Harris, J.K., Pace, N.R. and Brown, J.W. (1996) Structure and evolution of ribonuclease P RNA in Gram-positive bacteria, *Nucl. Acids Res.* **24**, 4775-4782.
56. Murphy, F.L. and Cech, T.R. (1993) An independently folding domain of RNA tertiary structure within the *Tetrahymena* ribozyme, *Biochemistry* **32**, 5291-5300.
57. Wang, Y.H., Murphy, F.L., Cech, T.R. and Griffith, J.D. (1994) Visualization of a tertiary structural domain of the *Tetrahymena* group I intron by electron microscopy, *J. Mol. Biol.* **236**, 64-71.
58. Guerrier-Takada, C. and Altman, S. (1993) A physical assay for and kinetic analysis of the interactions between M1 RNA and tRNA precursor substrates, *Biochemistry* **32**, 7152-7161.
59. Pan, T. (1995) Higher order folding and domain analysis of the ribozyme from *Bacillus subtilis* ribonuclease P, *Biochemistry* **34**, 902-909.
60. Loria, A. and Pan, T. (1997) Domain structure of the ribozyme from eubacterial ribonuclease P, *RNA* **2**, 551-563.
61. Zarrinkar, P.P., Wang, J. and Williamson, J.R. (1996) Slow folding kinetics of RNase P RNA, *RNA* **2**, 564-573.
62. Brown, R.S., Dewan, J.C. and Klug, A. (1985) Crystallographic and biochemical investigation of the lead(II)-catalyzed hydrolysis of yeast phenylalanine tRNA, *Biochemistry* **24**, 4785-4801.

63. Tallsjö, A., Svärd, S.G., Kufel, J. and Kirsebom, L.A. (1993) A novel tertiary interaction in M1 RNA, the catalytic subunit of *Escherichia coli* RNase P, *Nucl. Acids Res.* **21**, 3927-3933.
64. Mattsson, J.G., Svärd, S.G. and Kirsebom, L.A. (1994) Characterization of the *Borrelia burgdorferi* RNase P RNA gene reveals a novel tertiary interaction, *J. Mol. Biol.* **241**, 1-6.
65. Svärd, S.G., Mattsson, J.G., Johansson, K-E. and Kirsebom, L.A. (1994) Cloning and characterization of the RNase P RNA genes from two porcine mycoplasmas, *Mol. Microbiol.* **11**, 849-859.
66. Ziehler, W.A., Yang, J., Kurochkin, A.V., Sandusky, P.O. Zuiderweg, E.R.P. and Engelke, D.R. (1998) Structural analysis of the P10/11-P12 RNA domain of yeast RNase P RNA and its interaction with magnesium, *Biochemistry* **37**, 3549-3557.
67. Kirsebom, L.A. and Svärd, S.G. (1994) Base pairing between *Escherichia coli* RNase P RNA and its substrate, *EMBO J.* **13**, 4870-4876.
68. Svärd, S.G., Kagardt, U. and Kirsebom, L.A. (1996) Phylogenetic comparative mutational analysis of the base-pairing between RNase P RNA and its substrate, *RNA* **2**, 463-472.
69. Tallsjö, A., Kufel, J. and Kirsebom, L.A. (1996) Interaction between *Escherichia coli* RNase P RNA and the discriminator base results in slow product release, *RNA* **2**, 299-307.
70. Glemarec, C., Kufel, J., Földesi, A., Maltseva, T., Sandström, A., Kirsebom, L.A. and Chattopadhyaya (1996) The NMR structure of 31mer RNA domain of *Escherichia coli* RNase P RNA using its non-uniformly deuterium labelled counterpart (the 'NMR-window' concept), *Nucl. Acids Res.* **24**, 2022-2035.
71. Kufel, J. and Kirsebom, L.A. (1998) The P15-loop of *Escherichia coli* RNase P RNA is an autonomous divalent metal ion binding domain, *RNA* **4**, 777-788.
72. Hardt, W-D., Warnecke, J.M., Erdmann, V.A. and Hartmann, R.K. (1995) Rp-phosphorothioate modifications in RNase P RNA that interfere with tRNA binding, *EMBO J.* **14**, 2935-2944.
73. Harris, M.E. and Pace, N.R. (1995) Identification of phosphates involved in catalysis by the ribozyme RNase P RNA, *RNA* **1**, 210-218.
74. Lumelsky, N. and Altman, S. (1988) Selection and characterization of randomly produced mutants in the gene encoding for M1 RNA, *J. Mol. Biol.* **202**, 443-454.
75. Pan, T., Loria, A. and Zhong, K. (1995) Probing of tertiary interactions in RNA: 2'-hydroxyl-base contacts between the RNase P RNA and pre-tRNA, *Proc. Natl. Acad. Sci. USA* **92**, 12510-12514.

76. Loria, A. and Pan, T. (1997) Recognition of the T-stem-loop of a pre-tRNA substrate by the ribozyme from *Bacillus subtilis* ribonuclease P, *Biochemistry* **36**, 6317-6325.
77. Altman, S. (1989) Ribonuclease P: an enzyme with a catalytic RNA subunit, *Adv. Enzymol. Rela. Areas Mol. Biol.* **62**, 1-36.
78. Kikuchi, Y., Sasaki-Tozawa, N. and Suzuki, K. (1993) Artificial self-cleaving molecules consisting of a tRNA precursor and the catalytic RNA of RNase P, *Nucl. Acids Res.* **21**, 4685-4689.
79. Frank, D.N., Harris, M.E. and Pace, N.R. (1994) Rational design of self-cleaving pre-tRNA-ribonuclease P RNA conjugates, *Biochemistry* **33**, 10800-10808.
80. Frank, D.N. and Pace, N.R. (1997) *In vitro* selection for altered divalent metal specificity in the RNase P RNA, *Proc. Natl. Acad. Sci. USA* **94**, 143555-14360.
81. Davies, J., von Ahsen, U. and Schroeder, R. (1993) Antibiotics and the RNA world: A role for low-molecular-weight effectors in biochemical evolution. In *The RNA world* (Gesteland, R.F. and Atkins, J.F., eds.), pp. 185-204. Cold Spring Harbour Laboratory Press, New York.
82. Herrmann, T. and Westhof, E. (1998) Aminoglycoside binding to the hammerhead ribozyme: A general model for the interaction of cationic antibiotics with RNA, *J. Mol. Biol.* **276**, 903-912.

METAL ION - INDUCED CLEAVAGES IN PROBING OF RNA STRUCTURE

JERZY CIESIOŁKA
Institute of Bioorganic Chemistry, Polish Academy of Sciences, Noskowskiego 12/14, 61-704 Poznań, Poland

1. Principle

Certain metal ions induce hydrolytic degradation of RNA, and in some RNA molecules this process is exceptionally efficient and specific. The best known example, yeast $tRNA^{Phe}$, undergoes specific fragmentation in the D-loop in the presence of Pb^{2+} [1-3] and other ions: Eu^{3+}[4,5], Mn^{2+}[6] and Mg^{2+}[7,8]. It has been shown, based on X-ray analysis of yeast $tRNA^{Phe}$ crystals, that a Pb^{2+} ion induces the reaction acting from its tight metal ion-binding pocket [9,10]. This gave rise to an experimental approach that uses Pb^{2+} and other ions to localize some metal ion binding sites as well as to probe the structure of RNA molecules in the vicinity of the bound ions.

However, such highly efficient and specific cleavages are rather rarely observed. Other Pb^{2+}-induced cleavages are weaker and usually comprise several consecutive phosphodiester bonds. Most information on the specificity of hydrolysis has been obtained from studies on ribosomal 16S RNA [11] and 5S RNAs [12-14]. It turns out that cleavages occur preferentially in bulges, loops and other single-stranded RNA regions except those involved in stacking or other higher order interactions. Double-stranded RNA segments are essentially resistant to hydrolysis. Cleavages are also observed in paired regions destabilized by the presence of non-canonical interactions, bulges or other structural distortions. In general, it seems that flexibility of the polynucleotide chain determines its susceptibility to Pb^{2+}-induced hydrolysis [11-14].

It has been noted [15] that the mechanism proposed for the specific, Pb^{2+}-induced hydrolysis of yeast $tRNA^{Phe}$ [9,10,16] may account for all types of cleavages induced by metal ions. The simplified mechanism shown in Figure 1 is helpful in understanding the relation between RNA structure and susceptibility of a particular RNA region to hydrolysis.

J. Barciszewski and B.F.C. Clark (eds.), RNA Biochemistry and Biotechnology, 111–121.

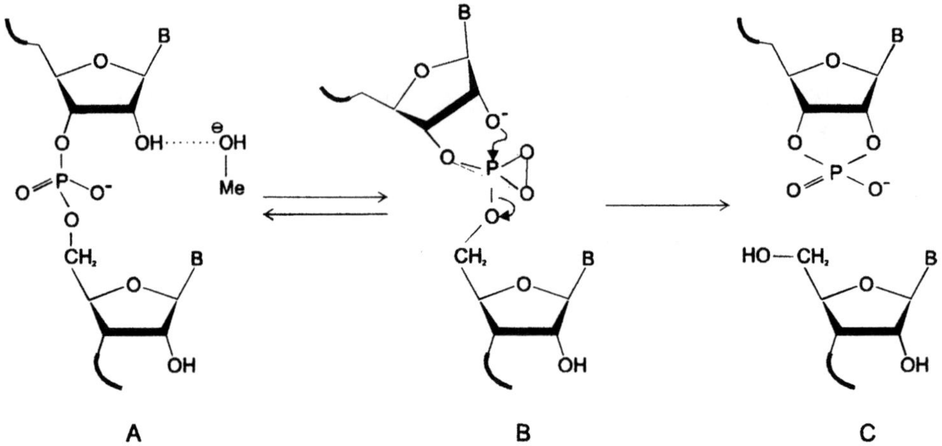

Figure 1. Mechanism of metal ion – induced RNA hydrolysis (discussed in the text).

An ionized metal ion hydrate abstracts a proton from the 2'-OH group of the ribose. An anionic 2'-O⁻ attacks a phosphorus atom and a pentacoordinated intermediate is formed. Subsequently, the phosphodiester chain is cleaved with the formation of 2',3'-cyclic phosphate and 5'-hydroxyl group. Since hydrolysis requires the presence of an ionized metal ion hydrate, at neutral pH the reaction occurs most effectively in the presence of Pb^{2+} (pK_a 7.2). For other ions: Eu^{3+}, Zn^{2+}, Mn^{2+}, Mg^{2+} and Ca^{2+} the reaction pH, time or temperature have to be increased (pK_a values of their hydrates are: 8.5, 9.6, 10.6, 11.4 and 12.6, respectively).

The cleavage efficiency of a particular phosphodiester bond in an RNA molecule depends on: i/ proper localization of the metal ion hydrate facilitating deprotonation of the 2'-OH group (Figure 1, transition A to B), ii/ sufficient conformational flexibility of the hydrolyzed region allowing the formation of pentacoordinated intermediate and subsequent breaking of the phosphodiester chain (transition B to C). Optimal distance and correct orientation of the bound metal ion hydrate seems to be of primary importance in case of RNAs undergoing highly specific hydrolysis. The cleavage of these RNAs occurs at relatively low concentration of Pb^{2+}, below 0.1 mM, at which hydrolysis of other phosphodiester bonds also takes place but is much slower. The cleavages of lower efficiency and specificity, above 0.5 mM Pb^{2+}, are most likely induced by ions acting from the solution or from their weak binding sites. The metal hydrates interact equally well with all accessible 2'-hydroxyl groups thus different rigidity/flexibility of the phosphates, hindering or facilitating conformational transitions, necessary for the reaction to occur, becomes a rate limiting factor [15].

The Pb^{2+} cleavage approach has been used in structural analysis of several RNAs and RNA complexes (Table 1). The experiments can be classified into three groups that aim to investigate (i) metal ion-binding sites (ii) RNA structure (iii) RNA - ligand interactions. Representative examples from each group will be discussed in the following chapters.

2. Metal ion – binding sites

The presence of a strong, highly specific metal ion-induced cleavage suggests that a tight metal ion binding site is created in the spatial RNA structure. Cleavage occurring in a particular RNA region does not implicate, however, the involvement of that region in direct coordination of the ion. Moreover, some tightly bound ions may not be able to induce specific hydrolysis. For instance, in yeast $tRNA^{Phe}$ a Pb^{2+} ion cleaves the D-loop but is bound in the TΨC-loop while the ion found in the anticodon loop does not induce specific cleavages [9,10,16]. A metal ion-binding pocket can usually accommodate different ions acting as a general metal ion binding site. In such cases, two additional experimental approaches can be used in order to confirm that a strong cleavage is induced by a tightly bound ion.

The first approach takes advantage of the fact that a metal ion-induced cleavage is suppressed if hydrolysis is performed in the presence of other ions competing for a common metal ion-binding site. This approach was applied in the studies of yeast $tRNA^{Phe}$ [17], HDV ribozyme [18], group I intron RNA [19], ribosomal 16S and 23S RNA [20] and *in vitro* selected Zn-binding RNA [21]. Quantitative analysis of the inhibition effect allowed to determine the K_d value corresponding to the binding of Mg^{2+} to yeast $tRNA^{Phe}$ [17].

The second approach relies on an observation that Pb^{2+}, Mg^{2+}, Mn^{2+}, and Eu^{3+} ions that bind in the D-TΨC region of yeast $tRNA^{Phe}$ induce strong cleavages at very similar positions of the RNA chain. Thus, it seems likely that also in other RNAs different metal ions occupying a common binding site are capable of inducing specific cleavages in a very similar RNA region. The cleavage reactions are performed at conditions taking into account different pK_a values of metal ion hydrates (discussed in the previous section). Beside Pb^{2+}, other ions Mg^{2+}, Mn^{2+}, Ca^{2+}, and Eu^{3+} are usually used in such experiments. This approach turned out to be successful in the studies of metal ion binding sites in several tRNA molecules [22-25], HDV ribozyme [26], RNase P RNA [27] and group I intron RNA [28].

TABLE 1. Examples of structural analysis of RNAs and RNA complexes by means of metal ion-induced cleavage approach.

RNA or RNA complex	Structural probe	Type of analysis	References
in vitro selected RNAs, aptamers and model oligonucleotides	Pb	ion binding sites RNA structure	[21][32] [15][21][40][42][43][44][45]
tRNAs, mutants, *in vitro* transcripts and fragments	Pb	ion binding sites	[2][3][16][17][29][46]
	Pb	RNA structure	[3][47]
	various Me	ion binding sites	[5][6][8][22][23][24][25][48][56]
	various Me	RNA structure	[22][23][24][25]
HDV ribozyme	Pb	ion binding sites	[18]
4.5S RNA	Pb	RNA structure	[33]
5S rRNA	Pb	RNA structure	[12][13][14][49]
U1 snRNA	Pb	RNA structure	[50]
RNase P RNA	Pb	ion binding sites	[30][31][51]
	various Me	ion binding sites	[27]
group I intron RNA	Pb	ion binding sites	[19]
	various Me	ion binding sites	[28]
10Sa RNA (tmRNA)	Pb	RNA structure	[52]
DMPK mRNA fragment	Pb	RNA structure	[34]
TfR mRNA fragment	Pb	RNA structure	[35]
SECIS mRNA fragment	Pb	RNA structure	[53][54]
CaMV 35 S RNA leader	Pb	RNA structure	[55]
16S rRNA fragment in 30S subunit and 70S rib.	Pb	RNA structure RNA - protein int. RNA - RNA int.	[11]
16S and 23S rRNA in 70S ribosome	Pb	ion binding sites	[20]
RNA aptamer - citrulline complex	Pb	RNA - aminoacid int.	[39]
RNA aptamer - antibiotic complex	Pb	RNA - antibiotic int.	[40][41]
Phe-tRNAPhe-EF-Tu:GTP complex	Pb	RNA - protein int.	[36]
HDV ribozyme-antibiotic complex	Pb	RNA - antibiotic int.	[18]
4.5S RNA - P48 protein complex	Pb	RNA - protein int.	[33]
5S rRNA - L18 protein complex	Pb	RNA - protein int.	[13]
RNase P RNA- pre-tRNA complex	Pb	RNA - RNA int.	[31][37][38]
TfR mRNA fragment-IRP1 complex	Pb	RNA - protein int.	[35]

Cleavages induced by different ions do not necessarily occur at exactly the same sites of an RNA chain but their positions usually differ by just one or two nucleotides. It is understandable when one takes into account the different coordination preferences of various metal ions, resulting in a slightly different arrangement of their hydrates in metal ion binding pockets.

The metal ion-induced cleavage method allows to identify similarities and differences in related RNA molecules in the regions involved in metal ion binding. Any structural changes introduced to an RNA by, for instance, point mutations or lack of modified nucleotides can also be detected. The RNA environment of tightly bound Pb^{2+} ions was probed in several mutants of yeast $tRNA^{Phe}$ [24,25,29] and RNase P RNAs from different sources [30,31], while Pb^{2+}, Mg^{2+} and Eu^{3+} were used in the studies of series of methionine [22] and phenylalanine [23] specific tRNAs. The Pb^{2+} ions were capable of identifying the long-range conformational effect that takes place in the D-loop of yeast $tRNA^{Phe}$ upon Y-base removal [3] and detecting changes in the tRNA structure lacking in modified nucleotides [24].

Finally, the Pb^{2+} ions were used for *in vitro* selection of several RNA molecules that undergo a highly efficient and specific cleavage reaction [32]. One of the selected motifs, the leadzyme, is now being extensively studied but those studies are beyond the scope of this article.

3. RNA structure

The susceptibility of several RNA molecules to Pb^{2+}-promoted hydrolysis was characterized and the cleavage patterns were used in the analysis of RNA structures (Table 1). It turns out, however, that Pb^{2+} patterns do not always correspond precisely to RNA secondary structure models. Experimental results are most consistent with predicted susceptibility of a particular RNA to hydrolysis assuming that cleavages occur in "flexible regions" of an RNA polynucleotide chain. However, taking into account that our knowledge on RNA conformational dynamics is still insufficient, the term "flexible regions of RNA" should be used cautiously while interpreting experimental data.

Recently, the susceptibility of several well-defined RNA secondary structure motifs: bulges, hairpin loops and single-stranded RNA stretches to Pb^{2+}-promoted hydrolysis has been characterized [15]. The studies show that the patterns of Pb^{2+} hydrolysis of single nucleotide bulge regions depend on the structural context of the adjacent base pairs. In general, pyrimidine flanking the bulged nucleotide facilitates hydrolysis while purine makes the bulge more resistant. This effect seems to correlate with the ability of the bulge to form stacking interactions with its neighbors. Hydrolysis of two- and three-nucleotide bulges depends only slightly on their nucleotide composition. In case of terminal loops hydrolysis usually increases with the loop size and strongly depends on its nucleotide composition. Particularly resistant to hydrolysis are stable tetraloops, most likely due to their high conformational rigidity. Most single-stranded RNA regions are cleaved by Pb^{2+} ions. Resistant to hydrolysis are, however, tracts of G residues and in most cases, the phosphodiester bonds present at the junction of unpaired RNA region and double-stranded helical stem. Presumably, this is an effect of increased conformational rigidity of these RNA regions due to extensive stacking interactions. The studies also clearly demonstrate that the same structural motifs present in different RNAs are characterized by essentially identically distributed series of cleavages. Thus, characteristic cleavages suggest the presence of particular RNA secondary structure motifs in RNAs of poorly defined structures. [15].

The Pb^{2+} cleavage method is very sensitive in detecting changes in conformation of an RNA polynucleotide chain. For instance, different conformational forms of 5S rRNAs from *E. coli* and rat liver were clearly distinguished [14], melting of 4.5S RNA [33] as well as DMPK mRNA fragment [34] were nicely followed at elevated temperatures.

4. RNA – ligand interactions

Remarkable reduction of cleavage intensities has been observed upon the formation of RNA complexes with proteins [11,13,33,35,36], other RNAs [31,37,38] or low molecular weight ligands - aminoacids [39] and antibiotics [18,40,41]. In RNA - protein complexes the shielding effect of a bound protein is, most likely, responsible for changes in cleavage intensities. This mapping method is obviously restricted to RNA regions susceptible to Pb^{2+}-induced hydrolysis, mainly bulges, loops and other single-stranded RNA stretches.

In some cases, binding of a ligand may result in changes in the Pb^{2+} cleavage pattern that correspond to unexpectedly large regions of the polynucleotide chain. This was especially evident in the case of *in vitro* selected RNA aptamers that bind antibiotics [41] and Zn ions [42]. The effect is probably caused by a loss of flexibility i.e. formation of a more rigid conformation of a large RNA fragment which is unstructured in the absence of the ligand. The ligand may stabilize one of the near-isoenergetic forms of RNA, while Pb^{2+} ions monitor the process of structure formation and not just the region of ligand binding [41,42].

There are several examples in which an enhancement of Pb^{2+} cleavages has been observed upon complex formation. For instance, the effect was seen in fragment of 16S rRNA in the naked form and in 30S ribosomal subunit and 70S ribosome [11], 5S rRNA and L18 protein [13], TfR mRNA fragment and IRP [35]. Moderately enhanced cleavages may suggest increased flexibility of the hydrolyzed RNA regions. The appearance of a very strong cleavage indicates that possibly a new strong metal ion binding site is created or a previously inactive ion has been rearranged in such a way that efficient metal ion-induced hydrolysis becomes feasible. In both cases the presence of a strongly bound ion needs to be verified by other methods, since it is conceivable that hydrolysis may also occur with no metal ion bound. Simply, the conformation of the phosphodiester bond could be changed upon ligand binding in a way that facilitates the formation of a reaction intermediate and polynucleotide strand scission (see Figure 1).

5. Conclusions

Metal ion-promoted hydrolysis of RNA can be successfully used in localizing some tight metal ion binding sites. The mechanism proposed for Pb^{2+} -induced cleavage reaction of yeast $tRNA^{Phe}$ is now being applied to the reactions induced by other ions and occurring in a variety of RNA molecules. The approach is one of the very few capable of identifying tight metal ion binding sites in RNA molecules. Results correspond well to those obtained with other methods and the simplicity of experiments enhances its attractiveness.

The past few years have shown that Pb^{2+} ions can be also applied in analysis of RNA higher order structure. This is an alternative to other limited digestion methods with the use of specific nucleases or chemical reagents. The specificity of Pb^{2+} -promoted hydrolysis is well characterized. Unlike enzymes or modifying reagents which monitor the accessibility of particular RNA regions to these probes, the Pb^{2+} ions cleave preferentially flexible regions of polyribonucleotide chains. Thus, better understanding of RNA conformational dynamics should further improve the correlation of cleavage patterns and RNA structure.

The interactions of several RNAs with proteins and other ligands have been probed by the Pb^{2+} method. Such interactions, in most cases, involve loops, bulges and other single-stranded RNA stretches. Since these regions are cleaved by Pb^{2+} and the ions do not show substantial base specificity, the Pb^{2+} approach turns out to be well suited to the analysis of RNA complexes.

Acknowledgements

This paper is dedicated to Professor Maciej Wiewiórowski on the occasion of his 80th birthday.
I thank Włodzimierz Krzyżosiak, in whose laboratory I started and I did most of my work on metal ion-induced RNA hydrolysis, for his contribution to this research. Volker A. Erdmann and Michael Yarus are thanked for their helpful suggestions on the use of lead ions in probing of RNA structure.
This work was supported by grant No. 6 P04A 001 12 from the State Committee for Scientific Research.

References

1. Werner, C., Krebs, B., Keith, G. and Dirheimer, G. (1976) Specific cleavage of pure tRNAs by plumbous ions, *Biochim. Biophys. Acta* **432**, 161-175.
2. Sampson, J. R., Sullivan, F. X., Behlen, L.S., DiRenzo, A.B. and Uhlenbeck, O.C. (1987) Characterization of two RNA catalyzed RNA cleavage reaction , *Cold Spring Harbor Symp. Quant. Biol.* **52**, 267-275.
3. Krzyżosiak, W. J., Marciniec, T., Wiewiórowski, M., Romby, P., Ebel, J. P. and Giege, R. (1988) Characterization of lead (II)-induced cleavages in tRNAs in solution and effect of the Y-base in yeast $tRNA^{Phe}$, *Biochemistry* **27**, 5771-5777.
4. Rordorf, B. F. and Kearns, D. R. (1976) Effect of europium(III) on the thermal denaturation and cleavage of transfer ribonucleic acids, *Biopolymers* **15**, 1491-1504.
5. Ciesiołka, J., Marciniec, T. and Krzyżosiak, W. J. (1989) Probing the environment of lanthanide binding sites in yeast $tRNA^{Phe}$ by metal ion-promoted cleavages, *Eur. J. Biochem.* **182**, 445-450.
6. Wrzesinski, J. Michałowski, D., Ciesiołka, J. and Krzyżosiak, W.J. (1995) Specific RNA cleavages induced by manganese ions, *FEBS Lett.* **374**, 62-68.
7. Wintermeyer, W. and Zachau, G. (1973) Mg^{2+} - katalysierte spezifische Spaltung von tRNA, *Biochim. Biophys. Acta* **299**, 82-90.
8. Marciniec, T., Ciesiołka, J., Wrzesinski, J. and Krzyżosiak, W. J. (1989) Specificity and mechanism of the cleavages induced in yeast $tRNA^{Phe}$ by magnesium ions, *Acta Biochim. Polon.* **36**, 115-122.
9. Brown, R. S., Hingerty, B. E., Dewan, J. C. and Klug, A. (1983) Pb(II)-Catalyzed cleavage of the sugar-phosphate backbone of yeast $tRNA^{Phe}$ - implications for lead toxicity and self-splicing RNA, *Nature* **303**, 543-546.
10. Rubin, J. R. and Sundaralingam, M. (1983) Lead ion binding and RNA chain hydrolysis in phenylalanine tRNA, *J. Biomol. Struct. Dyn.* **1**, 639-646.
11. Górnicki, P., Baudin, F., Romby, P., Wiewiórowski, M., Krzyżosiak, W. J., Ebel, J. P., Ehresmann, C. and Ehresmann, B. (1989) Use of lead(II) to probe the structure of large RNAs. Conformation of the 3' terminal domain of *E. coli* 16S rRNA and its involvement in building the tRNA binding sites, *J. Biomol. Struc. Dyn.* **6**, 971-984.
12. Brunel, C., Romby, P., Westhof, E., Ehresmann, C. and Ehresmann, B. (1991) Three - dimensional model of *Escherichia coli* ribosomal 5S RNA as deduced from structure probing in solution and computer modeling, *J. Mol. Biol.* **221**, 293-308.

13. Ciesiołka, J., Lorenz, S. and Erdmann, V. A. (1992) Structural analysis of three prokaryotic 5S rRNA species and selected 5S rRNA - ribosomal protein complexes by means of Pb(II)-induced hydrolysis, *Eur. J. Biochem.* **204**, 575-581.
14. Ciesiołka, J., Lorenz, S. and Erdmann, V. A. (1992) Different conformational forms of *Escherichia coli* and rat liver 5S rRNA revealed by Pb(II) - induced hydrolysis, *Eur. J. Biochem.* **204**, 583-589.
15. Ciesiołka, J., Michałowski, D., Wrzesinski, J., Krajewski, J. and Krzyżosiak, W.J. (1998) Patterns of cleavages induced by lead ions in defined RNA secondary structure motifs, *J. Mol. Biol.* **275**, 211-220.
16. Brown, R.S., Dewan, J.C. and Klug, A. (1985) Crystallographic and biochemical investigation of the lead(II) - catalyzed hydrolysis of yeast phenylalanine tRNA, *Biochemistry* **24**, 4785-4801.
17. Labuda, D., Nicoghosian, K. and Cedergren, R. (1985) Cooperativity in low-affinity Mg^{+2} binding to tRNA, *J. Biol.Chem.* **260,** 1103-1107.
18. Rogers, J., Chang, A.H., von Ahsen, U., Schroeder, R. and Davies, J. (1996) Inhibition of the self-cleavage reaction of the human hepatitis delta virus ribozyme by antibiotics, *J. Mol. Biol.* **259,** 916-925.
19. Streicher, B., von Ahsen, U. and Schroeder, R. (1993) Lead cleavage sites in the core structure of group I intron – RNA, *Nucleic Acids Res.* **21**, 311-317.
20. Winter, D., Polacek, N., Halama, I., Streicher, B. and Barta, A. (1997) Lead-catalysed specific cleavage of ribosomal RNAs, *Nucleic Acids Res.* **25**, 1817-1824.
21. Ciesiołka, J., Gorski, J. and Yarus, M. (1995) Selection of an RNA domain that binds Zn^{2+}, *RNA* **1**, 538-550.
22. Ciesiołka, J., Wrzesinski, J., Górnicki, P., Podkowiński, J. and Krzyżosiak, W. J. (1989) Analysis of magnesium, europium and lead binding sites in methionine initiator and elongator tRNAs by specific metal-ion-induced cleavages, *Eur. J. Biochem.* **186**, 71-77.
23. Marciniec, T., Ciesiołka, J., Wrzesinski, J. and Krzyżosiak, W.J. (1989) Identification of the magnesium, europium and lead binding sites in *E. coli* and lupine $tRNA^{Phe}$ by specific metal ion induced cleavages, *FEBS Lett.* **243**, 293-298.
24. Michałowski, D. Wrzesiński,J., Ciesiołka,J. and Krzyżosiak,W.J. (1996) Effect of modified nucleotides on structure of yeast $tRNA^{Phe}$. Comparative studies by metal ion-induced hydrolysis and nuclease mapping, *Biochimie* **78,** 131-138.
25. Michałowski, D., Wrzesinski, J. and Krzyżosiak, W.J. (1996) Cleavages induced by different metal ions in yeast $tRNA^{Phe}$ U59C60 mutants, *Biochemistry* **35**, 10727-10734.
26. Matysiak, M., Wrzesinski, J. and Ciesiołka, J. (1998) unpublished data
27. Kazakov, S. and Altman, S. (1991) Site - specific cleavage by metal ion cofactors and inhibitors of M1 RNA, the catalytic subunit of RNase P from *Escherichia coli, Proc. Natl. Acad. Sci. U.S.A.* **88**, 9193-9197.
28. Streicher, B. Westhof, E. and Schroeder, R. (1996) The environment of two metal ions surrounding the splice site of a group I intron, *EMBO J.* **15,** 2556-2564.
29. Behlen, L., Sampson, J. R., DiRenzo, A. B. and Uhlenbeck, O.C. (1990) Lead - catalyzed cleavage of yeast $tRNA^{Phe}$ mutants, *Biochemistry* **29**, 2515-2523.
30. Zito, K., Huttenhofer, A. and Pace, N. R. (1993) Lead - catalyzed cleavage of ribonuclease P RNA as a probe for integrity of tertiary structure, *Nucleic Acids Res.* **21**, 5916-5920.
31. Ciesiołka, J., Hardt, W-D., Schlegl, J., Erdmann, V. A. and Hartmann, R. K. (1994) Lead-ion-induced cleavage of RNase P RNA, *Eur. J. Biochem.* **219**, 49-56.
32. Pan, T. and Uhlenbeck O.C. (1992) *In vitro* selection of RNAs that undergo autolytic cleavage with Pb^{2+}, *Biochemistry* **31**, 3887-3895.

33. Lentzen, G., Moine, H., Ehresmann, C.H., Ehresmann, B. and Wintermeyer, W. (1996) Structure of 4.5S RNA in the signal recognition particle of *Escherichia coli* as studied by enzymatic and chemical probing, *RNA* **2,** 244-253.
34. Napierała, M. and Krzyżosiak, W.J. (1997) CUG repeats present in myotonin kinase RNA form metastable "slippery" hairpins, *J. Biol. Chem.* **272,** 31079-31085.
35. Schlegl, J., Gegout,V., Schläger, B., Hentze, M.W., Westhof, E., Ehresmann, C.H., Ehresmann, B. and Romby, P. (1997) Probing the structure of the regulatory region of human transferrin receptor messenger RNA and its interaction with iron regulatory protein-1, *RNA* **3,** 1159-1172.
36. Otzen, D.E., Barciszewski, J. and Clark, B.F.C. (1993) Altered lead(II)-cleavage pattern of free Phe-tRNA^{Phe} and Phe-tRNA^{Phe} in ternary complex with EF-Tu:GTP, *Biochem. Mol. Biol. Int.* **31,** 95-103.
37. Hardt, W.D., Schlegl, J., Erdmann,V.A. and Hartmann, R.K. (1993) Role of the D arm and the anticodon arm in tRNA recognition by eubacterial and eukaryotic RNase P enzymes, *Biochemistry* **32,** 13046-13053.
38. Hardt, W.D., Schlegl, J., Erdmann,V.A. and Hartmann, R.K. (1995) Kinetics and thermodynamics of the RNase P RNA cleavage reaction: analysis of tRNA 3'-end variants, *J. Mol. Biol.* **247,** 161-172.
39. Burgstaller, P., Kochoyan, M. and Famulok, M. (1995) Structural probing and image selection of citrulline-and arginine-specific RNA aptamers identify base positions required for binding, *Nucleic Acids Res.* **23,** 4769-4776.
40. Wallis, M.G., Streicher, B., Wank, H., von Ahsen, U., Clodi, E., Wallace, S.T., Famulok, M. and Schroeder, R. (1997) *In vitro* selection of a viomycin-binding RNA pseudoknot, *Chemistry & Biology* **4,** 357-366.
41. Wallace, S.T. and Schroeder (1998) *In vitro* selection and characterization of streptomycin-binding RNAs: Recognition discrimination between antibiotics. *RNA* **4,** 112-123.
42. Ciesiołka, J. and Yarus, M. (1996) Small RNA-divalent domains, *RNA* **2,** 785-793.
43. Szweykowska-Kulińska, Z., Krajewski, J. and Wypijewski, K. (1995) Mutations of *Arabidopsis thaliana* pre-tRNA^{Tyr} affecting pseudouridylation of U_{35} , *Biochim. Biophys. Acta* **1264,** 87-92.
44. Majerfeld, I. and Yarus, M. (1998) Isoleucine:RNA sites with associated coding sequences, *RNA* **4,** 471-478.
45. Welch, M., Majerfeld, I. and Yarus, M. (1998) 23S rRNA similarity from selection for peptidyl transferase mimicry, *Biochemistry* **36,** 6614-6623.
46. Deng, H.Y. and Termini, J. (1992) Catalytic RNA reactions of yeast tRNA^{Phe} fragments, *Biochemistry* **31,** 10518-10528.
47. Baron, C., Westhof, E., Böck, A. and Giegé, R. (1993) Solution structure of selenocysteine-inserting tRNA^{Sec} from *Escherichia coli.* Comparison with canonical tRNA^{Ser} , *J. Mol. Biol.* **231,** 247-292.
48. Ciesiołka, J., Marciniec, T., Dziedzic, P., Krzyżosiak, W.J. and Wiewiórowski, M. (1987) in *Biophosphates and Their Analogues: Synthesis, Structure, Metabolism and Activity* (Bruzik, K.S. and Stec, W.J., eds) pp. 409-414, Elsevier Science Publishers BV, Amsterdam.
49. Ciesiołka, J. and Krzyżosiak,W.J. (1996) Structural analysis of two plant 5S rRNA species and fragments thereof by lead-induced hydrolysis, *Biochem. Mol. Biol. Int.* **39,** 319-328.
50. Ziętkiewicz, E., Ciesiołka, J., Krzyżosiak, W. J. and Słomski, R. (1990) The secondary structure model of mouse U1 snRNA as determined from the results of Pb^{2+}-induced hydrolysis. In: *Nuclear Structure and Function,* (Harris, J.R. & Zbarsky, J.B., eds), pp. 453-457, Plenum Press, New York.

51. Tallsjö, A., Svärd, S.G., Kufel, J. and Kirsebom, L.A. (1993) A novel tertiary interaction in M1 RNA, the catalytic subunit of *Escherichia coli* RNase P, *Nucleic Acids Res.* **21,** 3927-3933.
52. Felden, B., Himeno, H., Muto, A., McCutcheon, J. P., Atkins, J. F. and Gesteland, R. F. (1997) Probing the structure of the *Escherichia coli* 10Sa RNA (tmRNA), *RNA* **3,** 89-103.
53. Walczak, R., Westhof, E., Carbon, P. and Krol, A. (1996) A novel RNA structural motif in the selenocysteine insertion element of eukaryotic selenoprotein mRNAs, *RNA* **2,** 367-379.
54. Walczak, R., Carbon, P. and Krol, A. (1998) An essential non-Watson-Crick base pair motif in 3'UTR to mediate selenoprotein translation, *RNA* **4,** 74-84.
55. Hemmings-Mieszczak, M., Steger, G. and Hohn, T. (1997) Alternative structures of the cauliflower mosaic virus 35 S RNA leader: Implications for viral expression and replication, *J. Mol. Biol.* **267,** 1075-1088.
56. Matsuo, M., Yokogawa, T., Nishikawa, K., Watanabe, K. and Okada, N. (1995) Highly specific and efficient cleavage of squid $tRNA^{Lys}$ catalyzed by magnesium ions, *J. Biol. Chem.* **270,** 10097-10104.

PROTEIN-DNA RECOGNITION

DANIELA RHODES
MRC Laboratory of Molecular Biology, Hills Road, Cambridge CB2 2QH, UK, rhodes@mrc-lmb.cam.ac.uk

1. Introduction

Understanding how proteins recognize DNA in a sequence-specific manner is central to understanding the regulation of transcription and other cellular processes. I will review the principles of DNA recognition that have emerged from the large number of high-resolution structures determined over the last 10 years. The DNA-binding domains of transcription factors exhibit surprisingly diverse protein architectures, yet all achieve a precise complemetarity of shape facilitating specific chemical recognition of their particular targets. Although general rules for recognition can be derived, the complex nature of the recognition mechanism precludes a simple recognition code. In particular, it has become evident that the structure and flexibility of DNA and contacts mediated by water molecules contribute to the recognition process. Based on the structural information it has proven possible to design proteins with novel recognition specificities. Despite this considerable practical success, the thermodynamic and kinetic properties of protein/DNA recognition remain poorly understood (reviewed in 1).

2. Reconciling physical chemistry with structure

Protein and DNA molecules will interact if there is a loss of Gibbs free energy on complex formation. The change in free energy (ΔG) during complex formation depends upon the change in both entropy (DS) and enthalpy (ΔH) such that $\Delta G = \Delta H - (T \times \Delta S)$. The enthalpy term arises from the many very short-range non-covalent interactions between protein and DNA. The entropy term depends upon the nature of the solvent at the interacting surfaces of the protein and DNA before and after complex formation. If a significant number of ordered water molecules are displaced on complex formation, then the entropy term can favour the interaction. It is clear therefore that for a favourable contribution to ΔG, both the enthalpy and entropy terms require the protein to have a surface that is highly complementary both in terms of shape and chemistry to that of its DNA target.

3. Shape recognition

One of the most striking features of protein/DNA complexes is the complementarity of shape between protein and DNA at the interface. This is achieved with a variety of DNA-binding architectures, or domains, such as the helix-turn-helix motifs, homeodomains, basic leucine zipper proteins and zinc-fingers [2]. These domains are precisely docked on DNA by numerous contacts to the sugar phosphate backbone made from amino acids of the scaffold. This docking allows the "reading head" to reach inside

J. Barciszewski and B.F.C. Clark (eds.), RNA Biochemistry and Biotechnology, 123–126.

the major groove, where the pattern of hydrogen bond donors and acceptors is unique for each base-pair, and interact specifically with the bases. Many DNA-binding proteins employ an α-helical element to interact with bases in the major groove, but this is by no means universal.

4. Chemical recognition

The same physical rules that determine protein and nucleic acid structure govern their specific and non-specific interactions. The forces involved include hydrogen bonds, van der Waals forces, hydrophobic interactions, global electrostatic interaction and salt bridge interactions. High resolution crystal structures of protein-DNA complexes have revealed the three dimensional networks of interactions that tie the two molecules together. Certain common themes have emerged. In particular arginine and glutamine residues have long side chains that are able to make bidentate contacts to individual bases. Indeed the most commonly occurring interaction is of arginine with the N7 and O6 of guanine. Glutamine (and asparagine) can interact similarly with the N7 and N6 of adenine. However, both these amino acids are seen to make a variety of other contacts to bases. Finally, the bulky 5'-methyl group of thymine is suited to making van der Waals interactions with the methyls of several amino acids, although it may also play an important role in sterically preventing incorrect binding.

5. How to be more specific

DNA binding domains are generally small and compact and the consequence of this is that one such domain is not able to make a sufficient number of contacts with the DNA to specify a unique target site, or bind with reasonable affinity. Several strategies have been employed to overcome this problem. The first is simply to add on arms or tails that recognise additional features of the DNA, particularly in the minor groove. The second is to double up on the recognition by forming either homo- or hetero-dimers, thus specifying a longer DNA sequence [3]. In the latter case this also vastly increases the recognition possibilities through a combinatorial approach. The third method of increasing specificity is to employ multiple DNA-binding domains, either by using tandem repeats of the same type of DNA-binding motif e.g. the zinc-finger motif [2], or by linking together different types of motif.

6. The role of DNA structure

It is now generally accepted that both the conformation and rigidity of DNA are determined by local base stacking. For any one stretch of DNA helix the width and depth of the major and minor grooves, the displacement and orientation of the base pairs relative to the helix axis, the helical periodicity and the global bend of the DNA are determined by the sequence of bases. Consequently, the structure and flexibility of the double helix is continuously variable and this must play a role in protein-DNA recognition. Proteins recognise DNA sequences by both "direct readout" of the base sequence and through "indirect readout" of the relative positions of the phosphate and sugar moieties that comprise the backbone of the DNA. In a number of complexes it is apparent that the flexibility of DNA is important for protein-DNA recognition. There are now many examples of protein-DNA complexes in which the DNA is significantly distorted. An example of this is seen in the structure of the TATA binding protein bound to its DNA target, in which the DNA has two 90° bends [4]. Clearly the structural properties of the TATAAAA sequence are an important factor in facilitating this distortion.

7. Recognizing more than one DNA-sequence

Most specific DNA-binding proteins do not recognise a unique DNA target, rather they recognize a family of related DNA sequences. Alternative sequences are recognized using different types of adaptations. In the simplest case, the protein rearranges the conformation of surface side chains so as to create a slightly different network of hydrogen bonds. This is exemplified by the ERDBD bound to a non-consensus target. In other cases the DNA structure and the relative position of the protein change.

8. The role of solvent in specificity

In most of the protein-DNA crystal structures determined to date there are ordered water molecules present at the protein-DNA interface (reviewed in 5). The number of water molecules and their role varies in the different complexes. Whereas this is partly a consequence of the resolution at which the structures have been solved, it also seems that the nature of different protein-DNA interfaces can differ significantly in this regard. In the structure of the trp repressor bound to its DNA target there are few direct amino acid to base contacts. However there are many well ordered water molecules that participate in a network of hydrogen bonds bridging protein with DNA. Significantly, many of these water molecules were found to reside in essentially identical positions on the DNA in the absence of protein. Thus it appears that the protein is specifically recognising not just the DNA, but also the associated water structure.

9. Is there a discernible recognition code?

The wealth of structural information on specific protein/DNA complexes provides a database for addressing the long-standing question of whether or not there is a recognition code, in some way analogous to the genetic code. The simple answer is that there is no such code because protein DNA recognition is complex. However, for each family of DNA-binding motifs there is a discernible pattern of contacts. Amongst all of the DNA-binding motifs known, the classical C_2-H_2 zinc-finger motif provides the best candidate for understanding the rules for recognition (reviewed in 6). The framework of the zinc-finger is very simple and its orientation with respect to the DNA is probably dominated by the amino acid to base contacts. These occur from 4 main positions in the zinc-finger: one immediately preceding the α-helix and the other 3 from within the α-helix [2]. The binding site for a zinc-finger spans 3 bases on one strand with a single base contact on the other strand. Generally there is one to one recognition (one amino-acid ---- one base).

10. Understanding the thermodynamics and kinetics

Recent experimental data suggest that the change in hydration when proteins bind to DNA is a key thermodynamic variable in protein-DNA interactions, and that in general as the two surfaces come together, the interaction energies associated with hydration forces increase exponentially with the number of waters displaced. So is the displacement of water molecules sufficient to explain protein-DNA binding affinities?

Experiments show that on the formation of protein-DNA complexes there is a large increase in heat capacity (reviewed in 5 and 7). This is similar to the increase in heat capacity when protein molecules undergo a transition from a denatured to a folded state. An example of this is found in the ERDBD which is a monomer in solution, but binds to DNA as a dimer. The dimer interface appears disordered by NMR analysis prior to binding to DNA, but becomes ordered upon DNA-binding [8]. In addition to the need to account for the thermodynamics of protein-DNA interactions, we need to understand the kinetic events of binding and release of proteins from their DNA targets. These are particularly important if we are to understand the processes by which genes are turned on and off.

11. Custom built DNA-binding proteins

The available structural information on has been of immense help in allowing us to design custom-built DNA-binding proteins that will recognise either designed, or naturally occurring DNA targets. This is clearly very important for both biotechnological and medical applications. The most successful and efficient strategy for designing proteins with genuinely novel DNA-binding specificity has been to use the architecture of the C_2-H_2 zinc finger and the phage display technique (reviewed in 6). Using this approach a library containing millions of zinc-finger motifs has been produced. The binding site to be recognized (a base triplet) is then used to select the appropriate zinc-finger. These may then be linked together to recognize a binding site of the desired sequence and length.

References

1. Rhodes D., Schwabe, J.W.R., Chapman, L. and Fairall, (1996) *L. Phil. Trans. R. Soc. Lond.* B **351**, 501-509.
2. Fairall, L., Schwabe, J.W.R., Chapman, L., Finch, J.T. and Rhodes, D. (1993) *Nature* **366**, 483-487.
3. Schwabe, J.W.R., Chapman, L., Finch, J.T. and Rhodes, D. (1993) *Cell* **75**, 567-578.
4. Kim, J.L., Nikolov,D.B., and Burley, S.K. (1993) *Nature* **365**, 520-527.
5. Schwabe, J.W.R. (1997) *Curr. Opin. in Struct. Biol.* **7**, 126-134.
6. Choo, Y. and Klug, A. (1997) *Curr. Opin. in Struct. Biol.* **7**, 117-125.
7. Ladbury J.E. (1995) *Structure* **3**, 635-639.
8. Schwabe, J.W.R., Chapman, L., Finch, J.T. Rhodes, D. and Neuhaus, D. (1993) *Structure* **1**,187-204.

SPECIFIC INTERACTION BETWEEN DAMAGED BASES IN DNA AND REPAIR ENZYMES

KOSUKE MORIKAWA
The department of structural biology, Biomolecular Engineering Research Institute (BERI), 6-2-3 Furuedai, Suita, Osaka 565-0874, Japan

1. Abstract

T4 endonuclease V is a DNA repair enzyme from bacteriophage T4 which catalyzes the first reaction step of the pyrimidine dimer specific base excision repair pathway. The crystal structure of the enzyme complexed with a duplex DNA substrate, containing a thymine dimer, has been determined at 2.75Å resolution. The atomic structure of the complex reveals the unique conformation of the DNA duplex, which exhibits a sharp kink with a 60° inclination at the central thymine dimer. This kink divides the duplex into two B-DNA regions, each of which makes extensive polar interactions with the basic concave surface of the enzyme. The adenine base complementary to the 5' side of the thymine dimer is completely flipped out of the DNA duplex and is trapped in a cavity on the protein surface. These structural features allow an understanding of the catalytic mechanism and implicate a general mechanism of how other repair enzymes recognize damaged DNA duplexes.

J. Barciszewski and B.F.C. Clark (eds.), RNA Biochemistry and Biotechnology, 127–138.

2. Introduction

Excision repair directly acts on damaged moieties within DNA duplexes to proceed reactions for their elimination. Although nucleotide excision repair (NER) is assumed to play a central role in this process, the structural analyses of its molecular machinery is not impressive, because of their complicated and unstable architectures. On the other hand, base excision repair (BER) involves simpler enzymes at least for the initial steps of the reaction pathways, although it remains unclear how these enzymes may interact with other molecules participating in the later reaction steps. In the initial step of BER, a DNA glycosylase directly excises the modified base moieties from DNA duplexes to produce apurinic-apyrimidinic (AP) sites. The phosphate backbone at the abasic site is subsequently cleaved, either by the AP lyase activity belonging to the same DNA glycosylase or by the actions of other AP endonucleases. From the viewpoint of structural biology, the BER enzymes are indeed good targets to clarify the mechanism by which repair enzymes specifically recognize lesions within DNA duplexes.

This report focuses, in particular, on the direct visualization of a damage recognition mechanism, which is based on the first successful X-ray structure determination of a representative BER enzyme, T4 endonuclease V, complexed with a DNA duplex containing a pyrimidine photodimer. This crystallographic study revealed an unexpected damage recognition mechanism that flips a base out of a DNA duplex. The structural features of the DNA deformation implied that this mechanism can be generalized to damage recognition by various repair enzymes. Since then, the crystal structures of some other BER enzymes have been determined and such implications for the repair mechanism are indeed becoming a reality.

3. Crystal structure of T4 endonuclease V in the DNA-free state

Endonuclease (endo) V from bacteriophage T4 is the most popular DNA repair enzyme that has been biochemically investigated for a long time [1]. This enzyme has high affinity for both cyclobutane-type pyrimidine dimers (PDs) and AP sites within DNA duplexes and two distinct catalytic activities; scission of the glycosyl bond at the 5' side of PD and subsequent cleavage of the phosphodiester bond at the 3'position of the abasic site through β-elimination. This enzyme binds only to double stranded DNA and scans nontarget sequences by electrostatic interaction to search for the damaged site [1].

The T4 eodo V protein purified from E. coli overproducing strain was crystallized [2] and the first crystal structure of DNA repair enzymes was determined at 1.6Å resolution [3] and afterwards refined at 1.45Å resolution [4]. The enzyme is composed of a single compact domain, although it exhibits the two distinct activities. The molecule, consisting of three α-helices and the connecting loops, shows a remarkably unique arrangements of α-helices (Figure 1). The NH2 terminal segments penetrates between the two major H1 and H3 third helices, preventing their direct contacts. The two residues at the terminus lies on the molecular surface, whereas the internal peptide groups all form hydrogen bonds with the surrounding atoms. This folding scheme is inexplicable with the close packing category for the assembly of α-helices. The molecular graphic display of the electrostatic potentials exhibits a positively charged concave surface with a dimension of about 50Å x 40Å. This positive area, contributed by about ten basic residues, is largely over the diameter of a B-DNA duplex and hence overall appears to be consistent with the proposal that the enzyme can scan non-target sequences [1].

In parallel with crystallographic study, the extensive mutation analyses were carried out to identify crucial residues for the glycosylase activity [5]. Various mutant enzymes were isolated using site directed mutagenesis and their activities were examined. Among them, E23Q (replacement of Glu by Gln), E23D and R3Q completely abolished the activity. Notably, the former two mutants maintained the full DNA substrate binding ability, although R3Q completely lost this ability. Two mutants, R22Q and R26Q, also exhibited the extremely low activity. In the crystal structure, Arg 3, Arg22, Glu23, and Arg26, were concentrated in a small area on the basic and concave surface, as though the negative charge of the Glu23 side chain floats in a sea of the positive charges. In combination with the mutational analysis [5], thus, the crystal structure allowed the identification of the glycosylase catalytic center [3], where Glu23 are surrounded by the three basic Arg3, Arg22, and Arg26 residues.

The crystal structures of the three active site mutants, E23Q, E23D and R3Q, all were determined at atomic resolution [4]. The E23Q and R3Q mutants hardly showed the significant conformational change, except for the substituted side chains and some water molecules in the close vicinity of the catalytic center. On the other hand, the E23D mutation induced a small, but significant, change in the backbone structure, such as an increased central kink of the H1 helix at the Pro25. However, the catalytic center showed no notable structural change that should be directly associated with the impairment of the glycosylase activity. These results suggest that the negatively charged side chain of Glu23 and the positively charged guanidino group are crucial for the glycosylase activity and the DNA substrate binding, respectively.

4. Crystal structure of T4 endonuclease V complexed with a DNA duplex

The crystallographic study of the active site mutants also revealed that the E23Q mutant is the most appropriate to cocrystallize with a DNA substrate duplex, because it retains the full substrate binding ability and exhibits no significant conformational change. After the extensive screening of synthetic DNA duplexes, we finally obtained the crystals of the mutant enzyme complexed with a DNA duplex,

ATCGCG**TT**GCGCT
 AGCGCAACGCGAT

which has a thymine dimer (bold letters) at the center.

The crystal structure of the complex [6, 7] was solved at 2.75Å resolution by the molecular replacement method using a 1.45Å structure of the DNA-free enzyme. The conformation of the enzyme in complex with the DNA coincide well with that of the DNA-free enzyme, except for a loop (residues 125-130) near the carboxyl terminus and side chains and a part (83-91) of a long loop which shows different side chain orientations. The rmsd values between the two structures are 0.39Å for all main chain atoms and 0.68Å for all the atoms including side chains.

The sharply kinked (60 degrees) DNA duplex is bound to the concave basic surface and this kink divides the duplex into two B-DNA regions (Figure 1). Half of the basic surface covers the L-strand through lots of direct and water-mediated hydrogen bonds with a sugar-phosphate backbone and likewise the remaining half does the R-strand (Figure 1 and 2). However, any interactions were not observed between DNA bases and protein atoms at all. As suggested before [8], the enzyme interacts with the thymine dimer in the minor groove (Figure 1).

5. Base flipping out

Compared with other protein-DNA complex involved in transcription, the most remarkable feature of the complex is that the adenine base complementary to the 5' side thymine of the photo-dimer is completely flipped out of the B-DNA duplex interior, while the dimer itself remains inside the duplex (Figure 1 and 3). The apparent hole created by the base flipping-out is further expanded by the concavo-convex distortion of adjacent two base pairs, although the base-pairs themselves are not disrupted. The adenine base flipped out is accommodated into a cavity on the enzymatic surface (Figure 3). Surprisingly, this adenine makes no polar interaction with protein atoms within the cavity,. Instead, it appears to be sandwiched between two layers consisting of protein atoms and water molecules through van-der Waals interactions. In fact, T4 endo V could efficiently cleave the modified DNA substrate duplexes, which contain guanine, thymine cytosine and even 5-phenyl uracil in place of the complementary adenine (Maeda, M. unpublished result). This finding supports the above hypothesis that no polar interactions in the cavity are prerequisite for the cleavage.

6. Other interactions with DNA

The sharp kink at the central thymine dimer is accompanied with striking deformation of the backbone structure. Most notably, the spacings between the adjacent two phosphates, both of which belong to the thymidine dimer becomes smaller by 1.5Å, as compared with 6.7Å in the B-DNA duplex. These two phosphates are recognized through seven direct polar interactions with five basic residues including Arg3. In particular, the two phosphates form four hydrogen bonds with the side chain of Arg3, whose replacement by Gln abolished the substrate binding ability. Likewise, the neighboring phosphate pair with a shortened spacing forms water-mediated hydrogen bonds with the enzyme.

7. Roles of base flipping-out

The crystal structure of the DNA substrate has not been determined in the absence of the enzyme yet. However, NMR analyses revealed that synthetic DNA duplexes containing PD maintains all of base pairs, although hydrogen bonds with PD are weakened [9-11]. These results imply that the remarkable DNA kink is induced by the binding of the enzyme. Presumably, the electrostatic force between the enzyme and the DNA backbone may crash the PD moiety which has weaker stacking interactions. Thus, the kink should be coupled with the deformation of the sugar-phosphate back bone around PD. I assume that the flipping-out of the adenine base even takes place in concert with the kink. Then, the roles of the base flipping-out partly would be to alleviate the local tension generated by the kink. In addition to no polar interaction between the base and protein atoms within the cavity, this hypothesis is consistent with the finding that a DNA duplex containing an abasic site is efficiently bound to the enzyme [12].

Another interesting feature of the base flipping is that the resultant hole of the DNA duplex is filled up by many protein atoms that all participate in the catalytic reaction of endo V. Thus, the apparent hole provides an empty space for the catalytically active residues, such as the NH_2 terminus, Arg22, Glu23 and Arg26, to access to a target for the catalytic reaction. The enclosure of the hole also may play an important role in the protection of the transition state from unfavorable contacts with the solvent molecules. In addition, the trapping of the flipped-out base into the cavity may contribute to preventing the slippage of the DNA substrate along the protein interface and thus to making its correct and intimate contact with the enzyme.

Biochemical studies reveled that an imino-covalent enzyme-substrate intermediate is formed between the a-amino terminus and the C1' atom of the 5' deoxyribose of PD [13, 14]. The configurations of the catalytic residues in the hole essentially supports the catalytic scheme [7, 12, 13] proposed on the basis of this biochemical evidence. The a-amino terminus is actually located within a distance from the C1' atom, which is capable of the covalent bond formation. The carboxyl side chain of Glu23 would stabilize the positively charged Schiff base intermediate [15] and donate a proton to a pyrimidine ring so as to make the N-glycosyl bond cleavable. Furthermore, it also may participate in β-elimination. The side chains of Arg22 and Arg26 appear to have structural roles in securing of the thymine ring and sugar moiety.

8. Comparison with other enzymes

A similar flipping-out base was also found in DNA methyltransferase (MT)-DNA complexes [16, 17], although this enzyme does not belong to DNA repair enzymes. In detail, however, the base-flipping out in these complexes are seriously different from that in T4 endo V. First, in the MT-DNA complex, a loop of the enzyme undergoes a large conformational change and penetrates into the hole within the DNA duplex, although the T4 endo V-DNA complex shows no conformational change of the enzyme. Secondly, the former complex shows a straight B-DNA duplex, in contrast with the sharp kink in the latter complex. Thirdly, in the PD recognition, only one adenine base complementary to PD is flipped out, whereas the flipped-out cytosine in the MT-DNA complex is the actual target for the catalytic reaction. Most recently, the crystal structure of the uracil-DNA glycosylase (UDG) complexed with a DNA duplex [18] has been reported. This repair enzyme cleaves the N-glycosyl bond at toxic uracil bases within DNA duplex to produce abasic sites. The complex structure revealed that the uracil base is flipped out of the straight B-DNA duplex and is completely buried into the enzyme through many polar interactions with protein atoms. This structural feature is very similar to that of the MT-DNA complex, although one side chain of the enzyme is inserted into the hole of the DNA duplex. In both of the MT-DNA and UDG-DNA complexes, many polar interactions are concentrated in the close vicinity to the target bases for the catalytic reaction. The T4 endo V-DNA complex contrasts with these complexes. In the former complex, the polar interactions are spread over the entire DNA duplex with the sharp kink, and the mechanism of the DNA deformation induced by binding with the enzyme already has been discussed above. Therefore, at the moment, it remains unsolved whether there are two distinct base flipping-out mechanisms.

9. Concluding remarks

The base flipping out is indeed an important and universal phenomenon in the biological world. Conceivably, this is the DNA recognition mechanism specific for enzymes, since simple DNA binding proteins like transcrptional reguratory factors have never shown the disruption of base pairs. Replication or transcrption process requires very large molecular machinery which consist of dozens proteins. In order for these super molecular complexes to achieve sophisticated catalytic reactions involving sequencial many steps, they need to contact with the large internal region of a DNA duplex which contains many reaction targets, and hence the DNA duplex must be unwound to expose them. On the other hand, the target for BER or MT is simply a small base, and then catalytic residues within enzymes are allowed to access to target bases, inside or outside the duplex case by case, only if they could be flipped out of the interior to exterior. It is prerequisite for this hypothesis that the base flipping-out should be frequently found in DNA relevant enzymes, in particular, involved in repair for mismatch base pair.

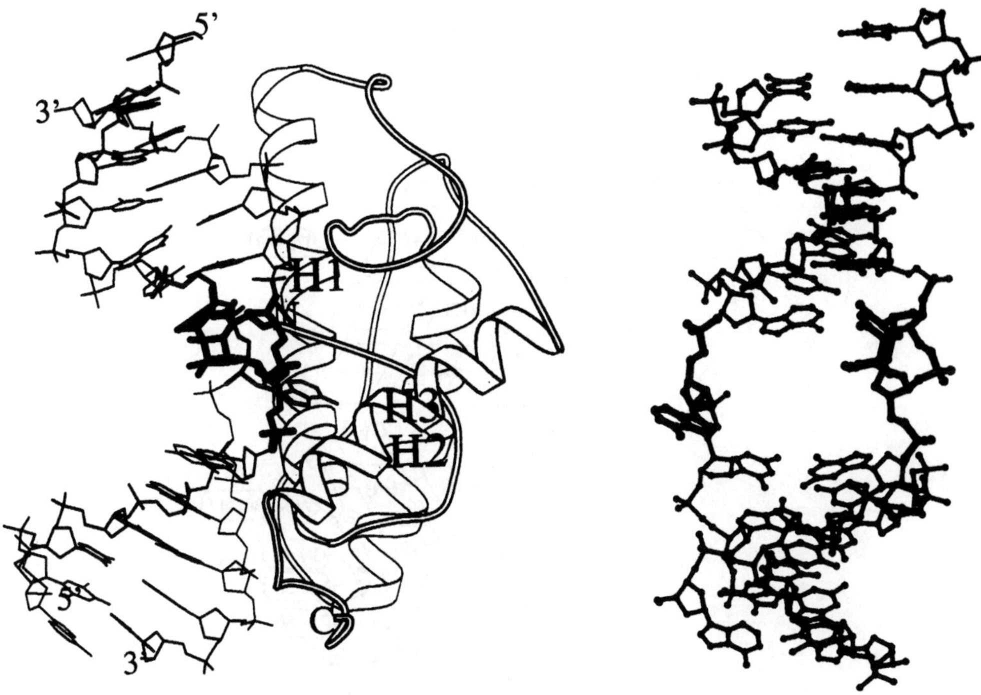

Figure 1. Structure of T4endoV-DNA substrate complex. Left: The protein and the DNA substrate containing a thymine-dimer at the center are shown by a ribbon model and a wire model. The thymine-dimer is denoted by bold lines. Note the flipped-out adenine base which lies behind the H1 helix. Right: Wire model of the DNA duplex in the complex (protein part not shown). The thymine-dimer and the complementary adenine on the 5' side are denoted by bold lines. Note the hole generated by the flipping-out of the adenine base and the distortions of the adjacent two base pairs. This hole is actually filled up by various chemical groups involved in the catalytic reaction.

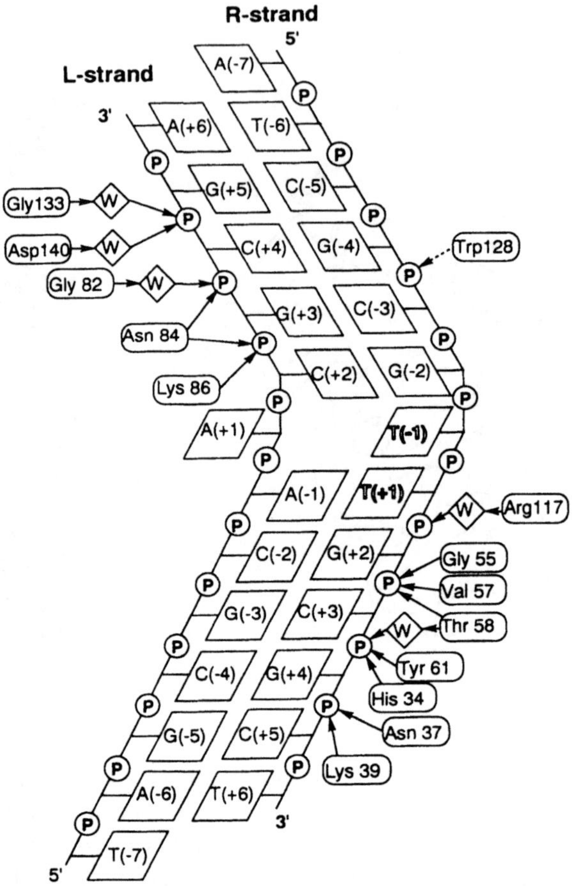

Figure 2. Schematic drawing of polar interactions between the enzyme and the DNA substrate. Open letters denote the thymine dimer. Polar interactions in the close vicinity of the thymine dimer are eliminated from the figure. Trp128 makes stacking interaction with the sugar ring of C(R-3). Note that direct read-out and water mediated indirect read-out are equally important for the enzyme-DNA substrate recognition.

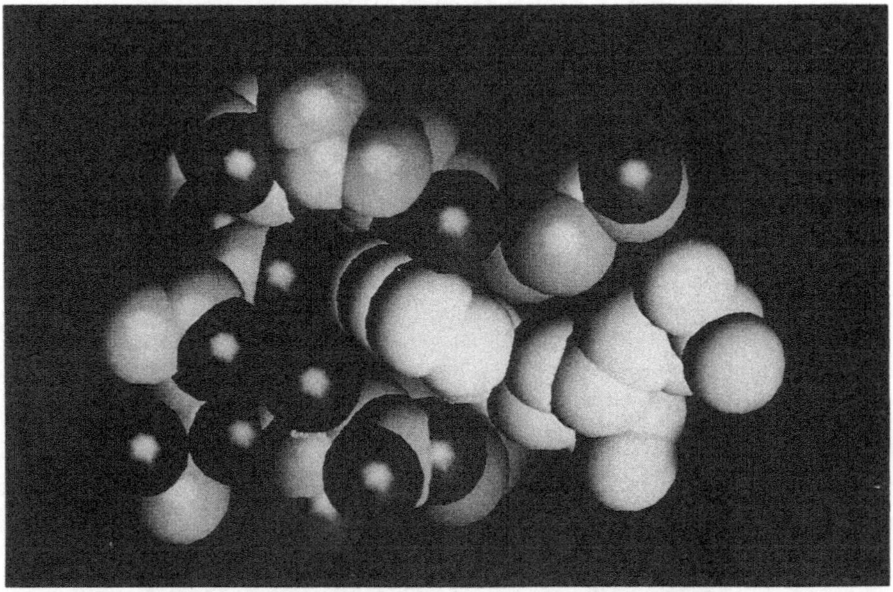

Figure 3. Flipped-out adenine base accommodated into the cavity on the protein surface. The adenine base (white) makes very intimate contacts with protein atoms.

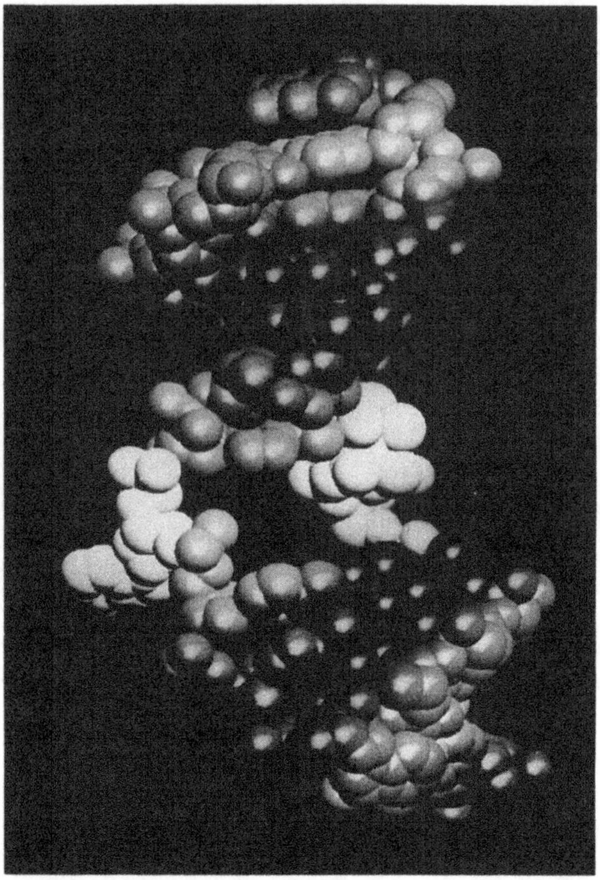

Figure 4. Space filling model representing the hole within the DNA duplex. The apparent space within the hole is actually occupied by protein atoms, which participate in the catalytic reaction.

References

1. Dodson, M.L., and Lloyd, R.S. (1989) Structure-function studies of the T4 endonuclease V repair enzyme, *Mutat. Res.* **218**, 49-65.

2. Morikawa, K. Tsujimoto, M., Ikehara, M., Inaoka, T., and Ohtsuka, E. (1988) Preliminary crystallographic study of pyrimidine dimer-specific excision-repair enzyme from bacteriophage T4, *J. Mol. Biol.* **202**, 683-684

3. Morikawa, K., Matsumoto, O., Tsujimoto, M., Katayanagi, K., Ariyoshi, M., Doi, T., Ikehara, M., Inaoka, T., and Ohtsuka, E. (1992) X-ray structure of T4 endonuclease V: an excision repair enzyme specific for a pyrimidine dimer, *Science* **256**, 523-526.

4. Morikawa, K., Ariyoshi, M., Vassylyev, D.G., Matsumoto, O., Katayanagi, K., and Ohtsuka, E. (1995) Crystal structure of a pyrimidine dimer-specific excision repair enzyme from bacteriophage T4: refinement at 1.45Å and X-ray analysis of three active site mutants, *J. Mol. Biol.* **249**, 360-375

5. Doi, T., Recktenwald, A., Karaki, Y., Kikuchi, M., Morikawa, K., Ikehara, M., Inaoka, T., Hori, N., and Ohtsuka, E. (1992). Role of the basic amino acid cluster and Glu-23 in pyrimidine dimer glycosylase activity of T4 endonucleae V, *Proc. Natl. Acad. Sci. USA* **89**, 9420-9424.

6. Vassylyev, D.G., Kashiwagi, T., Mikami, Y., Ariyoshi, M., Iwai, S., Ohtsuka, E. and Morikawa, K. (1995) Atomic model of a pyrimidine dimer excision repair enzyme complexed with a DNA substrate: structural basis for damaged DNA recognition, *Cell* **83**, 773-782.

7. Vassylyev, D.G., and Morikawa, K. (1997) DNA-repair enzymes, *Curr. Opin. Struc. Biol.* **7**, 103-109

8. Iwai, S., Maeda, M., Shirai, M., Shimada, Y., Osafune, T., Murata, T., and Ohtsuka, E. (1995) Reaction mechanism of T4 endonuclease V determined by analysis using modified oligonucleotide duplexes, *Biochemistry* **34**, 4601-4609.

9. Kemmink, J., Boelens, R., Koning, T.M.G., Kaptein, R., van der Marel, G.A., and van Boom, J.H. (1987) Conformational changes in the oligonucleotide duplex d(GCGTTGCG)·d(CGCAACGC) induced by formation of a *cis-syn* thymine dimer, *Eur. J. Biochem.* **162**, 37-43.

10. Kemmink, J., Boelens, R., Koning, T., van der Marel, G.A., van Boom, J.H., and Kaptein, R. (1987) ^{1}H NMR study of the exchangeable protons of the duplex d(GCGTTGCG)·d(CGCAACGC) containing a thymine photodimer, *Nucleic Acids Res.* **15**, 4645-4653

11. Lee, B.J., Sakashita, H., Ohkubo, T., Ikehara, M., Doi, T., Morikawa, K., Kyogoku, Y., Osafune, T., Iwai, S., and Ohtsuka, E. (1994) Nuclear magnetic resonance study of the interaction of T4 endonuclease V with DNA, *Biochemistry* **33**, 57-64.

12. Latham, K.A., Manuel, R.C., and Lloyd, R.S. (1995) The interaction of T4 endonuclease V E23Q mutant with thymine dimer- and tetrahydrofuran-containing DNA, *Jour. Bacteriol.* **177**, 5166-5168.

13. Schrock, R. D., III, and Lloyd, R.S. (1991) Reductive methylation of the amino terminus of endonuclease V eradicates catalytic activities, *J. Biol. Chem.* **266**, 17631-17639.

14. Dodson, M.L., Schrock III, R.D., and Lloyd, R.S. (1993) Evidence for an imino intermediate in the T4 endonuclease V reaction, Biochemistry **32**, 8284-8290.

15. Manuel, R.C., Latham, K.A., Dodson, M.L., and Lloyd, R.S. (1995) Involvement of glutamic acid 23 in the catalytic mechanism of T4 endonuclease V, *J. Biol. Chem.* **270**, 2652-2661

16. Cheng, X., Kumar, S., Posfai, J., Pflugrath, J.W., and Roberts, R.J. (1993) Crystal structure of the Hhal DNA methyltransferase complexed with S-adenosyl-L-methionine, *Cell* **74**, 299-307.

17. Kilmasaukas, S., Kumar, S., Roberts, R.J., and Cheng, X. (1994) Hhal methyltransferase flips its target base out of DNA helix, *Cell* **76**, 357-369.

18. Slupphaug, G., Mol, C.D., Kavil, B., Arvai, A.S., Krokan, H.E., and Tainer, J.A. (1996) Structure of human uracil-DNA glycosylase bound to DNA showa nucleotide flipping mechanism, *Nature* **384**, 87-92

TELOMERIC DNA RECOGNITION

DANIELA RHODES
MRC Laboratory of Molecular Biology, Hills Road, Cambridge CB2 2QH, UK; rhodes@mrc-lmb.cam.ac.uk

1. Introduction

Telomeres are the specialized protein-DNA complexes that cap linear eukaryotic chromosomes and are essential for the stable maintenance of chromosomes. The finding, about 20 years ago, that telomeric DNA consists of repeated sequence motifs opened the way for identifying both the specific proteins associated with telomeric DNA and the replication mechanism unique to telomeres. Telomeric DNA consists of species-specific short sequence motifs that typically contain clusters of three or four G-residues (e. g. TTGGGG in ciliates and TTAGGG in vertebrates). These sequence motifs are repeated in tandem, forming long double-stranded regions of telomeric DNA (reviewed in 1). The sequence and organization of telomeric DNA arises from the *de novo* addition of telomeric repeats by the telomerase, a reverse transcriptase containing an RNA template complementary to the G-rich strand. Since the telomerase is switched off in human somatic cells but is reactivated in cancer cells, telomere length has been implicated in both ageing and cancer (reviewed in 2).

The similarity between the telomeric DNA repeat sequences of widely divergent eukaryotes is likely to point towards some common, or conserved function. There is strong *in vivo* and *in vitro* evidence that the role of telomeric repeats is to recruit sequence-specific DNA-binding proteins and that such proteins are important for the structure and function of telomeres. Recently the three-dimensional structure of the DNA-binding domain of the yeast RAP1 in complex with a telomeric recognition site provided the first detailed picture of the recognition of the double-stranded telomeric DNA repeats [3]. I will review the insights this structure provides for understanding telomeric DNA recognition in general and also the possible emergence of a common protein fold for telomeric DNA recognition. This proposal arises from the discovery that other telomere binding factors, including the human TRF [4], contain a DNA-binding motif structurally related to RAP1.

2. Telomeric DNA recognition in *S. cerevisiae*

Of the proteins that bind to double-stranded telomeric DNA sequences, the best characterized in terms of its biological function is RAP1 (Repressor Activator Protein 1) from the budding yeast *S. cerevisiae*. Remarkably, RAP1 also functions as both an activator and repressor of transcription, hence its name (reviewed in 5). The interaction of RAP1 with the telomeric repeats is essential for chromosome stability, telomere length regulation, silencing of gene expression and the association of telomeres with the nuclear envelope. In silencing, the role of RAP1 is to recruit additional factors

J. Barciszewski and B.F.C. Clark (eds.), RNA Biochemistry and Biotechnology, 139–142.

(SIR2, SIR3, SIR4 and RIF1) to the telomeres, seeding the formation of transcriptionally repressed chromatin [5].

The 200-300 bp of *S. cerevisiae* telomeric DNA contain multiple copies of the RAP1 binding site $^{5'}$GGTGTGTGGGTGT$^{3'}$, which contains two tandem copies of the sequence $^{5'}$GGTGT$^{3'}$. RAP1 binds to DNA via a functionally independent and unusually large DNA-binding domain (DBD). The crystal structure [3] of the monomeric RAP1DBD (235 amino acid residues) in complex with an 18 base pair high affinity telomeric recognition site, shows that the protein has a bipartite structure consisting of two similar subdomains (Figure 1). Although the amino acid sequence of the protein gave no hint to this, the core of each domain consists of a bundle of three α-helices and hence is structurally closely related to both the homeodomain and Myb DNA-binding motif found in transcription factors.

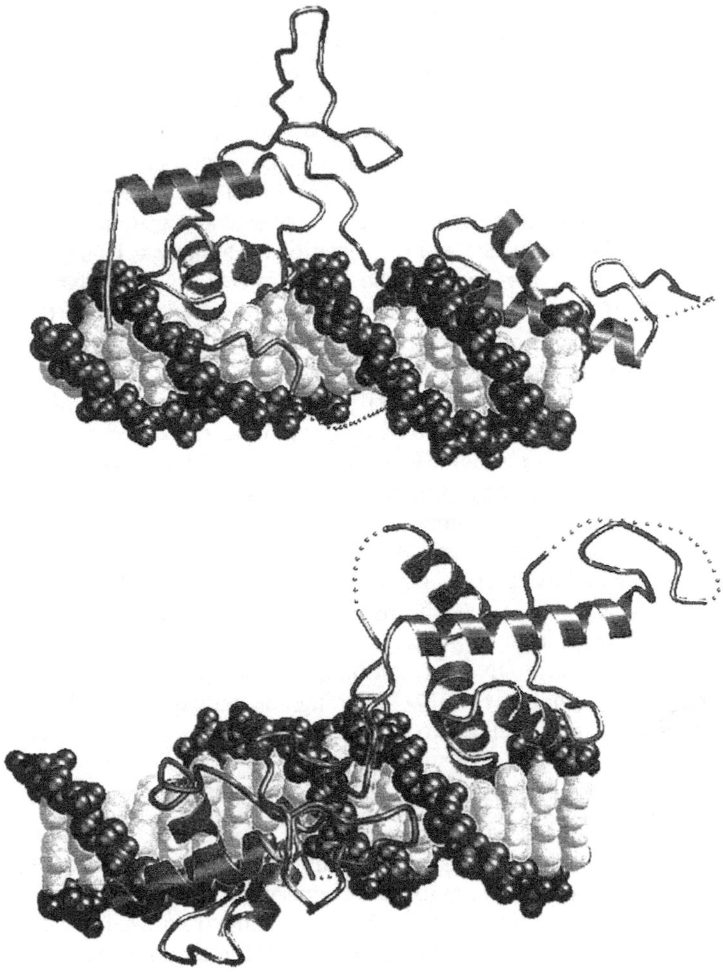

Figure 1.
Molscript representation of the DNA-binding domain of RAP 1 from *S. cerevisiae* in complex with a telomeric DNA binding site. Two views are shown.

The two domains contain additional structural elements and insertions to the three-helix bundles, and are connected by a loosely folded linker (Figure 1).

The most striking feature of the RAP1DBD structure is that the two domains are tandemly arranged on the DNA so that each is aligned on, and interacts with the intrinsic, tandemly repeated sequence GGTGT in the binding site (Figure 1). Each of the two domains make similar contacts to the repeated sequence motif: an N-terminal arm contacts a base in the minor groove and the third helix of the three-helix bundle, the "DNA-recognition helix", interacts with bases in the major groove of the DNA. Significantly, the clusters of guanines present in most telomeric repeats are contacted by the protein. The linker between the two domain crosses the minor groove between the two GGTGT repeats. The binding is further stabilized, as is common in protein/DNA complexes, by multiple interactions to the ribose-phosphate backbone.

3. Telomeric DNA recognition in other eukaryotes

In humans and other vertebrates the TTAGGG sequence motif is repeated in tandem for several thousand base pairs. Probing with nucleases shows that most of the double-stranded telomeric DNA region is packaged by histones, forming nucleosomes (reviewed in 6). The hunt for the first vertebrate telomeric protein led de Lange and colleagues to the identification of the human TTAGGG repeat binding factor (TRF1) [3]. Like RAP1, this protein is localized to the very ends of telomeric DNA and is involved in the regulation of telomere length. Although TRF1 and RAP1 appear to share some common functions at telomeres, they are not sequence homologues. However, and importantly for understanding how TRF1 may interact with telomeric target sites, its C-terminal region contains a single domain of about 50 amino acid residues which shows a significant sequence similarity to the DNA-binding motifs of the Myb proto-oncogenes. A Myb motif is also present in TAZ1, the recently cloned telomeric factor of the fission yeast *S. pombe* [6].

4. Telomeric DNA recognition by a conserved protein fold

The theme emerging from sequence comparison of proteins that bind to telomeric repeats and the structural information from RAP1 suggest that there might be a common structural motif for telomeric DNA recognition. It is instructive to note that in the absence of the structural information, the similarity between the DNA-binding motifs of these telomere binding proteins from evolutionarily distant species might not have been realized. The structural similarity between Myb motifs, homeodomains and RAP1DBD domains arises from the presence of a very similar bundle of three α-helices in all these DNA-binding motifs. The second and third helix of the three-helix bundles constitute the well known helix-turn-helix motif found in many prokaryotic and eukaryotic DNA-binding proteins. Although the amino acid sequence identity between these motifs is very low, a sequence alignment based on the structural information shows a conserved pattern of hydrophobic residues that form the structural core of the domains (Table 1).

```
                    Helix 1                    Helix 2           Helix 3

RAP1_1   HNKASFTDEEDEFILDVVRKN..PTRRTTHTLYDEISH.....VPNHTGNSIRHRFRVYLSKR
RAP1_2   SIKRKFSADEDYTLAIAVKKQFYRDLF....FFKHFAEE....HAAHTENAWRDRFRKFLLAY
MYB_2    LIKGPWTKEEDQRVIELVQKY.......GPKRWSVIAKH....LKGRIGKQCRERWH
MYB_3    VKKTSWTEEEDRIIYQAHKRL........GNRWAEIAKL....LPGRTDNAIKNHWNS
TRF1     RKRQAWLWEEDKNLRSGVRKY.......GEGNWSKILLHYKFN..NRTSVMLKDRWRTMKKLK
                   *     *                 *  *                *    *
```

Table 1. Amino acid sequence alignment based on structural information. Hydrophobic residues that form the core of the domains are indicated by (*)

Despite the close structural similarity Myb and RAP1 domains interact with DNA differently. At least two Myb motifs are required for DNA binding and the binding site of a single motif spans only 3 to 4 bp. In RAP1 the three-helix bundle of each domain is augmented by a short N-terminal arm that reaches into the minor groove where it makes specific contacts (Figure 1). `This has the effect of extending the length of the sequence that can be recognized from 3 or 4 bp by a Myb motif to 5 or 6 bp as seen for the RAP1 domains. Our recent model building and binding studies suggest that the Myb-domains of TRF1 do not bind to DNA like the Myb motifs of c-Myb, but bind to telomeric DNA just like the homeodomain-like domains of RAP1.

References

1. Rhodes, D. and Giraldo, R. (1995) *Curr. Opin. Struct. Biol.* **5**, 311-322
2. Harley, C. B. (1995) in *Telomeres* (eds. Blackburn, E.H.and Grider, C. W., Cold Spring Harbour Press Cold Spring harbour, New York) pp. 247-263
3. König, P., Giraldo, R., Chapman, L. and Rhodes, D. (1996) *Cell* **85**,125-136
4. Chong, L. *et al.* (1995) *Science* **270**,1663-1668
5. Shore, D. (1995) in *Telomeres* (eds. Blackburn, E.H.and Grider, C. W., Cold Spring Harbour Press Cold Spring harbour, New York) pp. 139-191
6. König, P.and Rhodes, D.(1997) *TIBS* **22**, 43-47

RECOGNITION OF ONE tRNA BY TWO CLASSES OF AMINOACYL-tRNA SYNTHETASE

M. IBBA, S. BUNJUN, H. LOSEY, B. MIN and D. SÖLL
Department of Molecular Biophysics and Biochemistry, Yale University, New Haven, CT 06520-8114, USA

Abstract

Lysyl-tRNA synthetases are unique amongst the aminoacyl-tRNA synthetases in being composed of two unrelated families. In most bacteria and all eukarya, the known lysyl-tRNA synthetases are subclass IIc-type aminoacyl-tRNA synthetases whereas some archaea and bacteria have been shown to contain an unrelated class I-type lysyl-tRNA synthetase. We have now examined substrate recognition by a bacterial (from *Borrelia burgdorferi*) and an archaeal (from *Methanococcus maripaludis*) class I lysyl-tRNA synthetase. The genes encoding both enzymes were able to rescue an *Escherichia coli* strain deficient in lysyl-tRNA synthetase, indicating their ability to functionally substitute for class II lysyl-tRNA synthetases *in vivo*. *In vitro* characterization revealed lysine activation and recognition to be tRNA-dependent, a phenomenon previously reported for other class I aminoacyl-tRNA synthetases. More detailed examination of tRNA recognition has shown that class I lysyl-tRNA synthetases recognize the same elements in $tRNA^{Lys}$ as their class II counterparts; specifically, the discriminator base (N73) and the anticodon serve as recognition elements. The implications of these results for the evolution of Lys-$tRNA^{Lys}$ synthesis and their possible indications of a more ancient origin for tRNA then aminoacyl-tRNA synthetases will be discussed.

Introduction

The accurate synthesis of aminoacyl-tRNAs is essential for faithful translation of the genetic code and is assumed to be one of the most highly conserved processes in biology. Recently, this dogmatic view has been called into question by the sequencing of a number of archaeal genomes; for example, the genomic sequence of *Methanococcus jannaschii* does not contain open reading frames (ORFs) encoding homologs of the asparaginyl-, cysteinyl-, glutaminyl- and lysyl-tRNA synthetases (1-3). The full complement of aminoacyl-tRNAs necessary for translation is not entirely formed by the aminoacyl-tRNA synthetases (AARS). In a significant number of cases, the AARSs serve to activate a non-cognate amino acid, while the generation of the correct aminoacyl-tRNA pair is subsequently brought about by a second protein. The use of such pathways for the formation of Gln-$tRNA^{Gln}$ (via Glu-$tRNA^{Gln}$) and Sec-$tRNA^{Sec}$ (via Ser-$tRNA^{Sec}$) is well documented in all the living kingdoms (4,5). It has also been found that in several Archaea

J. Barciszewski and B.F.C. Clark (eds.), RNA Biochemistry and Biotechnology, 143–148.

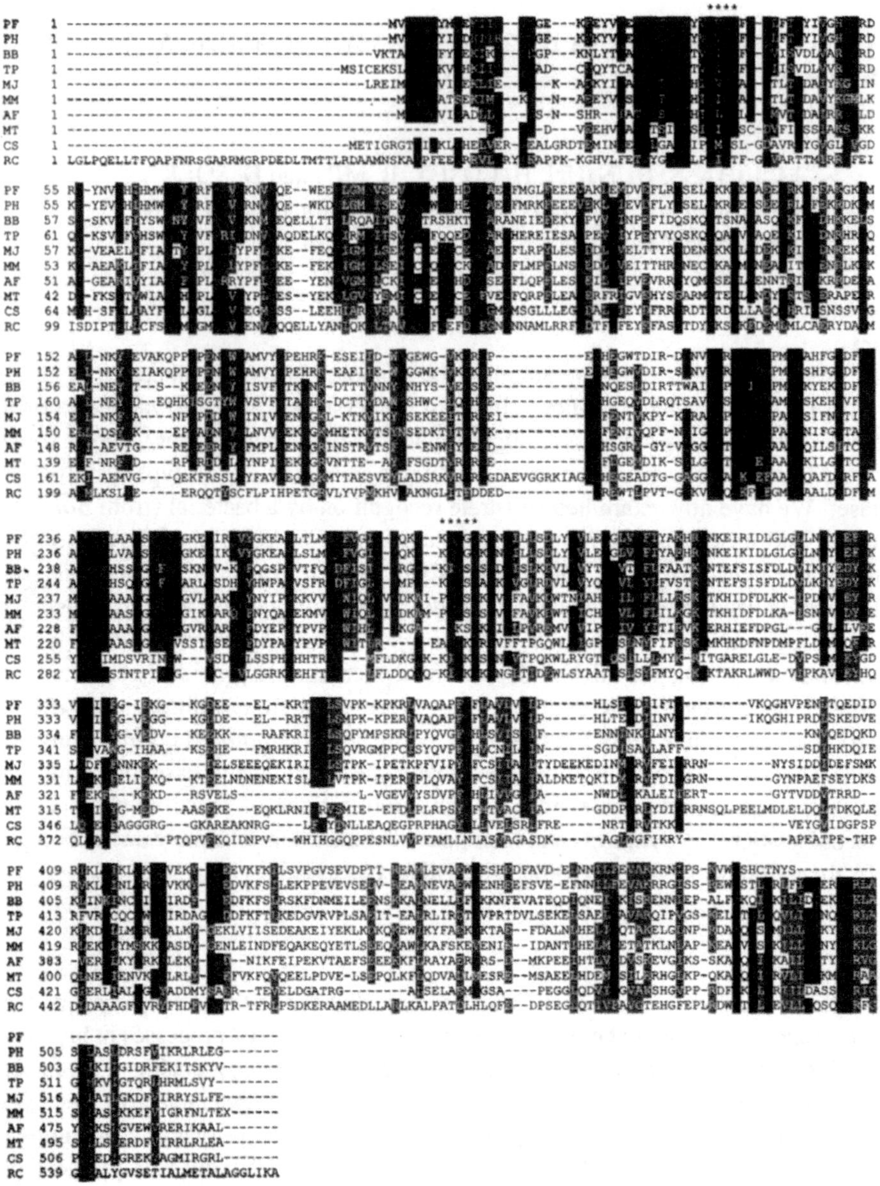

Figure. 1. Alignment of class I LysRS amino acid sequences. The signature sequences characteristic of class I aminoacyl-tRNA synthetases are indicated (*). The sequences were aligned using the CLUSTAL X program (11). The sequences shown are from *Pyrococcus furiosus* (PF), *Pyrococcus horikoshii* (PH), *Borrelia burgdorferi* (BB), *Treponema pallidum* (TP), *Methanococcus jannaschii* (MJ), *Methanococcus maripaludis* (MM), *Archaeoglobus fulgidus* (AF), *Methanobacterium thermoautotrophicum* (MT), *Cenarchaeum symbiosum* (CS) and *Rhodobacter capsulatus* (RC).

an additional aminoacyl-tRNA, Asn-tRNA^{Asn}, is also formed by transformation of a mischarged tRNA rather than by direct aminoacylation with asparaginyl-tRNA synthetase. Biochemical evidence indicates that aspartyl-tRNA synthetase initially synthesizes Asp-tRNA^{Asn}, which is subsequently converted to Asn-tRNA^{Asn} in a distinct tRNA-dependent transamidation reaction (6).

The use of two-step (indirect) aminoacylation pathways for the formation of Asn-tRNA^{Asn} and Gln-tRNA^{Gln} in some organisms circumvents the need for the enzymes which catalyze one step formation of these molecules, the asparaginyl- (AsnRS) and glutaminyl-tRNA synthetases (GlnRS) respectively. Consequently, it is not surprising that genes encoding these enzymes have not been found in the completed genomic sequences of organisms which employ one or both of the indirect pathways. However, in addition to lacking AsnRS and GlnRS, the genomic sequences of the euryarchaeons *Methanococcus jannaschii* and *Methanobacterium thermoautotrophicum* do not contain homologues of known cysteinyl- (CysRS) or lysyl-tRNA synthetases (1-3). While no adequate explanation yet exists for the apparent absence of CysRS, it has recently been shown that several members of the Archaea, including *M. jannaschii*, contain a functional lysyl-tRNA synthetase (LysRS) with no resemblance to known bacterial or eukaryal LysRSs or any other sequences in the public database (7). This is in contrast to all other AARSs which are highly conserved throughout the living kingdom. The high degree of conservation is exemplified by the invariant classification of AARSs into one of two classes defined by the presence of characteristic amino acid sequence motifs and topologically distinct nucleotide binding folds (8). This is not true of the recently identified archaeal LysRSs, which are class I-type AARSs and are thus easily distinguished from their known bacterial and eukaryal counterparts which are class II enzymes. While it was initially assumed that this novel type of LysRS was confined to certain Archaea, continued genomic sequencing efforts have suggested that it may also be found in some bacteria. This was confirmed by the cloning of a gene encoding a functional archaeal-type LysRS from the Lyme Disease Spirochete *Borrelia burgdorferi* (9). The spirochetes are a phylogenetically ancient bacterial group and the apparent existence of archaeal-type genes in these organisms raises a number of issues concerning their evolutionary origin and development.

Class I Lysyl-tRNA Synthetases

The aminoacyl-tRNA synthetases can be divided into two classes (I and II) of ten members each based on the presence of mutually exclusive amino acid sequence motifs (8). This division reflects structurally distinct topologies within the active site, class I aminoacyl-tRNA synthetases contain a Rossmann fold and class II a unique anti-parallel fold. It is generally assumed that an aminoacyl-tRNA synthetase of particular substrate specificity will always belong to the same class regardless of its biological origin, reflecting the ancient evolution of this enzyme family (10). The only known exceptions to this rule are the LysRSs which are composed of two unrelated families, namely class I enzymes in certain archaea and bacteria and class II enzymes in all other organisms (7, 9). The class I LysRSs are found in both kingdoms of archaea (Crenarchaotes and Euryarchaeotes), and two disparate bacterial taxa (spirochetes and α-proteobacteria). Comparison of known sequences shows that they all contain derivatives of the HIGH and

KMSKS sequence motifs characteristic of class I aminoacyl-tRNA synthetases (Fig. 1).

Archaeal Origin of Bacterial Class I LysRS

The level of sequence homology between bacterial and archaeal class I LysRSs suggest that they are more closely related than might be expected if they had evolved independently of each other. This is borne out by phylogenetic analyses which suggest that the bacterial examples originated in archaea from where they were acquired by horizontal gene transfer. Analysis of class I LysRS amino acid sequences by phylogenetic methods shows a relationship between the archaeal examples, which recapitulates phylogenies deduced from small subunit rRNA sequences. The crenarchaeal and euryarchaeal kingdoms are distinct, with the *Pyrococci* forming a coherent group within the latter. The bacterial class I LysRS proteins do not group together instead branching with the pyrococcal sequences in the case of the spirochetes, and *C. symbiosum* in the case of *R. capsulatus* (Fig. 2).

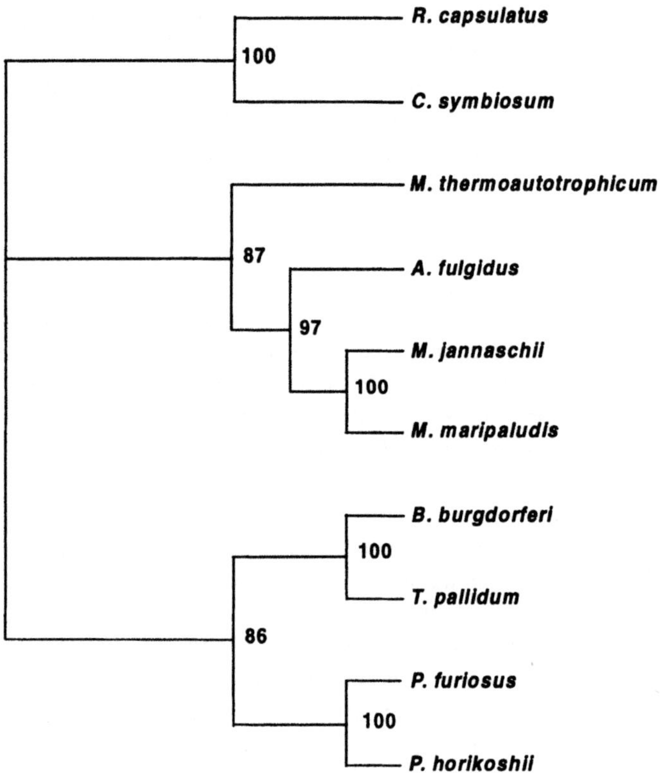

Figure 2. Phylogenetic analysis of class I lysyl-tRNA synthetases constructed using the maximum likelihood method implemented in the program PUZZLE 4.0 (12). Numbers represent the percentage occurrence of nodes during 1000 puzzling steps.

This separation of the bacterial sequences, none of which are deeply rooted in the class I LysRS phylogeny, indicates that they have arisen relatively recently following two separate horizontal gene transfer events from archaea. The absence of class I LysRS sequences from the genomic sequences of more deeply rooted bacteria such as *Aquifex aeolicus* and *Thermotoga maritima* indicates that the class I LysRS was absent during the early evolution of bacteria, supporting an archaeal origin for the contemporary bacterial examples of this protein.

tRNA Recognition by Class I Lysyl-tRNA Synthetases

The mechanism of $tRNA^{Lys}$ recognition by class I LysRSs was investigated *in vivo* and *in vitro*. It was shown by complementation of an appropriate deletion strain (14) that both an archaeal and a bacterial class I LysRS-encoding gene could functionally replace the class II LysRS-encoding genes of *E. coli*. *In vitro* characterization of class I LysRSs revealed lysine activation and recognition to be tRNA-dependent, a phenomenon previously reported for other class I aminoacyl-tRNA synthetases. A more detailed examination of tRNA recognition was then undertaken using a series of *E. coli* $tRNA^{Lys}$ variants (15) transcribed *in vitro*. This set of tRNAs was used to determine steady-state kinetic parameters for the aminoacylation reaction using the class I LysRS of *B. burgdorferi*. Comparison of these results with those for the class II LysRS of *E. coli* shows that both types of lysyl-tRNA synthetase recognize the same elements in $tRNA^{Lys}$. as their class II counterparts. Both classes of LysRS recognize the discriminator base (N73) and the anticodon, with nucleotides U35 and U36 being of particular importance. The observation that class I and class II LysRSs recognize the same identity elements in $tRNA^{Lys}$ *in vitro* correlates with the finding that one can functionally replace the other *in vivo*.

Conclusions

The discovery that lysyl-tRNA is synthesized by unrelated class I and class II aminoacyl-tRNA synthetases has revealed an unexpected degree of diversity in this family of enzymes. Preliminary investigations have now begun to indicate that despite the obvious differences between the class I and class II lysyl-tRNA synthetases, they recognize the same identity elements in $tRNA^{Lys}$. Further studies are now needed in order to understand the molecular basis by which substrate conservation has led to the functional convergence of these divergent lysyl-tRNA synthetases.

Acknowledgements

Work in the authors laboratory was supported by grants from the National Institute for General and Medical Scinces to DS.

References

1. Doolittle, R.F. (1998) Microbial genomes opened up. *Nature* **392**, 339-342.
2. Ibba, M., Curnow, A.W. and Söll, D. (1997) Aminoacyl-tRNA synthesis: divergent routes to a common goal. *Trends Biochem. Sci.* **22**, 39-42.
3. Dennis, P.P. (1997) Ancient ciphers: translation in Archaea. *Cell* **89**, 1007-1010.
4. Curnow, A.W., Hong, K.W., Yuan, R., Kim, S.I., Martins, O., Winkler, W., Henkin, T.M. and Söll, D.

(1997) Glu-tRNA^{Gln} amidotransferase: a novel heterotrimeric enzyme required for correct decoding of glutamine codons during translation. *Proc. Natl. Acad. Sci. USA* **94**, 11819-11826.

5. Baron, C. and Böck, A. (1995) The selenocysteine-inserting tRNA species: structure and function, in D. Söll and U.L. RajBhandary (eds.), *tRNA, Structure, Biosynthesis, and Function*, ASM Press, Washington, D.C., pp. 529-544.
6. Curnow, A.W., Ibba, M. and Söll, D. (1996) tRNA-dependent asparagine formation. *Nature* **382**, 589-590.
7. Ibba, M., Morgan, S., Curnow, A.W., Pridmore, D.R., Vothknecht, U.C., Gardner, W., Lin, W., Woese, C.R. and Söll, D. (1997) A euryarchaeal lysyl-tRNA synthetase: resemblance to class I synthetases. *Science* **278**, 1119-1122.
8. Arnez, J.G. and Moras, D. (1997) Structural and functional considerations of the aminoacylation reaction. *Trends Biochem. Sci.* **22**, 211-216.
9. Ibba, M., Bono, J.L., Rosa, P.A. and Söll, D. 1997. Archaeal-type lysyl-tRNA synthetase in the Lyme disease spirochete Borrelia burgdorferi. *Proc. Natl. Acad. Sci. USA* **94**, 14383-14388.
10. Brown, J. R. (1998) Aminoacyl-tRNA synthetases: evolution of a troubled family, in J. Wiegel and M.H.W. Adams (eds), *Thermophiles: The keys to molecular evolution and the origin of life?* Taylor and Francis, London, pp 217-230.
11. Thompson, J.D., Gibson, T.J., Plewniak, F., Jeanmougin, F. and Higgins, D.G. (1997) The CLUSTAL_X windows interface: flexible strategies for multiple sequence alignment aided by quality analysis tools. *Nucleic Acids Res.* **25**, 4876-4882.
12. Strimmer, K., and von Haeseler, A. (1996) Quartet puzzling: A quartet maximum likelihood method for reconstructing tree topologies. *Mol. Biol. Evol.* **13**, 964-969.
13. Deckert, G., Warren, P.V., Gaasterland, T., Young, W.G., Lenox, A.L., Graham, D.E., Overbeek, R., Snead, M.A., Keller, M., Aujay, M., Huber, R., Feldman, R.A., Short, J.M., Olsen, G.J. and Swanson, R.V. (1998). The complete genome of the hyperthermophilic bacterium *Aquifex aeolicus*. *Nature* **392**, 353-358.
14. Chen, J., Brevet, A., Lapadat-Tapolsky, M., Blanquet, S. and Plateau, P. (1994) Properties of the lysyl-tRNA synthetase gene and product from the extreme thermophile *Thermus thermophilus*. *J. Bacteriol.* **176**, 2699-2705.
15. Tamura, K., Himeno, H., Hasegawa, T., and Shimizu, M. (1992) *In* vitro study of *E.coli* tRNA^{Arg} and tRNA^{Lys} identity elements. *Nucleic Acids Res.* **20**, 2335-2339.

FUNCTIONAL STRUCTURES OF CLASS-I AMINOACYL-tRNA SYNTHETASES

Osamu NUREKI[1,2], Shun-ichi SEKINE[1,2], Atsushi SHIMADA[1], Takashi NAKAMA[1], Shuya FUKAI[1], Dmitry G. VASSYLYEV[2], Ikuko SUGIURA[3], Sachiko KUWABARA[3], Masaru TATENO[2], Masayoshi NAKASAKO[2], Dino MORAS[4], Michiiko KONNO[3] and Shigeyuki YOKOYAMA[1,2]

[1]Department of Biophysics and Biochemistry, Graduate School of Science, University of Tokyo, 7-3-1 Hongo, Bunkyo-ku, Tokyo 113-0033, Japan, [2]The Institute of Physical and Chemical Research (RIKEN), 2-1 Hirosawa, Wako-shi, Saitama 351-0198, Japan, [3]Department of Chemistry, Faculty of Science, Ochanomizu University, 2-1-1 Otsuka, Bunkyo-ku, Tokyo 112-8610, Japan, [4]Institut de Génétique et de Biologie Moléculaire et Cellulaire, CNRS, BP 163, 67404 Illkirch Cedex, France

1. Introduction

Aminoacyl-tRNA synthetase (aaRS) catalyzes the two-step aminoacylation reaction: the amino acid and the ATP form an aminoacyl-AMP in the first step, and the aminoacyl moiety is transferred to the 3' adenosine of the tRNA in the second step.

$$\text{aaRS} + \text{aa} + \text{ATP} \rightarrow \text{aaRS}\bullet\text{aa-AMP} + \text{PP}_i \quad (1)$$

$$\text{aaRS}\bullet\text{aa-AMP} + \text{tRNA}^{\text{aa}} \rightarrow \text{aaRS} + \text{aa-tRNA}^{\text{aa}} + \text{AMP} \quad (2)$$

The twenty aminoacyl-tRNA synthetases are divided into two classes I and II (each consisting of 10 members), on the basis of the motifs for ATP binding [1]. As the catalytic-site architectures are largely different between classes I and II of aminoacyl-tRNA synthetases, these two classes must have

J. Barciszewski and B.F.C. Clark (eds.), RNA Biochemistry and Biotechnology, 149–158.

evolved from separate ancestor enzymes [2]. Class I consists of methionyl-, isoleucyl-, valyl-, leucyl-, cysteinyl-, arginyl-, glutaminyl-, glutamyl-, tyrosyl, and tryptophanyl-tRNA synthetases (MetRS, IleRS, ValRS, LeuRS, CysRS, ArgRS, GlnRS, GluRS, TyrRS, and TrpRS, respectively). These ten class-I synthetases are characterized by the characteristic ATP-binding motifs, His-Ile-Gly-His (HIGH) and Lys-Met-Ser-Lys (KMSK) [1].

Crystal structures have been determined for *Bacillus stearothermophilus* TyrRS [3], *Escherichia coli* GlnRS (bound with $tRNA^{Gln}$) [4], *E. coli* MetRS [5], *Thermus thermophilus* GluRS [6], *B. stearothermophilus* TrpRS [7], *T. thermophilus* IleRS [8], and *Saccharomyces cerevisiae* ArgRS [9]. These synthetases share a catalytic Rossmann fold and an intervening domain between the NH_2- and COOH-terminal halves.

2. Three Subclasses of Class-I Aminoacyl-tRNA Synthetases

Class-I synthetases are subdivided into three subclasses (classes Ia, Ib, and Ic) from the primary structures; class Ia consists of MetRS, IleRS, ValRS, LeuRS, and CysRS; class Ib consists of GlnRS, GluRS, and ArgRS; class Ic consists of TyrRS and TrpRS [2]. GlnRS, GluRS, and ArgRS require the cognate tRNA as a cofactor for the aminoacyl-AMP formation (the first step of aminoacyl-tRNA synthesis), while other synthetases can catalyze this reaction in the absence of tRNA. This was one of the reasons for classification of GlnRS, GluRS, and ArgRS to the same subclass (class Ib). Recently, however, the crystal structure of yeast ArgRS was determined [9]. We have also determined the crystal structure of *T. thermophilus* ArgRS (Shimada et al., unpublished results), which is similar to the yeast ArgRS structure. With respect to the catalytic-domain and anticodon-binding-domain structures, ArgRS is much more similar to MetRS and IleRS than to GlnRS and GluRS. Therefore, ArgRS belongs to class Ia rather than to class Ib.

The GlnRS and GluRS structures [4, 6] are compared in Figure 1 (upper and lower, respectively). The NH_2-terminal halves of these two class-Ib synthetases are highly homologous to each other. In contrast, the COOH-terminal half of GlnRS is rich in β-structure, while that of GluRS is composed of α-helices. The COOH-terminal halves are the anticodon-binding domains [4, 10]. Therefore, these two enzymes might have evolved, one from the other, by replacement of the entire anticodon domains possibly through exon shuffling [6]. It is also suggested that the GlnRS gene was horizontally transferred from eukaryotes to bacteria, in rather late evolution

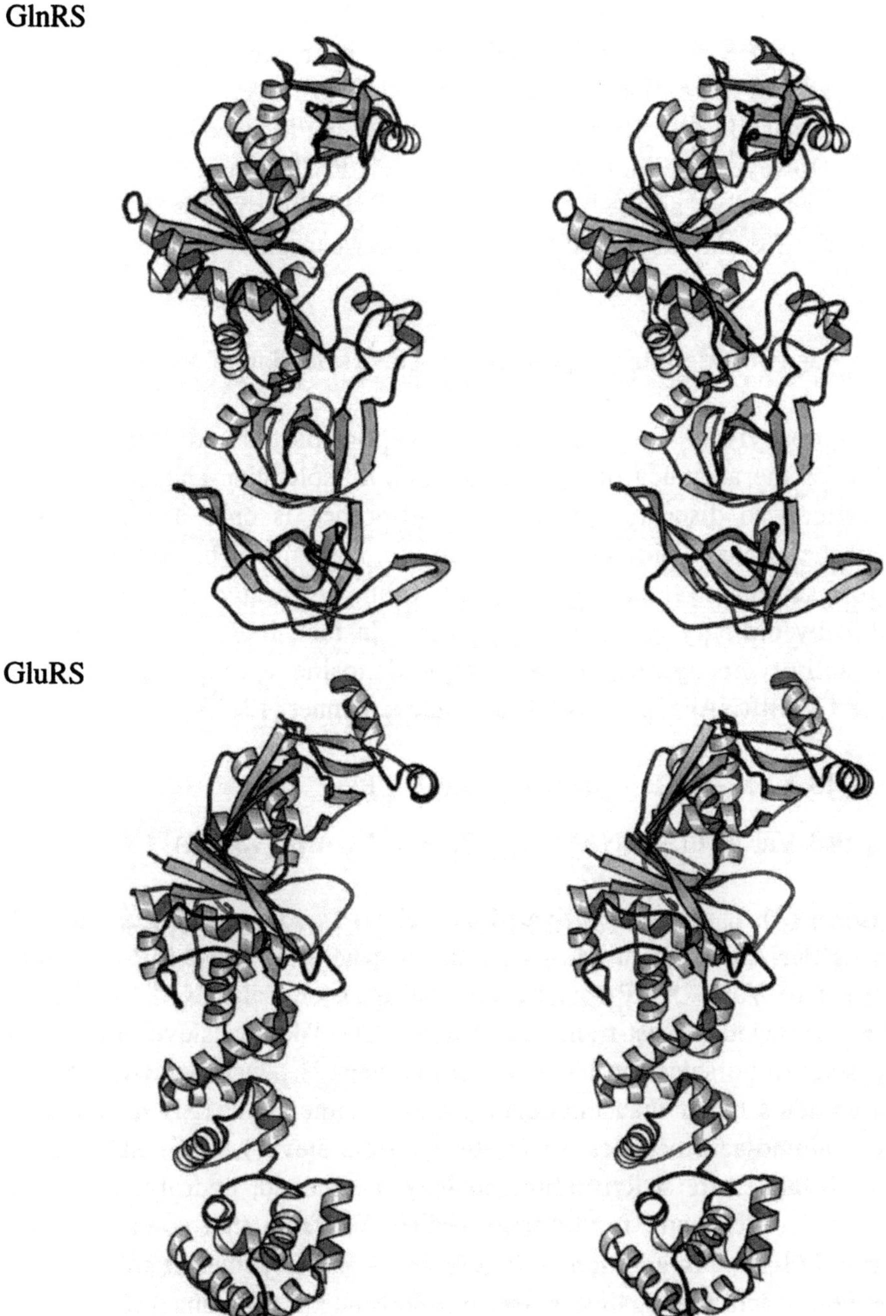

Figure 1. Crystal structures of *E. coli* GlnRS (from the $tRNA^{Gln}$-bound form) [4] and *T. thermophilus* GluRS [6] (stereoviews).

[11]. On the other hand, crystal structures of the two class-Ic synthetases, TyrRS and TrpRS, are highly homologous to each other [7]. TyrRS and TrpRS (class Ic) are dimeric, which is essential for the enzymatic activities. In contrast, class-Ia and -Ib synthetases are monomeric, except that MetRS is a dimer. However, the monomeric form of MetRS lacking the dimerization domain is as active as the wild-type MetRS in the dimeric form. This indicates a different enzymatic mechanism between class-Ia/Ib and class-Ic synthetases.

3. IleRS has two catalytic sites for double-sieve selection of amino acids

Some aminoacyl-tRNA synthetases are responsible for editing activities essential for the accuracy of translation [12]. L-isoleucine and L-valine are most difficult to discriminate, as their difference is only one methylene group in the aliphatic side chains. Therefore, it seems to be impossible for IleRS to discriminate L-isoleucine from L-valine so strictly (the error rate of 1/40,000) by ordinary one step recognition. In fact, IleRS has an essential editing activity to hydrolyze both valyl-adenosine monophosphate (Val-AMP) and Val-$tRNA^{Ile}$ in a $tRNA^{Ile}$-dependent manner [12, 13].

$$\text{IleRS} + \text{Val} + \text{ATP} \rightarrow \text{IleRS}\bullet\text{Val-AMP} + \text{PP}_i \quad (3)$$

$$\text{IleRS}\bullet\text{Val-AMP} + \text{tRNA}^{\text{Ile}} \rightarrow \text{IleRS} + \text{Val} + \text{AMP} + \text{tRNA}^{\text{Ile}} \quad (4)$$

In equation (3), L-valine is activated as well as L-isoleucine. Equation (4) shows the tRNA-dependent editing by IleRS, where the Val-AMP is directly hydrolyzed to Val + AMP (pre-transfer editing), or Val-$tRNA^{Ile}$ is formed and then deacylated (post-transfer editing). The "double sieve" model for two-step substrate selection has been proposed [12]. In this model (Figure 2), amino acids larger than the cognate L-isoleucine are strictly excluded by the first, amino-acid activation site (the "coarse sieve"), and smaller ones, such as L-valine, are strictly eliminated by the second, hydrolytic site (the "fine sieve"). The site for editing is different from that for amino-acid activation [13-15]. In addition to IleRS, valyl-tRNA synthetase (ValRS) has similar editing activities against an isosteric substrate, L-threonine [12].

$$\text{ValRS} + \text{Thr} + \text{ATP} \rightarrow \text{ValRS}\bullet\text{Thr-AMP} + \text{PP}_i \quad (5)$$

$$\text{ValRS}\bullet\text{Thr-AMP} + \text{tRNA}^{\text{Val}} \rightarrow \text{ValRS} + \text{Thr} + \text{AMP} + \text{tRNA}^{\text{Val}} \quad (6)$$

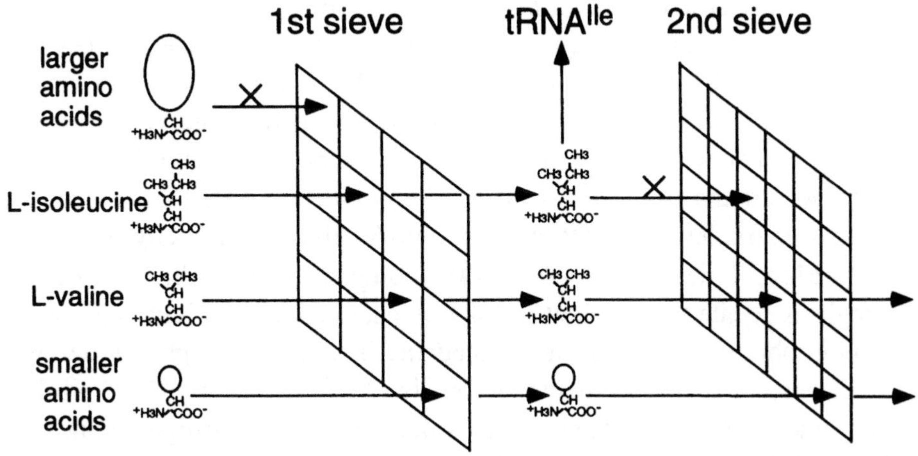

Figure 2. "Double-sieve" mechanism for two-step selection of L-isoleucine by IleRS.

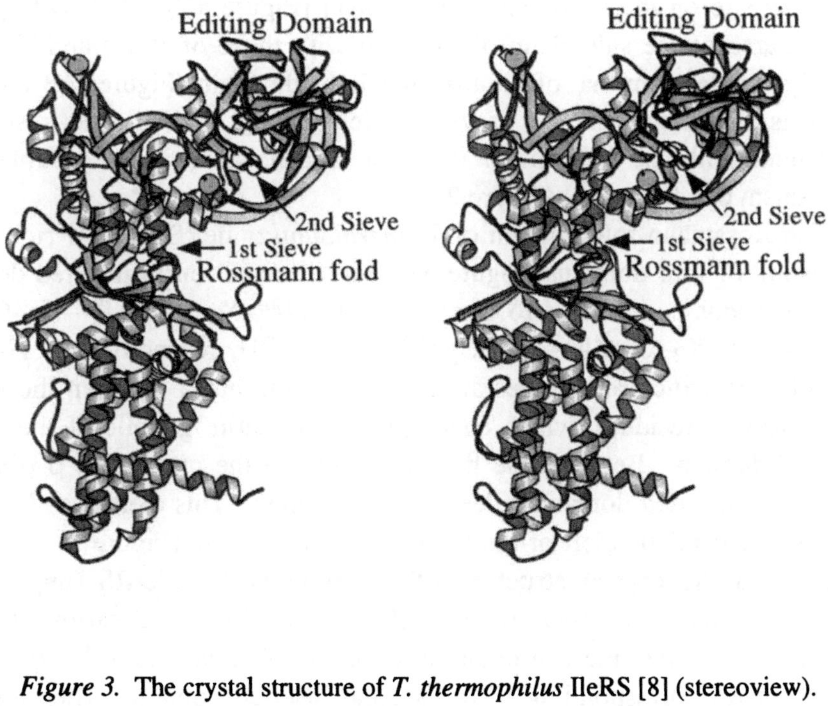

Figure 3. The crystal structure of *T. thermophilus* IleRS [8] (stereoview).

The crystal structure of *T. thermophilus* IleRS (120 kDa) at a resolution of 2.5 Å was determined, together with those of the complexes of IleRS with L-isoleucine and L-valine at resolutions of 2.8 Å [8]. The IleRS structure exhibits the Rossmann-fold domain at the center, β-rich intervening domains at the top, and an α-rich cylindrical domain at the bottom (Figure 3).

The Rossmann-fold domain of *T. thermophilus* IleRS has a deep cleft with two characteristic ATP-binding motifs, His^{54}–Val^{55}–Gly^{56}–His^{57} and Lys^{591}–Met^{592}–Ser^{593}–Lys^{594}. In the complex with IleRS, a L-isoleucine molecule is located at the bottom of this catalytic cleft (Figures 3 and 4A). The hydrophobic side chain of L-isoleucine is bound, through van der Waals interactions, in a pocket made by Pro^{46}, Trp^{518}, and Trp^{558} , while the amino and carboxyl groups of L-isoleucine form hydrogen bonds with Asp^{85} and Gln^{554}, respectively (Figure 4A). These residues are completely conserved among IleRSs. Steric hindrance prevents the side chains of L-leucine and larger amino acids from binding into this pocket of IleRS. Figure 4B shows how L-isoleucine fits into the pocket of the Rossmann-fold domain of IleRS by space-filling presentation.

In the L-valine•IleRS complex, an L-valine molecule is actually bound to the same site on the Rossmann-fold domain (Figure 4C). The hydrophobic contact area of the side chain of L-valine with those of Pro^{46} and Trp^{558} is slightly smaller than that of the L-isoleucine side chain (Figures 4B and 4C), which is consistent with the previous calculation [12]. All these results are consistent with the concept of the first, coarse sieve in the double-sieve mechanism of editing [12] (Figure 2).

The Rossmann-fold domain of *T. thermophilus* IleRS has a β-rich insert consisting of four domains (Figure 3). The first and fourth inserted domains have zinc ions coordinated to Cys^{181}–X_2–Cys^{184}–X_{204}–Cys^{389}–X_2–Cys^{392} and Cys^{461}–X_2–Cys^{464}–X_{37}–Cys^{502}–X–Cys^{504}, respectively. The first zinc-binding structure includes an unusually large insert between the second and third Cys residues, which encompasses the entire globule of the second inserted domain. Between the β-barrel core and the protruding β-ribbon of the second inserted domain, a deep cleft is formed. This cleft is as large and deep as the catalytic cleft of the Rossmann-fold domain (Figure 3). We have determined the crystal structure of *T. thermophilus* MetRS (unpublished results). The *T. thermophilus* MetRS has a zinc-coordination structure similar to the first zinc-coordination structure of *T. thermophilus* IleRS, but has no insert corresponding to the second inserted domain. Therefore, the second inserted (Figure 5) domain is characteristic to the IleRS structure.

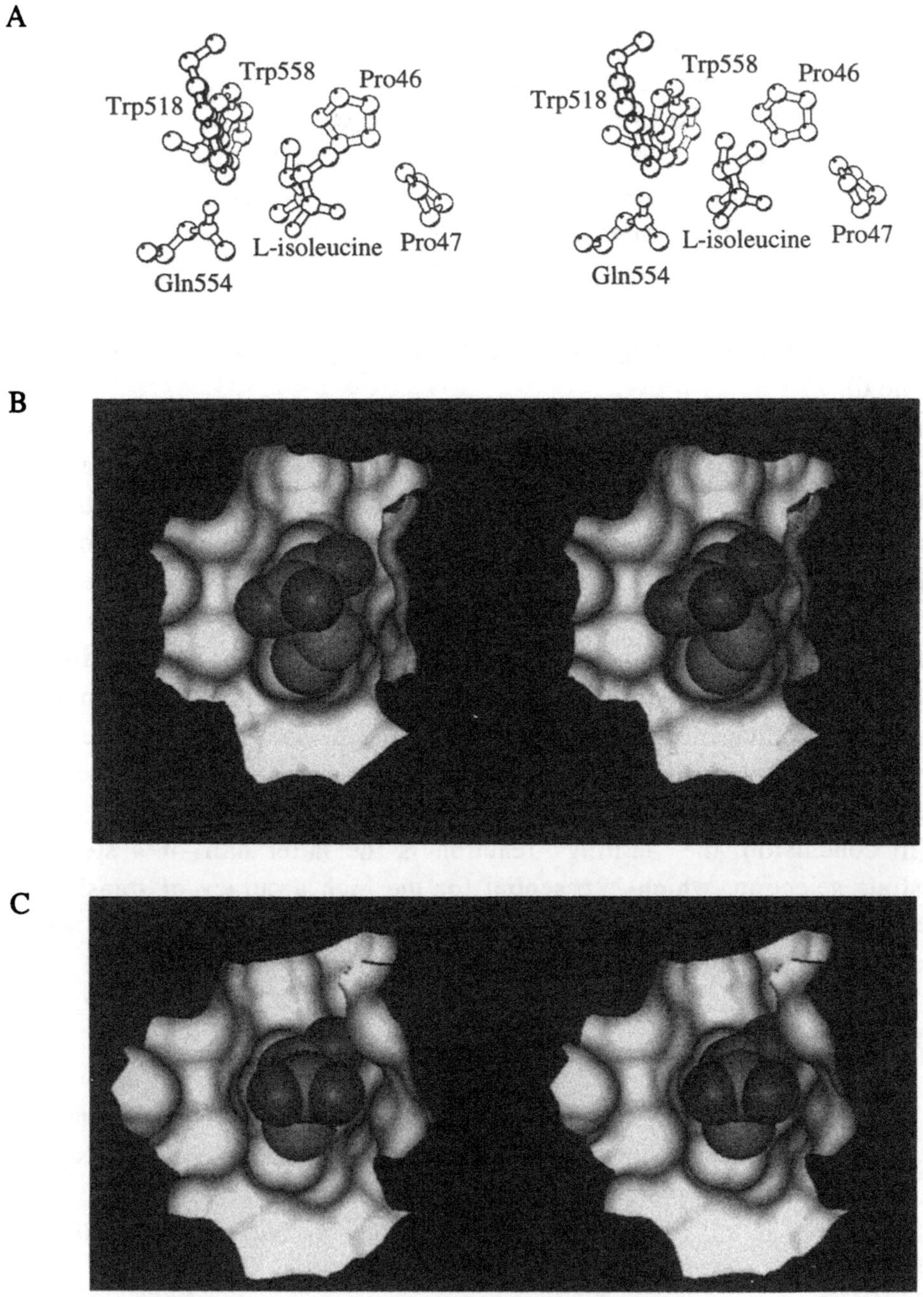

Figure 4. L-Isoleucine (A and B) and L-valine (C) bound in "the first, coarse sieve" on the Rossmann-fold domain of IleRS [8] (stereoviews).

A fragment, corresponding to the first and second inserted domains, cloned from *E. coli* IleRS retains the specific Val-tRNA^{Ile} editing activity [15]. In the L-valine•IleRS complex, a second L-valine molecule was identified at the bottom of the deep cleft in the second inserted domain [8] (Figures 5 and 6A). In contrast, in the L-isoleucine•IleRS complex, no L-valine molecule was observed in this second pocket [8]. Therefore, the second pocket on the second inserted domain is specific for L-valine, and indicated to be "the second, fine sieve" of the double-sieve mechanism (Figure 2) [12]. The space made by the conserved Trp^{232} and Tyr^{386} residues is just large enough for L-valine, and is too small for L-isoleucine (Figure 6).

The mutation deleting residues 219–265, including the Trp^{232} of the L-valine-specific pocket, of IleRS completely abolished the Val-tRNA^{Ile} editing activity [8]. This mutant has nearly the same activity for Ile-tRNA^{Ile} formation as that of the wild-type IleRS. In contrast, the activity of this mutant IleRS to mischarge tRNA^{Ile} with L-valine is drastically higher than that of the wild-type IleRS. Surprisingly, the K_M value for L-valine is the same as that for L-isoleucine in aminoacyl-tRNA formation. Accordingly, the discrimination between L-isoleucine and L-valine primarily depends on the editing. Mutagenesis analyses made on the *E. coli* IleRS revealed that Thr and Asn residues in the strictly-conserved Thr-rich sequences near the L-valine-specific pocket are important for the editing activity [8].

In conclusion, the "editing" reaction is the latter half of a stepwise substrate selection, which is essential for the high accuracy of translation. The ATP consumption with regard to L-valine is non-productive and is solely for substrate selection, which demonstrates the high cost of accuracy.

4. References

1. Eriani, G., Delarue, M., Poch, O., Gangloff, J., Moras, D. (1990) Partition of tRNA synthetases into two classes based on mutually exclusive sets of sequence motifs, *Nature* **347**, 203-206.
2. Cusack, S. (1995) Eleven down and nine to go, *Nature Struct. Biol.* **2**, 824-831.
3. Brick, P.,, Bhat, T. N., Blow, D. M. (1989) Structure of tyrosyl-tRNA synthetase refined at 2.3 Å resolution. Interaction of the enzyme with the tyrosyl adenylate intermediate, *J. Mol. Biol.* **208**, 83-98.
4. Rould, M. A., Perona, J. J., Söll, D., Steitz, T. A. (1989) Structure of *E.*

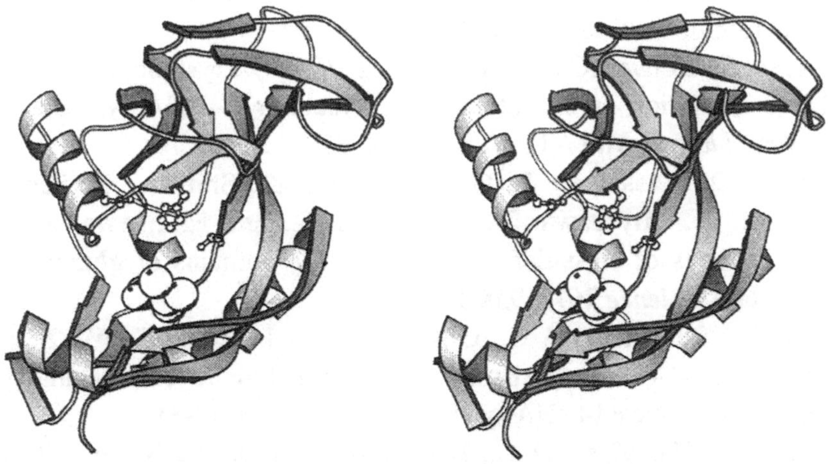

Figure 5. The editing domain of IleRS bound with L-valine [8] (stereoview)

A

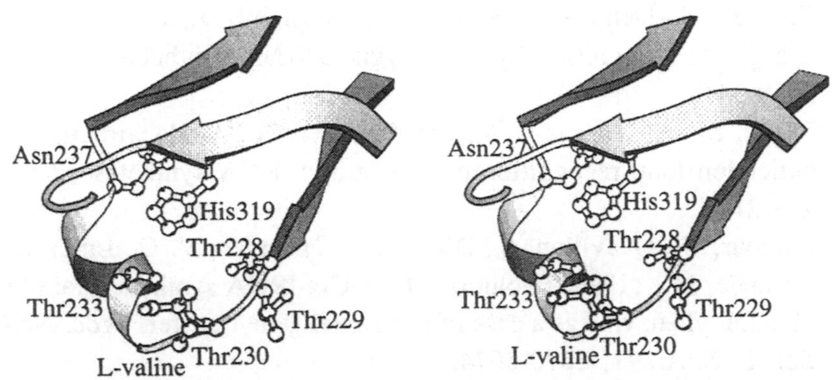

B

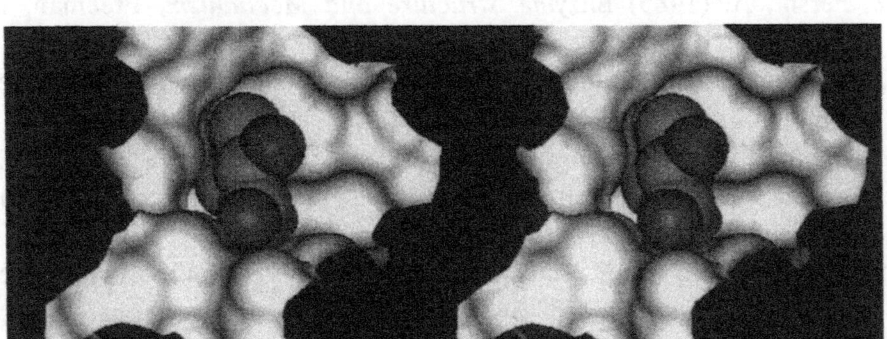

Figure 6. L-valine bound in "the second, fine sieve" on the editing domain of IleRS [8]. Conserved amino acid residues (A) and space-filling model (B) (stereoviews).

coli glutaminyl-tRNA synthetase complexed with tRNA^{Gln} and ATP at 2.8 Å resolution, *Science* **246**, 1135-1142.
5. Brunie, S., Zelwer, C., and Risler, J. L. (1990) Crystallographic study at 2.5 Å resolution of the interaction of methionyl-tRNA synthetase from *Escherichia coli* with ATP, *J. Mol. Biol.* **216**, 411-424.
6. Nureki, O., Vassylyev, D. G., Katayanagi, K., Shimizu, T., Sekine, S., Kigawa, T., Miyazawa, T., Yokoyama, S., and Morikawa, K. (1995) Architectures of class-defining and specific domains of glutamyl-tRNA synthetase, *Science* **267**, 1958-1965.
7. Doublié., S., Bricogne, G., Gilmore, C., and Carter, C. W., Jr. (1995) Tryptophanyl-tRNA synthetase crystal structure reveals an unexpected homology to tyrosyl-tRNA synthetase, *Structure* **3**, 17-31.
8. Nureki, O., Vassylyev, D. G., Tateno, M., Shimada, A., Nakama, T., Fukai, S., Konno, M., Hendrickson, T. L., Schimmel, P., and Yokoyama, S. (1998) Enzyme structure with two catalytic sites for double-sieve selection of substrate, *Science* **280**, 578-582.
9. Cavarelli, J., Delagoutte, B., Eriani, G., Gangloff, J., and Moras, D. (1998) L-arginine recognition by yeast arginyl-tRNA synthetase, *EMBO J.* **17**, 5438-5448.
10. Rould, M. A., Perona, J. J., and Steitz, T. A. (1991) Structural basis of anticodon loop recognition by glutaminyl-tRNA synthetase, *Nature* **352**, 213-218.
11. Lamour, V., Quevillon, S., Diriong, S., N'Guyen, V. C., Lipinski, M., and Mirande, M. (1994) Evolution of the Glx-tRNA synthetase family: the glutaminyl enzyme as a case of horizontal gene transfer, *Proc. Natl. Acad. Sci. U. S. A.*, 91, 8670-8674.
12. Fersht, A. (1985) *Enzyme Structure and Mechanism,* Freeman, New York.
13. Schmidt, E. and Schimmel, P. (1994) Mutational isolation of a sieve for editing in a transfer RNA synthetase, *Science* 264, 265-267.
14. Schmidt, E. and Schimmel, P. (1995) Residues in a class I tRNA synthetase which determine selectivity of amino acid recognition in the context of tRNA, *Biochemistry* **34,** 11204-11210.
15. Lin, L., Hale, S. P., and Schimmel, P. (1996) Aminoacylation error correction, *Nature* **384,** 33-34.
16. Hale, S. P., Auld, D. S., Schmidt, E., and Schimmel, P. (1997) Discrete determinants in transfer RNA for editing and aminoacylation, *Science* **276**, 1250-1252.

AMINOACYLATION OF tRNA INDUCES A CONFORMATIONAL SWITCH ON THE 3′-TERMINAL RIBOSE

ANDREAS SCHLOSSER, BERND BLECHSCHMIDT, BARBARA NAWROT[#] MATHIAS SPRINZL*
Laboratorium für Biochemie der Universität Bayreuth, D-95440 Bayreuth, Germany
[#]Polish Academy of Sciences, Centre of Molecular and Macromolecular Studies, Department of Bioorganic Chemistry, 90-363 Lodz, Sienkiewicza 112, Poland
** Author for correspondence. Tel.: +49-921-552420; Fax: +49-921-552432*
E-mail: mathias.sprinzl@uni-bayreuth.de

Abstract
The 3′-terminal adenosine-76 in tRNA is an important recognition element for the interaction of tRNA or aminoacyl-tRNA with proteins and nucleic acids. The aromatic adenine-76 is usually located in a lipophilic binding pocket of the tRNA binding proteins. The ribofuranose ring of nucleosides is not planar and can adopt a C3′-*endo*-C2′-*exo* (north sugar, N) or C2′-*endo*-C3′-*exo* (south sugar, S) conformation. This sugar pucker defines the position of the nucleobase attached to C1′ either in the axial or equatorial position. The conformation of the ribose residues in the RNA is determined by the electronic nature of the sugar ring and its substituents as well as by the chemical nature and stacking interactions of the nucleobase. Using a fluorescent analogue of tRNA, tRNA-CCF, we demonstrate, that the aminoacylation of the tRNA results in a conformational change of the 3′-terminal nucleobase. The increase of the fluorescence intensity of formycin in tRNA-CCF[1] upon aminoacylation is explained by partial destacking of the 3′-terminal base moiety. Several analogues of aminoacyl-tRNA which were modified on 3′-terminal ribose were tested in their ability to interact with elongation factor Tu·GTP complex. The results indicate that the aminoacylation of the ribose in the tRNA triggers a conformational N⇆S ribose switch which determines the position of adenine-76 for recognition by the elongation factor Tu.

1. Introduction

[1] *Abbreviations*: EF-Tu, elongation factor Tu; F, formycin or 7-amino-(β-D-ribofuranosyl)pyrazolo-(4,3-d)pyrimidine; $tRNA^{Val}$-CCF, valine-specific tRNA with formycin instead of terminal adenosine in position 76;

J. Barciszewski and B.F.C. Clark (eds.), RNA Biochemistry and Biotechnology, 159–168.

Transfer ribonucleic acid, tRNA, is composed of 70 to 90 nucleotides folded to a cloverleaf secondary and an L-shaped tertiary structure. Several thousand tRNA sequences have been determined so far ([18] Sprinzl et al., 1998). All tRNAs possess an anticodon sequence located in the anticodon loop which translates the genetic code and is recognized by several aminoacyl-tRNA synthetases ([2] Arnez & Moras, 1997). The core region of tRNA involves the interacting D-and T-loops and adjacent helical regions. Aminoacylation of tRNA takes place on the aminoacyl domain composed of the aminoacyl stem and NCCA single stranded 3′-end. The aminoacyl residue is attached to the ribose of the 3′-terminal adenosine by the aminoacyl-tRNA synthetase. Identity elements for interaction with aminoacyl-tRNA synthetases are often located on the aminoacyl stem. Similarly the discriminator base, N-73, which connects the aminoacyl stem with the invariant CCA-end serves as a recognition element for aminoacyl-tRNA synthetases. Besides this function, N-73 connects the single-stranded CCA end with the rigid body of the tRNA. Biochemical analyses and structures of tRNAs in complex with aminoacyl-tRNA synthetases ([2] Arnez & Moras, 1997) and aminoacyl-tRNA in complex with elongation factor Tu and a GTP analogue ([10] Nissen et al., 1995) suggest that the fixed geometry of tRNA, which defines the distance between the anticodon and adenosine-76 is important for effective recognition of tRNA by aminoacyl-tRNA synthetase and is a prerequisite for correct translation of the genetic code ([3] Barciszewski et al., 1994).

The aminoacyl residue in aminoacyl-tRNA acts as a discriminator in different recognition events. Uncharged elongator tRNAs bind to aminoacyl-tRNA synthetases and dissociate from this enzymes as aminoacyl-tRNAs. On the other hand the GTP form of the elongation factor Tu binds aminoacyl-tRNA with much higher affinity than not aminoacylated tRNA ([4] Janiak et al. 1990). Protein factors involved in initiation, such as initiation factor 2 ([12] Petersen et al., 1979), or Met-tRNA transformylase ([16] Schmitt et al., 1996), also recognize only aminoacylated initiator tRNA. This raises the question about structural differences between aminoacylated and non-aminoacylated tRNA which are recognized by these proteins.

The main problem which hampers the structural investigations of aminoacyl-tRNA by X-ray structure analysis or NMR spectroscopy is the chemical instability of the aminoacyl ester bond. ([17] Schuber & Pinck, 1974). Chemical probing, light-scattering and other physical or chemical methods did not identify any large structural changes in the tertiary structure of tRNA upon aminoacylation. Only subtle structural differences which influence the interaction of magnesium ions or binding of complementary oligonucleotides to the 3′-terminal region of tRNA were identified as consequence of tRNA aminoacylation. Light scattering experiments suggested that a Mg^{2+} binding site near the CCA end of tRNA may be altered by aminoacylation ([13] Potts et al., 1977, [14] Potts et al., 1981). The interpretations of these experiments were later discussed by ([1] Antosiewicz & Porschke, 1989) who argued against a major change of the tRNA shape caused by aminoacylation. However, these authors could not exclude subtle conformational changes localized on the 3′-terminus of aminoacyl-tRNA. In addition, more recent NMR studies confirmed the presence of a metal ion binding site in the aminoacyl domain of several tRNAs ([7] Limmer et al., 1993, Ott et al. 1993). This suggests that the aminoacylation-dependent conformational change of the terminal adenosine may influence a metal-ion binding site as already reported by ([14] Potts et al. 1981).

2. Structure of aminoacyl-adenosine

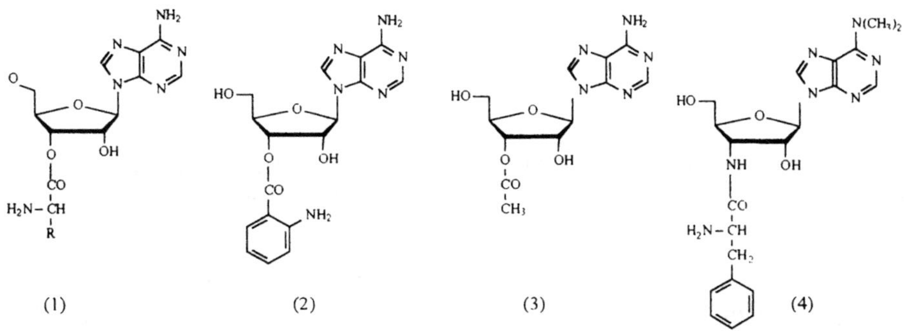

Figure 1. Structure of (1) 3′-O-aminoacyl-adenosine (isomeric 2′-O-aminoacyl-adenosine is formed by fast spontaneous transacylation from (1)), (2) 3′-O-anthraniloyl-adenosine, (3) 3′-O-acetyl-adenosine, (4) puromycine.

Detailed X-ray or NMR structures of adenosine with a natural α-amino acid attached to the 2′-or 3′-hydroxyl group are not known. The NMR measurements performed with several aminoacyl-adenosines (Fig. 1a) revealed a fast transacylation between the 2′- and 3′-hydroxyls of the ribose ([20] Taiji et al. 1983). Similar to the rate of aminoacyl hydrolysis, the rate of this transacylation is slightly dependent on the chemical nature of the aminoacyl residue. X-ray structures were determined only for the chemically stable 3′-O-acetyl-adenosine ([15] Rao & Sundaralingam, 1970), puromycine ([19] Sundaralingam & Arora, 1972) and 3′-O-antraniloyl-adenosine ([9] Nawrot et al., 1997) (Fig 1). The structures of 3′-O-anthraniloyl-adenosine and 3′-O-acetyl-adenosine are different from the structure of the antibiotic puromycine. Most significantly, the ribose ring has a C2′-*endo*-C3′-*exo* (S) conformation in the case of 3′-O-anthraniloyl-adenosine and 3′-O-acetyl-adenosine as compared to a C3′-*endo*-C2′-*exo* (N) conformation of puromycine. A prevalent C2′-*endo*-C3′-*exo* ribose conformation was also found for 3′-O-anthraniloyl-adenosine in solution by NMR spectroscopy ([9] Nawrot et al., 1997). In contrast, puromycine has C3′-*endo*-C2′-*exo* conformation both in crystal as well as in solution a ([5] Jardetzky, 1963).

The sugar ring pucker is extremely important for recognition of aminoacyl-tRNA by proteins since it defines the orientation of the glycosidic bond in the 3′-

terminal adenine residue, which is equatorial for the C2′-*endo*-C3′-*exo* conformation, but axial for the C3′-*endo*-C2′-*exo* conformation (Fig. 2).

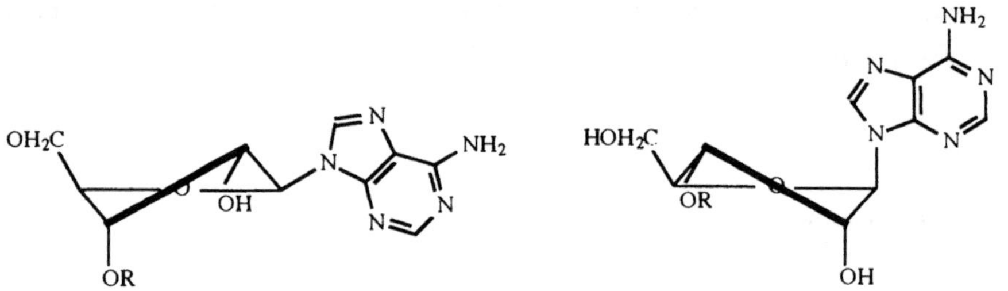

Figure 2. Conformation of the ribofuranose ring determines the equatorial (left) or axial (right) location of the nucleobase in ribonucleosides. Left: south sugar (S), C2′-*endo*-C3′-*exo*. Right: north sugar (N), C3′-*endo*-C2′-*exo*.

Interesting is the comparison of the interaction 3′-O-anthraniloyl-adenosine and puromycine with EF-Tu•GTP. Whereas puromycin does not bind to EF-Tu•GTP, 3′-O- anthraniloyl-adenosine interacts with EF-Tu•GTP with a dissociation constant in the μM range ([7] Limmer et al., 1998).

Substrate properties of the two analogues reflect the requirement of EF-Tu for the C2′-*endo*-C3′-*exo* sugar conformation which is provided only by 3′-O-anthraniloyl-adenosine. In a regular tRNA structure, in which the terminal adenosine is involved in stacking interactions with the adjacent nucleobases of the NCCA end, a stabilization of the C2′-*exo*-C3′-*endo* conformation of the terminal ribose is expected ([21] Varani & Ramos 1997, [22] Viani-Puglisi et al., 1994, [6] Limmer et al., 1993). Recently, the three dimensional structure of the ternary complex composed of yeast Phe-tRNA, *Thermus aquaticus* EF-Tu and a GTP analogue was determined by X-ray crystallography ([10] Nissen et al., 1995). This is the only known crystal structure of an aminoacyl-tRNA. The aminoacyl-adenosine residue of Phe-tRNA is located near the interface of domains I and II of EF-Tu. The adenine base of the 3′-terminal residue-76 is not stacked to the cytosine-75 as would be expected for an A-type RNA. Instead it is placed in a hydrophobic pocket of the protein. Thus, a destacking of adenine-76 from cytosine-75 has to take place first, before the adenine ring is positioned to its binding site on EF-Tu. Such structural change can be either a consequence of an induced fit during interaction with the protein or is intrinsically determined by the aminoacylation of tRNA. To answer this question we investigated the structural alterations of the 3′-terminal nucleoside in the tRNA upon aminoacylation.

3. Aminoacylation of tRNA and hydrolysis of the aminoacyl residue from aminoacyl-tRNA are accompanied by a conformational change which can be detected by fluorescence spectroscopy.

To study the effect of aminoacylation and amino acid hydrolysis by fluorescence spectroscopy the 3'-terminal adenosine of tRNA was replaced by the fluorescent adenosine analogue formycine (Fig. 3) ([8] Maelicke et al., 1974). The local structural changes caused by aminoacylation were then monitored by measurement of the

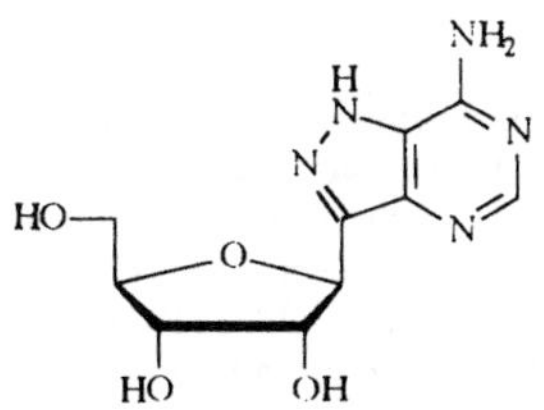

Figure 3. Structure of formycine (7-amino-(β-D-ribofuranosyl)pyrazolo-(4,3-d)pyrimidine)).

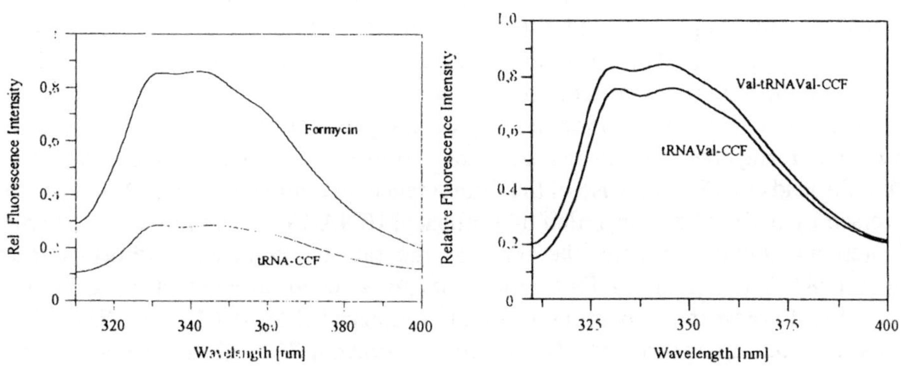

Figure 4. Fluorescence emission spectra of, left, formycine and $tRNA^{Val}$ -CCF, (formycine was formed by RNase A treatment of $tRNA^{Val}$ -CCF and the difference in fluorescence intensity reflects the quenching caused by incorporation of formycin to tRNA) and right, $tRNA^{Val}$ -CCF and Val- $tRNA^{Val}$ -CCF (the difference in the fluorescence intensity reflects the effect of aminoacylation). Excitation was at 293 nm. Spectra were measured at 20°C in 100 mM Tris/HCl (pH 8.6), 100mM NaCl and 10 mM $MgCl_2$.

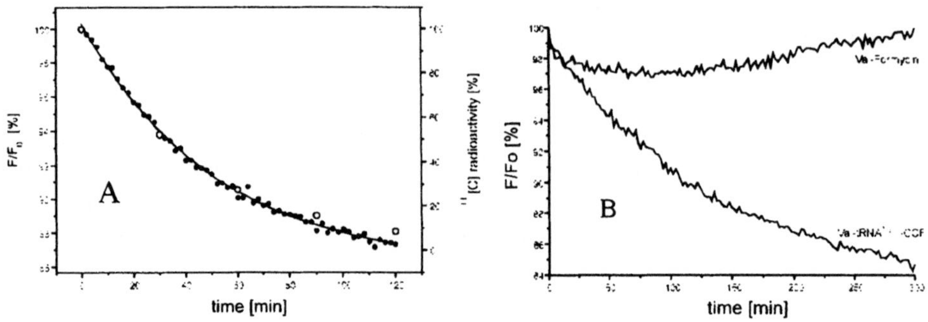

Figure 5. (A) change of the fluorescence intensity during incubation of Asp-tRNA^{Asp}-CCF at 30°C in a buffer containing 100 mM Tris/HCl (pH 8.6), 100mM NaCl and 10 mM $MgCl_2$ (●). Excitation wavelength was at 293 nm, emission was measured continuously at 342 nm. Change of the fluorescence intensity follows the rate of [^{14}C]Asp-tRNA^{Asp} -CCF hydrolysis as determined by measurement of trichloroacetic acid-precipitable radioactivity (✐). (B) Change of the fluorescence intensity of Val-tRNA^{Val}-CCF due to hydrolysis of aminoacyl residue (under conditions defined in A) before (Val-tRNA^{Val}-CCF) and after (Valyl-formycin) treatment with RNase A.

formycine fluorescence. Formycine has a fluorescence excitation maximum at 293 nm and an emission maximum at 342 nm. No other naturally occurring nucleotides with a fluorescence in this spectral region were present in applied tRNAs. Incorporation of formycine into the single-stranded 3'-terminus of tRNA leads to a strong decrease in formycine fluorescence. The relative quantum yield of free formycine is 0.052 as compared to the relative quantum yield of 0.012 for formycine in tRNA-CCF ([8] Maelicke et al. 1974). This fluorescence quenching was attributed to the stacking of the terminal formycine base to the adjacent cytidine-75 residue. Aminoacylation of tRNA-CCF leads to 15% increase in the fluorescence quantum yield (Fig 4), which can be reversed by hydrolytic deacylation of aminoacyl-tRNA-CCF. The rate of the change in fluorescent signal is exactly the same as the rate of aminoacyl hydrolysis from aminoacyl-tRNA-CCF (Fig. 5). Furthermore, it can be demonstrated that the difference between the fluorescence quantum yield of aminoacyl-tRNA-CCF and tRNA-CCF depends on the interaction of the terminal formycin-76 with cytidine-75 since aminoacyl-formycine, which is formed by RNase cleavage of aminoacyl-tRNA-CCF and in which the rest of the tRNA molecule is missing, does not significantly change its fluorescence intensity upon hydrolysis of the ester bond (Fig. 5). Thus, the fluorescence change upon aminoacylation of tRNA-CCF (Fig. 4) is determined essentially only by the interaction of the formycin-76 with the neighboring cytidine-75. Aminoacylation of the ribose on the 2'- or 3'-hydroxyl group results in an unstable structure which allows a rapid transacylation of the aminoacyl residue. Spectroscopic analysis of this dynamic system by NMR or fluorescence can provide only an average signal of both 2′- and 3′-aminoacyl-adenosine derivatives ([20] Taiji et al., 1983). This may be the reason why the fluorescence change observed as a result of aminoacylation of tRNA-CCF is substantially smaller (13 %) than the fluorescence increase observed by RNase cleavage of tRNA-CCF (68 %) (Fig. 4).

4. Interaction of EF-Tu•GTP with aminoacyl-tRNAs modified on the 3′-terminal ribose

As discussed above, the aminoacylation of the 3′-position of adenosine changes the conformation of the ribose and induces a destacking of the terminal formycine-76 from the rest of tRNA. Analogous structural change in the native tRNA-CCA may facilitate the binding of the adenosine ring into its binding pocket on EF-Tu. However, in order to interact with EF-Tu•GTP the aminoacyl residue must be linked

Table 1: Effect of modification of the 3′-terminal ribose in aminoacyl-tRNA on the equillibrium dissociation constant for the interaction with EF-Tu.GTP. The Kd values wer de-termined by a fluorescence assay ([11] Ott, et al., 1989)

Number	Aminoacyl-tRNA-	K_d [nM]
1	-CCA	1.7
5	-CC3′dA	12,0
6	-CC2′dA	4.2
7	-CC3′NH2A	123.0
8	-CCAoxi-red	183.0

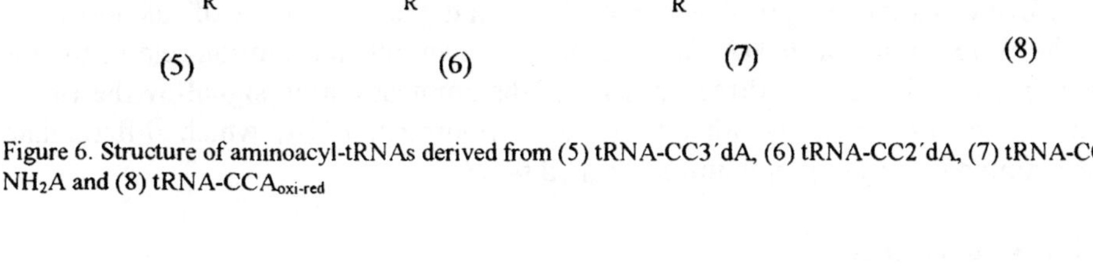

(5) (6) (7) (8)

Figure 6. Structure of aminoacyl-tRNAs derived from (5) tRNA-CC3′dA, (6) tRNA-CC2′dA, (7) tRNA-CC3′-NH_2A and (8) tRNA-$CCA_{oxi\text{-}red}$

to the 3′-position of the ribose by an ester bond. The tRNA, in which the aminoacyl residue is attached via an amide bond (an analogue of puromycin), is expected to be a weak substrate for EF-Tu•GTP. This prediction was tested by measurement of equillibrium dissociation constants, K_d, for aminoacyl-tRNA analogues shown in Fig. 6. The results are presented in Table 1. As expected aminoacyl-tRNA-CC3′NH_2A, has more then 50 times higher dissociation constant for the binding to EF-Tu•GTP than the native aminoacyl-tRNA-CCA (Table 1.) The presence of the 3′-NH- in the terminal adenosine of tRNA-CC3′NH_2A is, however, not in conflict with the structure of the phenylalanyl-tRNA·EF-Tu•GTP since there are no interactions with the protein involving this atom ([10] Nissen et al. 1995). Therefore, the only plausible explanation for the reduced capability of aminoacyl-tRNA-CC3′NH_2A to bind EF-Tu•GTP is, in analogy with the properties of puromycin in this reaction, the C3′-endo sugar conformation in the 3′-terminal ribose. In this context it was also interesting to determine the dissociation constants of isomeric aminoacyl-tRNA-CC2′dA and aminoacyl-tRNA-CC3′dA for the reaction with EF-Tu•GTP. As determined by X-ray structure analysis of aminoacyl-tRNA•EF-Tu•GppNHp ternary complex the 2′-OH group of the aminoacylated terminal adenosine interacts with the carboxylate of an conserved glutamic acid residue ([10] Nissen et al. 1995). Although, in deoxy analogues (5) and (6) the 2′OH group of the 3′-terminal adenosine is not available for such interaction the affinity for EF-Tu•GTP is not strongly reduced (only 2 to 7 fold). This suggests that the conformational switch of the ribose and the recognition of the adenine residue are more important for efficient binding then the direct recognition of the substituents on the ribofuranose ring. Thus, the large, 10,000 fold, difference between the affinity of aminoacyl-tRNA and tRNA for EF-Tu•GTP is not caused by direct interaction of aminoacyl residue with the protein but is a result of an indirect recognition of the change in the ribose pucker induced by aminoacylation. In agreement with this suggestion the affinity of aminoacyl-$tRNA_{oxi\text{-}red}$, in which the ribofuranose ring is destroyed by periodate oxidation and the ribose switch can not be activated, for EF-Tu•GTP is drastically decreased (Table 1).

In the structure of the aminoacyl-tRNA•EF-Tu•GppNHp ternary complex ([10] Nissen et al. 1995) the adenine moiety does not interact with the adjacent cytidine. Instead it binds to a hydrophobic pocket of the protein, and the aminoacyl residue is placed to a protein cleft. Specific interactions of the aminoacyl residue with aminoacyl side chains of EF-Tu were not identified. Thus, it is reasonable to assume that the conformation of the 3'-terminal ribose and the destacking of adenosine-76 from the 3'-terminus of tRNA are the most important recognition elements for elongation factor Tu. This indirect transfer of the aminoacylation signal by the ribose switch may be used also by other proteins or ribosomal RNA, which differentiate between aminoacylated and not aminoacylated tRNA.

5. Acknowledgements

The work was supported by the Human Frontier Science Program, RG-369/93, the *Deutsche Forschungsgemeinschaft*, (Sp-243/5) and Fonds der Chemischen Industrie. We thank Heike Rütthard for help with the preparation of the manuscript.

6. References

[1] Antosiewicz, J. and D. Porschke. 1989. Effect of aminoacylation on tRNA conformation. *Eur.Biophys.J.* 17:233-235.
[2] Arnez, J.G. and D. Moras. 1997. Structural and functional considerations of the aminoacylation reaction. *Trends.Biochem.Sci.* 22:211-216.
[3] Barciszewski, J., M. Sprinzl, and B.F.C. Clark. 1994. Aminoacyl-tRNAs. Diversity before and unity after interaction with EF-Tu:GTP. *FEBS Lett.* 351:137-139.
[4] Janiak, F., V.A. Dell, J.K. Abrahamson, B.S. Watson, D.L. Miller, and A.E. Johnson. 1990. Fluorescence characterization of the interaction of various transfer RNA species with elongation factor Tu·GTP: evidence for a new functional role for elongation factor Tu in protein biosynthesis. *Biochem.* 29:4268-4277.
[5] Jardetzky, O. 1963. Proton magnetic resonance of purine and pyrimidine derivatives. X. The conformation of puromycin. *J.Am.Chem.Soc.* 85:1823-1825.
[6] Limmer, S., H.-P. Hofmann, G. Ott, and M. Sprinzl. 1993. 3'-terminal end (NCCA) of tRNA co-determines the structure and stability of the aminoacyl acceptor stem. *Proc.Natl.Acad.Sci.USA* 90:6199-6202.
[7] Limmer, S., Vogtherr, M., Nawrot, B., Hillenbrand, R., and Sprinzl, M. 1998. Specific recognition of a minimal model of aminoacylated tRNA by the elongation factor Tu of bacterial protein biosynthesis. 36, 2485-2489.
[8] Maelicke, A., M. Sprinzl, F. von der Haar, T.A. Khwaja, and F. Cramer. 1974. Structural studies on phenylalanine transfer ribonucleic acid from yeast with the spectroscopic label formycin. *Eur.J.Biochem.* 43:617-625.
[9] Nawrot, B., W. Milius, A. Ejchart, S. Limmer, and M. Sprinzl. 1997. The structure of 3`-O-anthraniloyladenosine, an analogue of the 3`-end of aminoacyl-tRNA. *Nucleic Acids Res.* 25:948-954.
[10] Nissen, P., M. Kjeldgaard, S. Thirup, G. Polekhina, L. Reshetnikova, B.F.C. Clark, and J. Nyborg. 1995. Crystal structure of the ternary complex of Phe-$tRNA^{Phe}$, EF-Tu, and a GTP analog. *Science* 270:1464-1472.
[11] Ott, G., L. Arnold, and S. Limmer. 1993. Proton NMR studies of manganese ion binding to tRNA-derived acceptor arm duplexes. *Nucleic Acids Res.* 21:5859-5864.
[12] Petersen, H.U., T. Roll, M. Grunberg-Manago, and B.F.C. Clark. 1979. Specific interaction of initiation factor IF2 of *E. coli* with formylmethionyl-$tRNA_f^{met}$. *Biochem.Biophys.Res.Comm.* 91:1068-1074.
[13] Potts, R., M.J. Fournier, and N.C.J. Ford. 1977. Effect of aminoacylation on the conformation of yeast phenylalanine tRNA. *Nature* 268:563-564.
[14] Potts, R.O., N.C.J. Ford, and M.J. Fournier. 1981. Changes in the solution structure of yeast phenylalanine transfer ribonucleic acid associated with aminoacylation and magnesium binding. *Biochem.* 20:1653-1659.
[15] Rao, S.T. and M. Sundaralingam. 1970. Stereochemistry of nucleic acids and their constituents. XIII. The crystal and molecular structure of 3'-O-acetyladenosine. Conformational analysis of nucleosides and nucleotides with *syn* glycosidic torsional angle. *J.Am.Chem.Soc.* 92:4963-4970.

[16] Schmitt, E., S. Blanquet, and Y. Mechulam. 1996. Structure of crystalline *Escherichia coli* methionyl-tRNA$_f^{Met}$ formyltransferase: Comparison with glycinamide ribonucleotide formyltransferase. *EMBO J.* 15:4749-4758.
[17] Schuber, F. and M. Pinck. 1974. On the chemical reactivity of aminoacyl-tRNA ester bond. I - Influence of pH and nature of the acyl group on the rate of hydrolysis. *Biochimie* 56:383-390.
[18] Sprinzl, M., Horn, C., Brown, M., Ioudotitch, A., and Steinberg, S. Compilation of tRNA sequences and sequences of tRNA genes. *Nucl.Acids Res.* 26, 148-153. 1998.
[19] Sundaralingam, M. and S.K. Arora. 1972. Crystal structure of the aminoglycosyl antibiotic puromycin dihydrochloride pentahydrate. Models for the terminal 3'-aminoacyladenosine moieties of transfer RNA's and protein-nucleic acid interactions. *J.Mol.Biol.* 71:49-70.
[20] Taiji, M., S. Yokoyama, and T. Miyazawa. 1983. Transacylation rates of (aminoacyl)adenosine moiety at the 3'-terminus of aminoacyl transfer ribonucleic acid. *Biochem.* 22:3220-3225.
[21] Varani, G. and A. Ramos. 1997. Structure of the acceptor stem of *Escherichia coli* tRNA^{Ala}: role of the G3·U70 base pair in synthetase recognition. *Nucleic Acids Res.* 25:2083-2090.
[22] Viani Puglisi, E., J.D. Puglisi, J.R. Williamson, and U.L. RajBhandary. 1994. NMR analysis of tRNA acceptor stem microhelices: Discriminator base change affects tRNA conformation at the 3' end. *Proc.Natl.Acad.Sci.USA* 91:11467-11471.

POINT MUTANTS OF ELONGATION FACTOR TU FROM *E. COLI* IMPAIRED IN BINDING AMINOACYL-tRNA

C. R. KNUDSEN, F. MANSILLA, G. N. PEDERSEN AND B. F. C. CLARK
Division of Biostructural Chemistry, Institute of Molecular and Structural Biology, Aarhus University, Gustav Wieds Vej 10C, DK-8000 Århus C.

Abstract

Protein engineering is a powerful technique, which can be used to establish the relationship between the structure and function of a protein. The protein engineering process is cyclic and runs through the following steps: design of substitution based on structural information, site-directed mutagenesis, expression and purification of mutant protein and finally a functional and structural characterisation.

We have used the translational elongation factor EF-Tu from *E. coli* as a model for carrying out protein engineering. EF-Tu bound to its cofactor GTP transports aminoacylated tRNA to the mRNA-programmed ribosome. Codon-anticodon interaction triggers GTP-hydrolysis by EF-Tu, which undergoes a conformational change forcing the factor to leave the ribosome. For many years, only the structure of EF-Tu in its inactive, GDP-bound form was known. Information about the mode of binding aa-tRNA came from modification and cross-linking studies supplemented with protein engineering studies. Recently, the structure of the ternary complex, EF-Tu:GTP:aa-tRNA, was solved in our lab allowing a structural interpretation of our mutational studies. Our mutants impaired in binding aa-tRNA could be split into two groups: (1) those for which a residue directly involved in binding was substituted and (2) those of which the substitution was situated outside the region involved in binding aa-tRNA. Examples from the different groups will be given. Furthermore, examples of various techniques applied to characterise the interactions between EF-Tu and aa-tRNA will be given.

When the structure of the ternary complex was compared with that of another elongation factor EF-G it was possible to envisage a simplification of the entire translation process. The concept of structural macromolecular mimicry between RNA and protein was proposed and gave rise to suggestions about G-protein molecular evolution an to common structural elements and activities among translation factors at initiation, elongation and termination steps. The current status is briefly summarised.

1. On the structure and function of EF-Tu

Elongation factor Tu (EF-Tu) is a guanine-nucleotide binding protein with the characteristic features of cycling between an active, GTP-bound form and an inactive, GDP-bound form [1, 2]. In the GTP-form, EF-Tu has a high affinity for aa-tRNA with

J. Barciszewski and B.F.C. Clark (eds.), RNA Biochemistry and Biotechnology, 169–193.

dissociation constants in the nanomolar range [3]. The affinity for deacylated tRNA is several orders of magnitude lower [4] and the GDP-form of EF-Tu has negligible affinity for aa-tRNA [5]. The ternary complex, EF-Tu·GTP·aa-tRNA binds to the A-site of the mRNA-programmed ribosome. If the triplet of the tRNA anticodon matches the A-site exposed codon, hydrolysis of the GTP-molecule bound to EF-Tu is triggered [6]. GTP-hydrolysis induces a conformational change of EF-Tu causing the factor to loose its affinity for aa-tRNA and the ribosome from which it dissociates in complex with GDP. The newly delivered aa-tRNA remains bound to the ribosome and the amino acid attached to the tRNA is incorporated in the nascent polypeptide chain. EF-Tu·GDP is recycled by the action of its guanine nucleotide exchange factor, EF-Ts, which catalyses the exchange of GTP for GDP (Figure 1).

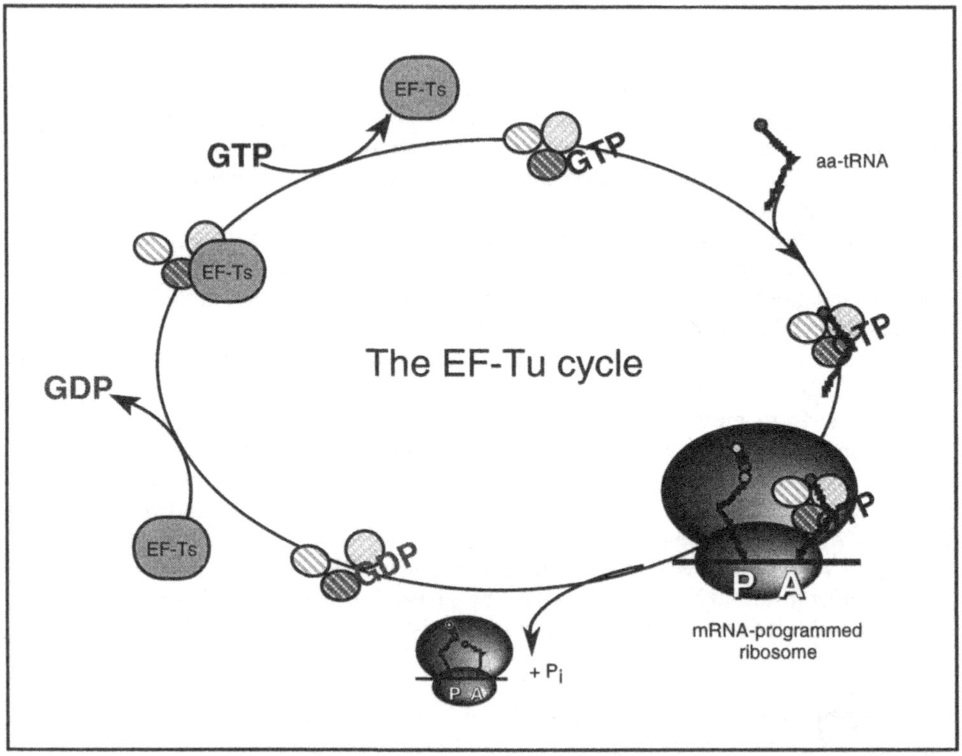

Figure 1. A simplified representation of the EF-Tu elongation cycle.

The cycle of EF-Tu is structurally well-characterised since the structures of the following complexes are available: Intact and proteolytically modified EF-Tu·GDP from *E. coli* [7, 8, 9] and *Thermus aquaticus* [8], EF-Tu·GDPNP (where GDPNP is a non-hydrolysable GTP-analogue) from *Th. thermophilus* [10] and *Th. aquaticus* [11], EF-Tu·GDPNP·Phe-tRNA^{Phe} consisting of EF-Tu from *Th. aquaticus* and tRNA^{Phe} from yeast [12] and EF-Tu·EF-Ts from *E. coli* [13] and *Th. thermophilus* [14]. EF-Tu is composed of three domains of which domain 1 is the guanine-nucleotide-binding domain or G-domain. This domain shares large structural homology with the G-domains of other G-binding proteins [15] and possesses the characteristic switch regions, which undergo large conformational changes during the elongation cycle. Domains 2 and 3, which are pure β-barrel domains, form another structural unit. When EF-Tu cycles between its GDP and GTP-bound forms, these two domains move as a rigid body relative to domain 1. This movement is induced by structural rearrangements internally in domain 1 involving the two switch regions in particular. The rearrangements expand from domain 1 throughout the entire molecule resulting in a rotation of domain 1 of 90° relative to domains 2 and 3. The GDP-form has an open structure, with a characteristic hole in the middle, whereas the GTP-form is compact and closed. In this form, EF-Tu is able to bind aa-tRNA - an event, which causes no major conformational changes.

1.1 THE STRUCTURE OF THE TERNARY COMPLEX, EF-TU·GTP·AA-TRNA.

The first macromolecular component of the ternary complex, EF-Tu, was described in the previous section. The second, the tRNA, consists of two double-helical segments almost perpendicular to each other. One segment, the acceptor helix, consists of the T-stem and acceptor stem and ends with the conserved 3'-CCA sequence, to which the amino acid is attached via an ester bond to a hydroxyl group of the terminal ribose ring. The other helical segment is composed of the D-stem and the anticodon stem and terminates in the anticodon loop [16].

The ternary complex has an elongated, corkscrew-like structure, where the anticodon helix of the tRNA, which is exposed, forms the screw and the handle consists of EF-Tu·GTP and the acceptor helix of the tRNA [12, 17] (Figure 2).

EF-Tu interacts with three distinct parts of the aa-tRNA: the T-stem, the 5'-end and the 3'-CCA-Phe-end. Surprisingly few direct contacts are established between aa-tRNA and EF-Tu. All three domains of EF-Tu are in contact with the aa-tRNA, and the two switch regions 1 and 2, comprising the effector region and the helix B, respectively, play important roles in the binding. None of the interactions, except for the recognition of A76 of the common CCA-motif at the 3'-end, are base specific but are mediated via interactions with the phosphates and 2'-OH groups of the backbone. This is in contrast to the tRNA synthetases, which requires a high degree of tRNA specific recognition to function properly.

1.1.1 *A detailed description of interactions within the ternary complex*

In the following, the structure of the ternary complex will be described in more detail. The residue numbers refers to EF-Tu from *Th. aquaticus*, and the identity of the corresponding residues in EF-Tu from *E. coli* will be given in brackets if differing. The *E. coli* residue number is obtained by subtracting 1, 11 or 12 residues from the *Th. aquaticus* number for residues in the regions 39-180, 191-260 and 261-405, respectively. The residues from 181 to 190 are unique to *Th. aquaticus* EF-Tu.

The 3'-CCA-Phe end is bound in a narrow cleft formed between domains 1 and 2. The phenylalanine is docked into a pocket where it stacks on His67(66). The pocket is lined with the side chains of Phe229(218), Asp227(216), Glu226(215) and Thr239(228). The pocket can accommodate any of the 20 naturally occurring amino acids. The amino group of the ester bond can form hydrogen bonds to the main chain CO of Asn285(273) and the main chain NH of His273(Phe261). The carbonyl group of the ester bond can form hydrogen bonds with the main chain NH of Arg274(262), while the side chain of this residue interacts with the phosphate of A76. The ester bond is

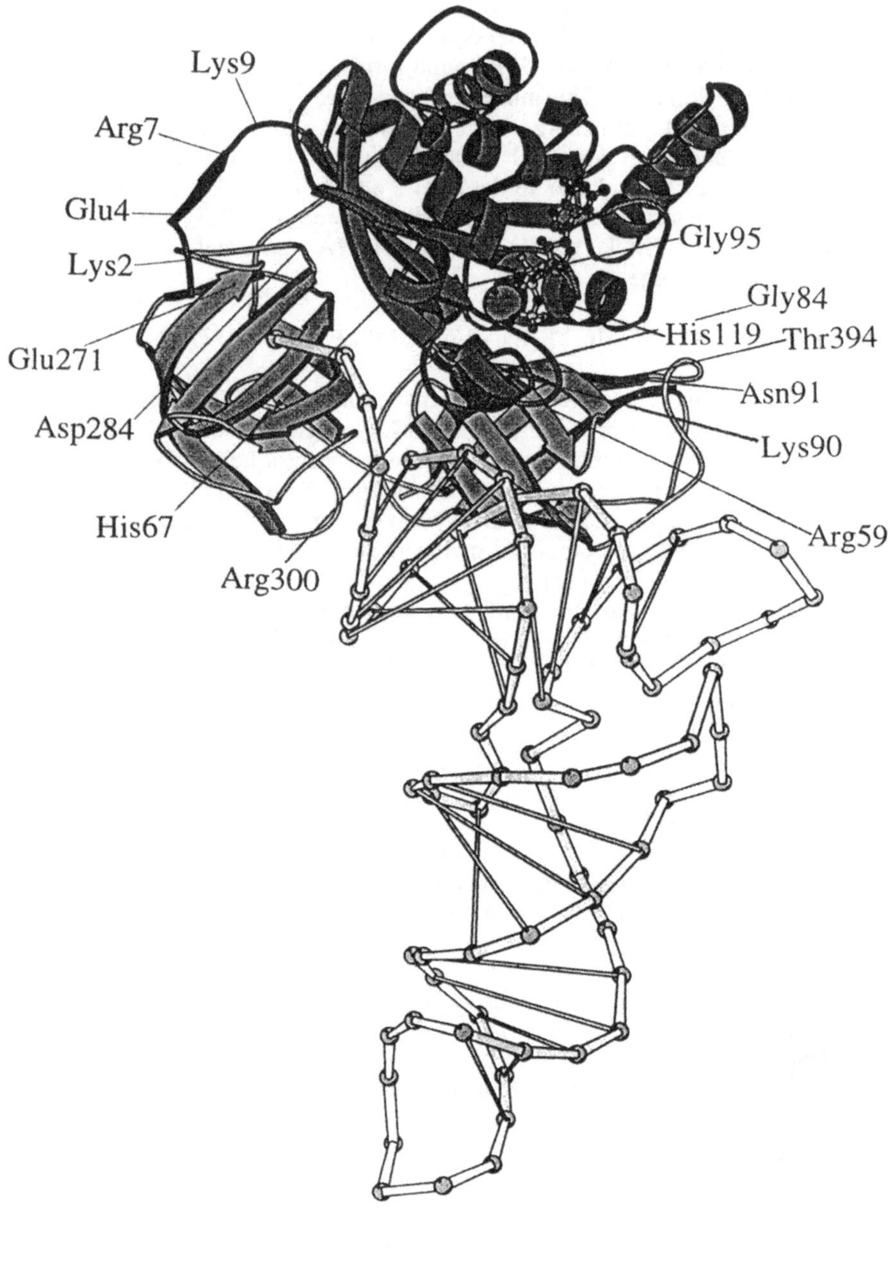

Figure 2. The structure of the ternary complex, EF-Tu·GDPNP·Phe-tRNA^{Phe}, composed of EF-Tu from *Th. aquaticus* and yeast Phe-tRNA^{Phe}. Mutagenesis studies of the indicated residues have been carried out in our laboratory. The numbering refers to EF-Tu from *Th. aquaticus* although EF-Tu from *E. coli* has been used as the mutagenesis template.

made to the 3'-OH of the terminal ribose, while the 2'-OH can form a hydrogen bond with the side chain of Glu271(259).

Aminoacylation favours the unstacking of A76, which is docked in a pocket formed by two protruding loops of domain 2. The conserved Glu271(259) stacks with the adenine on one side, whereas a hydrophobic platform made up by the conserved Val237(226), Leu289(277) and Ile231(220) is formed on the other side. The phosphates of positions 74 and 75 are in contact with Lys52(Asn51). The three bases of A73, C74 and C75 are not in contact with EF-Tu but stacks with each other and point away from the protein. The identity of these bases seems to be unimportant for the formation of the ternary complex [4].

Helix A", the C-terminal part of helix B and two loops between β-strands e2 and f2, and b3 and c3 form a cavity for the 5'-end of the tRNA, which is thereby in contact with all three domains of EF-Tu. The phosphate of G1 forms a salt bridge with the conserved Arg300(288) and the ribose is in contact with the conserved Lys90(89) and Asn91(90). Also the ribose of C2 and the phosphate of G3 make contact with the protein.

One side of the double stranded part of the T-stem is aligned in parallel with the β-barrel structure of domain 3 and forms contacts with the main and side chains of this domain. Surprisingly, domain 3 only exhibits a low degree of conservation of its solvent exposed residues. However, several core residues are invariant ensuring the maintenance of the overall domain fold.

The structures of the two free macromolecular components of the ternary complex are almost the same as when bound in the ternary complex. The most pronounced shift observed in EF-Tu is that of helix A", which is part of the switch 1 region. The single stranded 3'-end of the tRNA has a helical curvature induced upon binding to EF-Tu. The angle and twist between the acceptor and T-stem are slightly distorted resulting in a 16Å displacement of the position of the 3'-end. The anticodon helix is bent. Some of these discrepancies are caused by crystal packing effects.

There is, however, no doubt about the physiological relevance of this structure. The crystal structure has been shown to be similar to the structure found in solution [18], and more recently the crystal structure of a ternary complex consisting of *E. coli* Cys-tRNA and *T. aquaticus* EF-Tu has been solved. This structure is very similar to that of the ternary complex with yeast Phe-tRNA, although the crystal packing is very different [19].

2. Macromolecular mimicry in translation

A comparison of the structures of the ternary complex and another elongation factor, EF-G, which promotes the translocation of tRNAs relative to the mRNA, revealed a striking structural similarity [20] (Figure 3).

EF-G, which is composed of five structural domains, displays an elongated structure corresponding to that of the ternary complex, EF-Tu·GTP·aa-tRNA [21, 22]. Domains 1 and 2 of EF-G and EF-Tu are closely related and their relative positions are the same in the EF-Tu·GTP and EF-G·GDP complexes. Domains 3, 4 and 5 of EF-G mimic the tRNA part of the ternary complex. Domain 3 is shaped like the acceptor stem, domain 4 like the anticodon helix and domain 5 like the T-stem. The folds of domains 4 and 5 also resemble those of the ribosomal proteins S5 and S6, respectively.

Domain 4 is elongated and presents an unusual left-handed cross-over connection between two β-strands in a βαβ motif.

Figure 3. A comparison of the structures of the ternary complex, EF-Tu·GDPNP·Phe-tRNA^{Phe} (left) and EF-G (right).

The structural relationship between EF-G·GDP and the ternary complex suggests a functional relationship implying a common mechanism of interacting with the ribosome. The ribosomal binding site for the two factors may be the same or partially overlapping. This idea is also supported by earlier studies [23-27]. More specifically, it has been suggested that the hydrolysis of GTP by EF-G induces a conformational change of the factor, which creates a pocket on the ribosome prepared for the subsequent binding of the ternary complex [28, 29]. Likewise the triggering of the GTPase activities of EF-G and EF-Tu may proceed via the same pathway. Unfortunately, the crystallisation of EF-G·GTP has turned out to be very difficult, so we are left with speculations about the nature of the conformational changes taking place in EF-G upon GTP-hydrolysis.

At first it may seem strange why the structures of the inactive form of EF-G and the active form of EF-Tu resemble one another. However, the two factors catalyse reverse transitions of the ribosome. EF-Tu·GTP·aa-tRNA drives the ribosome from the post-translocational state into the pre-translocational state, while EF-G promotes the opposite transition. Therefore, the GTP-forms of both EF-Tu and EF-G stabilise a transition state of the ribosome lying in-between the pre- and post-translocational states.

The initiation and termination processes of protein biosynthesis also involve G-binding proteins. Prokaryotic translational initiation involves three factors IF1, IF2 and IF3 of which IF2 binds guanine nucleotides and functions as a carrier of fMet-tRNA^{fMet} to the 30S ribosomal P-site in analogy with the function of EF-Tu. The GTPase centre of IF2 is activated upon formation of the 70S initiation complex. Cross-linking and binding studies have shown that IF2, EF-G, EF-Tu and release factors (RFs) bind to overlapping sites on the ribosome [30] The ternary complex, EF-Tu·GTP·aa-tRNA, binds to the A-site during decoding of the mRNA, EF-G is positioned in the A-site during translocation and the release factor complex between RF3 and RF1/RF2 fills the A-site in response to a stop-codon leading to the termination of the translation process. IF1 and IF2 locate the initiator fMet-tRNA^{fMet} in the P-site. During this process, the initiation factors need to prevent the initiator tRNA from interacting with the A-site. This could be achieved by the docking of one or more initiation factor domains into the site, while the initiator tRNA is delivered to the P-site. The high homology between the nucleotide-binding domains of EF-Tu, EF-G, IF2 and RF3 is well-known. Apart from this, manual homology searches [31] reveal that domain 4 of IF2 shares homology with domains 2 of EF-G and EF-Tu indicating that this domain provides a site for the binding of the 3'-end of the initiator tRNA. The C-terminal part of domain 4 of IF2 finds a counterpart in domain 3 of EF-G, while domain 3 of IF2 is homologous to the C-terminal part of domain 4 and the entire domain 5 of EF-G. Finally, IF1 shares homology with the N-terminal region of domain 4 of EF-G. In this way, IF2 - like EF-G - possesses a proteinaceous domain mimicking tRNA. Actually all domains of EF-G show homology to domains in IF1 and 2. The tasks of both EF-G and the initiation factors are to position a tRNA in the ribosomal P-site, and the structural similarities strongly suggest that this is achieved via similar functional mechanisms.

The bacterial release factors RF1 and RF2 respond to stop-codons on the mRNA in a codon specific manner, and their activities are stimulated by the guanine-nucleotide binding release factor, RF3. The codon-specific release factors RF1 and RF2 share homology with domain 4 of EF-G and have therefore been proposed to possess a tRNA-like domain, which accounts for the binding of the factor to the ribosomal A-site and mimic the anticodon for pairing with the stop codon [32]. Site-directed mutagenesis of two amino acids equivalent to residues in domain 4 of EF-G resulted in a dominant lethal phenotype or abnormal termination in response to sense codons [32]. According to this model, RF3 should mimic EF-Tu or part of EF-G. However, a recent study of the eukaryotic release factors eRF1 and eRF3 corresponding to the prokaryotic

RF1/RF2 and RF3, respectively, showed that eRF1 can function without forming a stable complex with eRF3. This indicates that eRF3 plays other roles than just simple mimicry of EF-Tu/EF-G [33].

The ribosomal release factor (RRF) is crucial for the recycling of ribosomes and acts in the splitting of the termination complex. This requires the interaction with a ribosome having an empty A-site and a P-site occupied by deacylated tRNA. It has been speculated that also the RRF has a tRNA-mimicry domain for binding to the A-site of the ribosome [34].

Recently, the structures of different functional states of the complete ribosome obtained by electron cryomicroscopy (cryo-EM) [35–39] has allowed the placing of the concept of macromolecular mimicry in a ribosomal context. The A, P, and E binding sites for the tRNAs have been identified resulting in the location of the decoding site on the 30S subunit and the peptidyl transfer site [35, 37, 39]. A site for the binding of EF-Tu has been identified below the L7/L12 stalk on the large ribosomal subunit [36]. Domain 1 of EF-Tu is in contact with the 50S subunit, while domain 2 is in contact with the 30S subunit. Interestingly, all translation factors possess these two domains [40] in support of the theory of the existence of a common GTPase centre on the ribosome. The conservation of the fold of domain 2 has been found to be very extensive in the case of EF-Tu and EF-G. Even the specific pocket of domain 2 of EF-Tu involved in the binding of the terminal A-base of the tRNA is mimicked very closely in EF-G [19].

More recently, also the position of EF-G corresponding to the post-translocational stage could be mapped on the ribosome using cryo-EM [41]. The factor contacts both the small and the large ribosomal subunits in a manner correlating nicely with the map obtained by hydroxyl radical foot printing of the ribosomal RNAs directed from 18 different surface positions of EF-G [42]. Domain 5 of EF-G seems to contact the base of the L7/L12 stalk, while domain 1 contacts the ribosome via a structural element protruding from the stalk. The latter contact is very similar to that observed for EF-Tu. The common interaction with L7/L12 could well be the source of triggering the GTPase activities of the G-binding translation factors. Like the anticodon helix of the ternary complex, domain 4 of EF-G reaches into a cleft of the 30S subunit in close proximity to the decoding centre in accordance with the crucial role implicated for particularly this domain in translocation catalysis [43]. Also mutation studies have ascribed an important role to this domain, which is the counterpart of the anticodon helix in the ternary complex [44, 45]. Furthermore, the homologous, mammalian EF-2 is inactivated by diphtheria toxin via ADP ribosylation of the tip of domain 4. It seems that the conformation of the loop is more important than the actual amino acid sequence consistent with its proposed ability to serve as an anticodon mimic.

It has been suggested that domain 4 of EF-G and the anticodon helix of tRNA use similar mechanisms to carry out their different tasks [42]. Hydrolysis of GTP induces a conformational change that is transmitted to the ribosome via domain 4 or the anticodon helix. Thereby an unlocked state of the ribosome is generated. In this ribosomal transition state, the tRNA-binding contacts are relaxed and the movement of substructures within the ribosome allows new contacts to be established involving a rotational movement of the two bound tRNA molecules [46].

Further, domains 2 of EF-Tu and EF-G seem to be oriented similarly on the ribosome. Domain 2 of EF-G contacts the S4/S12 region of the small ribosomal subunit. This direct contact, however, cannot be observed in the structure of the ternary complex on the ribosome. The protein S4 and the S4 region of 16S rRNA have been suggested to influence the accuracy of the translation process - the latter via interactions with EF-Tu [42, 47].

The discovery of macromolecular mimicry among translation elongation factors has provided new insight into the elongation process in particular, but has also suggested an even broader consequence involving the proceeding and succeeding processes of translation, initiation and termination, respectively, as well. If we assume an RNA world in early evolution then the ternary complex is an early stage where RNA function could be replaced by a protein. Then EF-Tu could be an early progenitor for translation factors and even for GTP-binding proteins.

3. Classical protein engineering

Protein engineering is most often used to shed light on the relation between the structure and function of a protein. These functions include ligand binding, enzymatic activity, folding or stability. A basic knowledge about the design of proteins is thereby obtained and can be used to improve the features of a protein of interest. The classical protein engineering cycle (Figure 4) runs through a number of steps, which in principle can be repeated infinitely. Structural and functional knowledge of the protein of interest (or related proteins) are important tools when planning a mutagenesis strategy that can solve a particular problem. This includes two decisions: which amino acid to mutate and what to substitute it with. The alignment of the protein of interest with related proteins can be used as a guideline for the identification of important residues [48]. Residues that are directly involved in a common function such as catalysis will be among the most conserved. However, also residues with a structural role will show a high degree of conservation. The potential danger of misinterpretations is obvious. Conversely, the nonconserved residues may play a role in functions unique to each protein such as substrate binding and specificity.

Two sources of information can be used when trying to choose a suitable substitution. Natural evolution has used *in vivo* point-mutations to engineer stable protein structures for specific functions. A similar fold is maintained for functionally related molecules regardless of the accumulation of natural mutations. Considerable variation can be tolerated in external loops, while protein cores are maintained as well-packed regions with topographically equivalent side chains in each species. Overall, the surface should appear hydrophilic in character, while the core is stabilised by hydrophobic interactions [49]. Another source of information comes from other protein engineering studies, but the principles for "safe" mutations deduced from these two types of studies are similar.

Basically, two kinds of mutations can be introduced into a protein [50]: (1) substitution of one amino acid by an isosteric residue of different function (f.ex. Glu by Gln), or (2) replacement of one amino acid by another of identical function but different structure (f.ex. Glu by Asp). The first kind of substitution examines function while keeping structure constant, whereas the second kind explores how function depends on structure. In general, glycines and prolines should not be mutated or introduced, since these residues often have a great influence on protein structure.

It pays to spend some time on the design phase. Mutants are easy to produce, but the results of the subsequent functional characterisation can turn out to be laborious or even impossible to evaluate in a sensible manner, if the mutants have not been designed properly.

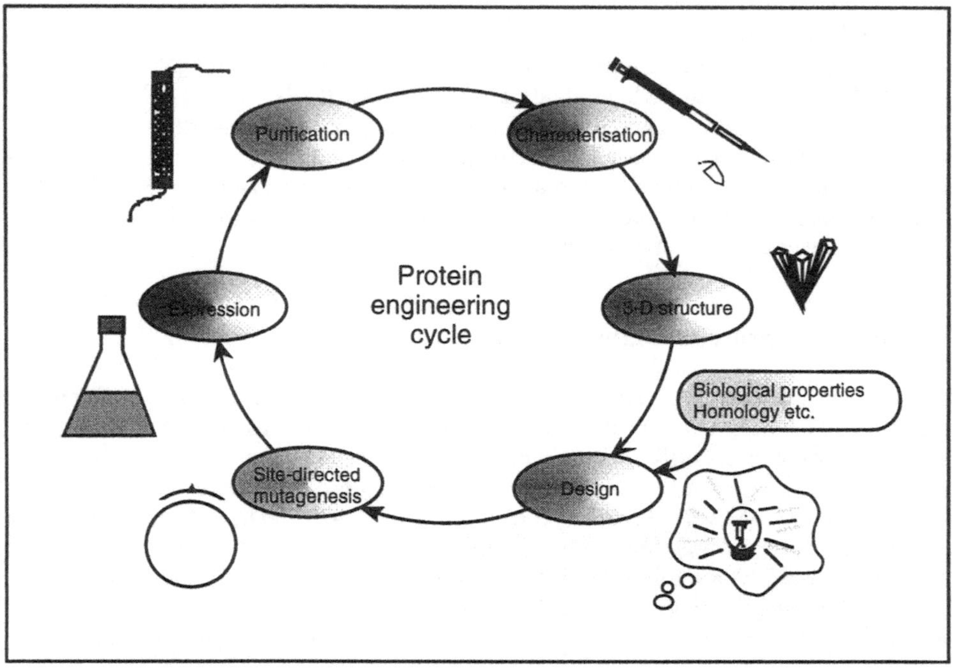

Figure 4. The classical protein engineering cycle.

The "charged-to-alanine scanning mutagenesis"-technique is also employed for the identification of functionally important regions. This technique involves the progressive replacement of each of the charged amino acids in the protein with alanine. This primary screen should be followed by a more thorough investigation of the contributions of the neighbouring polar and neutral amino acids in the regions identified [51].

After having decided on a mutagensis strategy, the mutation(s) is(are) introduced using site-directed mutagenesis. The engineered gene is then expressed and the resulting protein purified. It is advantageous to choose a generally applicable scheme such as a fusion-gene approach to avoid the need of individual optimisation for each mutant [52]. The wild-type protein should be expressed and purified in exactly the same manner as the mutant protein to serve as a reference in the functional characterisation.

The purified protein can be characterised in various ways depending on the interests of the experimentor. The first task is to ensure that the overall structure of the mutant protein is preserved. Often, the stable expression and solubility of a mutant protein can be taken as evidence of its structural integrity. The measurement of the thermostability of the mutant protein using for example calorimetry or the recording of various spectra in solution are fast methods suitable for providing the necessary information about the overall structure of the new mutant. If no global or long-range structural distortions can be observed, one can proceed to the functional assays carried out to access the role of the substituted amino acid in terms of catalytic activity, involvement in ligand-binding etc. The ultimate characterisation of a mutant is of course the determination of its three-dimensional structure. This structure allows a complete evaluation of the functional characterisation in structural terms, and serves as the starting point for a new round of improvements in the protein-engineering cycle.

4. Protein engineering of elongation factor Tu

Since the appearance of the first structure of the G-domain of EF-Tu·GDP lacking the effector region [53], the protein has been subjected to structure-function studies using protein engineering. The first studies focussed specifically on the binding of GDP/GTP and the mechanism of GTP-hydrolysis [54, 55]. Most of these studies used the isolated G-domain as a model. This was done to have a simplified way of purifying the overexpressed mutant protein, which would otherwise have chromatographic characteristics resembling those of the endogenously expressed wild-type EF-Tu so much that the separation of the two proteins would be impossible or very time consuming using conventional techniques. Furthermore, the structures of domains 2 and 3 were not very well-defined and their functions not clear. The use of the truncated EF-Tu model did of course limit the features of EF-Tu that could be studied. The study of the binding of aa-tRNA and interaction with EF-Ts or the ribosome was out of reach. Even the studies of guanine nucleotide binding and intrinsic GTP hydrolysis that was carried out was somewhat speculative since it was clear that these features were regulated by the presence of domains 2 and 3 [56, 57].

The need for an improved purification system allowing the separation of full-length wild-type and mutant EF-Tu was obvious and as the fusion protein technology allowing affinity chromatography had started to emerge, the strategy was clear. We developed a system for the purification of EF-Tu mutants with an N-terminal fusion to glutathione-S-transferase (GST). A factor Xa recognition sequence was incorporated between the two fusion partners enabling the removal of the GST-handle [58]. Other groups have used a thrombin site for the release of the native protein [59], or used a hexa-His-handle as the purification tag, which turned out not to affect the functions of EF-Tu significantly and could therefore be left on the factor in the succeeding *in vitro* characterisation [60].

Our studies of aa-tRNA-binding was initiated before any structural information about the ternary complex became available. The guidelines for mutant design were chemical modification and cross-linking studies and later on the structure of EF-Tu in its GTP-conformation was used to deduce potential sites for the binding of aa-tRNA. In other cases, aa-tRNA-binding was not the aim for design of mutants. However, by chance even some of these mutants have provided information about this feature of EF-Tu.

Crystallographic structures are important tools when we want to understand how macromolecules carry out their function. Structural studies, however, cannot stand alone but need to be supplemented with biochemical studies, which can explain and expand the information embedded in a three-dimensional structure. Structures of macromolecules need to have a very high resolution before the positions of hydrogen bonds, salt bridges, water molecules etc. can be determined with certainty. The water molecules are important since most reactions performed by macromolecules take place in aqueous solution and often specifically co-ordinated water molecules play important functional roles. One should also bear in mind that crystal structures are static pictures of highly dynamic systems and essential information about the protein mobility is

TABLE 1. List of mutants made in our laboratory, which are affected with respect to their abilities to form a ternary complex. Residue numbers refer to *E. coli* EF-Tu.

Residue	Mutated to	Aim	Role in aa-tRNA binding	Impairment*	Ref.
Lys2	Ala, Glu	Role of N-terminal region; chemical modification; ternary complex formation	Indirect, stabilise the conformation of residues 5–9. Dynamic role?	+(+), +(+)	[81]
Lys4	Ala, Glu	Role of N-terminal region; chemical modification; ternary complex formation	Indirect, stabilise the conformation of residues 5–9. Dynamic role?	-, +(+)	[82]
Arg7	Ala, Glu	Role of N-terminal region; ternary complex formation	Indirect, forms salt bridge with Glu272; stabilise the domain 1/2 interface	+++(+), +++(+)	[81]
Lys9	Ala, Glu	Role of N-terminal region; ternary complex formation	Indirect, forms salt bridge with Glu70; stabilisation	+(+), +++(+)	[82]
Arg58	Ala, Glu	GTP-hydrolysis; ternary complex formation	Indirect, coordination of Glu54 and Tyr87. Contribution of a positively charged environment	++, +++(+)	[71]
His66	Ala	cross-linking; chemical modification; ternary complex formation	Direct, part of the aminoacyl docking site; stacks with the Phe-ring	+	[68, 69]
Gly83	Ala	Switch mechanism; N-terminal cap of helix B in the GDP-form	Indirect, helix B is involved in positioning of the major groove between the acceptor stem and the T-stem	+++	[75]
Lys89	Ala, Glu	Solvent exposed, conserved residue positioned in helix B	Direct, docking site for the 5'-end; positioned between the phosphates of G1 and G2	+, ++	[69]
Asn90	Ala, Glu	Solvent exposed, conserved residue positioned in helix B	Direct, docking site for the 5'-end; forms hydrogen bond with 2'-OH of G1	++, ++	[69]
Gly94	Ala	Switch mechanism; C-terminal cap of helix B in the GDP-form	Indirect, helix B is involved in positioning of the major groove between the acceptor stem and the T-stem	+(+)	[74, 75]
His118	Ala, Glu	cross-linking; chemical modification; ternary complex formation	Indirect, dynamic role?	++, +++	[68, 69]
Glu259	Ala, Asp, Gln, Tyr	Structure confirmation	Direct, main chain CO forms hydrogen bond to amino ester NH; side chain makes hydrogen bond with terminal 2'-OH and stacks on one side of A76	++, ++, -, ++++	[80]
Glu272	Arg	Confirmation of Arg7-Glu272 salt bridge	Indirect, salt-bridge formation with Arg7; stabilisation of the domain 1/2 interface	+++(+)	[84]
Arg288	Ala, Ile, Lys, Glu	Structure confirmation; allosteric regulation	Direct, forms salt bridge with the phosphate of G1	++, +, -, +++	[76]
Thr382	Ser, Ala, Asp, Tyr	Role of phosphorylation	Indirect, stabilisation of the domain 1/3 interface	-, ++(+), ++(+), +++(+)	[85]

*Degree of impairment: - not impaired; + moderate; ++ pronounced; +++ severe; ++++ abolished
A bracket indicates that the result can be dependent on the method of analysis.

concealed. Finally, packaging contacts are introduced upon crystallisation of the macromolecule and some of these might distort the protein and introduce artificial contacts with no physiological relevance. Thus there are many reasons why biochemical studies using structural data as the starting point should not be neglected.

In the following, a number of mutants characterised in our lab will be described and placed in a structural context. The mutational analysis is performed on EF-Tu from *E. coli*, thus the residue identity and numbers used below refers to this system. Any differences from the *Th. aquaticus* EF-Tu, which is our main source of structural information, will be indicated in brackets. The mutants are named as follows: the first amino acid of the name refers to the original amino acid, the number gives its position in the primary structure and the last amino acid gives the identity of the introduced amino acid. A schematic overview of the mutants characterised in our lab can be found in table 1 and figure 2. Residues mentioned in the explanation of a specific mutant will be highlighted in bold font, if they have also been subjected to mutational analysis in a separate study described elsewhere. This description is proceeded by a short introduction to the methods, which can be applied to characterise the interaction between EF-Tu and aa-tRNA.

4.1 METHODS

The aim of this section is to focus on techniques used by us and others to characterise the binding of aa-tRNA to EF-Tu. In connection to this a few words need to be said about other assays used to evaluate the structural integrity of the mutants in order to render probable that no long range effects have been introduced along with the point mutations.

Already upon expression and purification of a mutant of EF-Tu we get an indication whether or not the overall structure of the mutant is maintained as structural rearrangements often lead to denaturation, insolubility resulting in the formation of inclusion bodies and an increased susceptibility to proteolytic degradation. Next we test the stability of the mutants towards thermal and/or chemical denaturation. Finally, the mutant's ability to bind guanine nucleotides, hydrolyse GTP and/or take part in protein synthesis *in vitro* are taken as strong indications that the global structure of the mutant is comparable to that of the wild type.

The strength of the interaction between aa-tRNA and EF-Tu have been studied in different laboratories using various methods. These comprise fluorescence studies [61, 62], gel filtration experiments [63], membrane filter assays [64], band shift assays [65, 66], methods based on EF-Tu's ability to protect the labile aminoacyl bond [3, 67] etc. Here I will focus mainly on the methods applied in our lab.

The formation and stability of the EF-Tu·GTP·aa-tRNA complex are studied by following the reaction EF-Tu·GTP·aa-tRNA ↔ EF-Tu·GTP+aa-tRNA. If EF-Tu·GTP and aa-tRNA are mixed and applied on a non-denaturing polyacrylamide gel, which is concomitantly stained with Coomassie Blue, a mobility shift can be observed for EF-Tu·GTP bound to aa-tRNA as the negative charges of the phosphate backbone of the tRNA make the complex migrate faster than the free protein. If a fixed amount of EF-Tu is incubated with varying amounts of aa-tRNA, a titration curve can be obtained allowing the estimation of an apparent dissociation constant, K_d, for the complex. No absolute value can be obtained as the method operates under non-equilibrium conditions causing a dissociation of the ternary complex during electrophoresis. However, the method is adequate for comparisons of different mutants with wild-type EF-Tu. Also weak interactions can be observed by this method even if no distinct band is visible at the position of the ternary complex. In these cases, a faint interaction will be seen as a smearing of the band representing binary complex, which is released gradually during electrophoresis as dissociation of the ternary complex is favoured in the gel. Zone-interference gel electrophoresis is a similar though more exact method [65]. A pre-formed ternary complex is subjected to electrophoresis under equilibrium conditions, as the macromolecular complex is surrounded by zones containing different concentrations of aa-tRNA. The measurement of migration distances allows the determination of K_d values, which are closer to the absolute value.

Two assays take advantage of the ability of EF-Tu to protect the labile aminoacyl bond against spontaneous hydrolysis [67] and enzymatic hydrolysis by RNaseA [3]. Both methods can be used to determine an apparent dissociation rate constant, k_{-1}, whereas the determination of K_d may be impossible, if the affinity of the mutant EF-Tu·GTP complex for aa-tRNA is too low. A ternary complex is formed in the presence of phosphoenolpyruvate and pyruvate kinase, which forces EF-Tu into its GTP conformation. Then the dissociation of the ternary complex is induced by either increasing the temperature form 0 to 20°C (non-enzymatic hydrolysis) or by addition of RNaseA (enzymatic hydrolysis). The resulting data are analysed according to the first-order rate law $dC/dt=-k_{-1}\cdot t$ by plotting $-\ln(C_t/C_0)$ versus t, where C_t and C_0 denote the concentrations of ternary complex at times t and zero, respectively. k_{-1} is found as the slope. These methods are non-equilibrium methods, as the equilibrium is driven towards the right owing to the continuous loss of aa-tRNA. The disappearance of the ternary complex is monitored by the loss of TCA-precipitable, radiolabelled aa-tRNA. The aa-tRNA synthetase should be removed from the charged aa-tRNA by phenol extraction prior to use in the spontaneous hydrolysis assay, as it will otherwise affect the deacylation by recharging the tRNA. The RNaseA protection assay is the faster of the two assays and the determined values of k_{-1} are closer to the true values although overestimated due to the constant removal of aa-tRNA by the RNase. RNaseA specifically attacks pyrimidine nucleotides by cleaving the 3'-adjacent phosphodiester bond Pyp/N. In this particular case, the attack of C74 and C75 are followed. The amount of RNaseA to add is critical and has to be determined carefully. Two assays are carried out in parallel at different concentrations of RNaseA - one with EF-Tu·GTP and another without. The best RNaseA concentration to use is found by the following criteria: (a) The blank curve should decrease very rapidly to a constant value, (b) the wild-type EF-Tu curve should be linear and (c) the difference between the blank and the wild-type EF-Tu curves should be as large as possible. The "k_{-1} value" provided for free aa-tRNA corresponds to the maximum measurable dissociation rate constant. The limit is set by the rate at which the aa-tRNA is digested by RNaseA. The same applies to the spontaneous hydrolysis assay, where the limit is set by the rate of spontaneous hydrolysis of the aminoacyl bond.

The gel electrophoresis techniques are generally applicable depending on the availability of material and a suitable detection method (staining, autoradiography, immunoblotting etc.). Also the protection methods can be utilised to follow an interaction between a protein and its ligand. This requires that a measurable protection is established during the interaction, for example against heat or urea-induced protein denaturation or enzymatic or chemical conversion of the substrate.

4.2 MUTANTS OF *E. COLI* EF-TU

His66 and His118 [68, 69]. A number of modification and cross-linking studies indicated that His66(67) and His118(119) were involved in the binding of aa-tRNA. His118 is invariant and His66 is extremely well-conserved amongst procaryotes. His66 was therefore mutated to alanine, and His118 was changed to alanine or glutamic acid. All mutants were found to be hampered in their ability to bind aa-tRNA. A similar result was reported in a study of the mutant His118Gly [70]. However, the His118-mutants, which behave similarly despite the different nature of their substitutions were impaired to a much larger extent than the His66 mutant. All other features examined were like those of the wild type except for the thermostabilities, which were slightly reduced.

How does the behaviour of the mutants fit with the structural data? His66 is positioned in the narrow cleft formed between domains 1 and 2 upon GTP-binding. It is part of the pocket into which the amino acid docks and in the case of Phe-tRNA^{Phe} it is stacked onto the amino acid. His118 is much more curious. The aa-tRNA is at least 16Å away from His118, which is completely buried in the interface between domains 1 and 3. Even in the GDP-form, the solvent-accessible surface area is only 0.5 Å^2 and the residue therefore hardly available for interaction. It is likely that His118 is temporarily exposed during the rearrangement of the domain 1/3 interface taking place upon switching between the GDP and GTP forms and thereby play a role in the structural transition, which is of course important for the binding of aa-tRNA. This suggests an indirect role of His118 in aa-tRNA binding.

Arg58 [71]. Arg58(59) located in the so-called effector region, was proposed to be involved in the hydrolysis of GTP and to stabilise an orthoester anion intermediate formed by the aminoacyl bond of aa-tRNA upon ternary complex formation [72]. Arg58 was therefore mutated to either alanine or glutamic acid to test these suggestions.

The effects of the mutations on GDP/GTP-binding and GTP hydrolysis were only slight or absent. The stabilisation of an eventual orthoester anion intermediate could not be confirmed since the removal of the positive charge as in the mutant Arg58Ala had only a minor effect on aa-tRNA binding. Later on also the structure of the ternary complex rejected this theory. The mutant Arg58Glu was affected much more significant with respect to aa-tRNA binding. The structure of the ternary complex shows that Arg58 is in contact with the conserved Glu54(55), which forms a hydrogen bond with the 2'-OH of C2, and with the conserved Tyr87(88), which is positioned over G1 and C2 together with the conserved Ile62(63). This co-ordination is lost upon mutation of Arg58 causing a destabilisation of the ternary complex, which is probably intensified by the introduction of a negative charge at this position. This effect may amplify and lead to a more general problem of positioning the acceptor helix and the aminoacyl bond correctly on EF-Tu. Others have demonstrated that also the invariant Thr61(62) of the effector region plays a role in the binding of aa-tRNA. K_d is decreased 40 times upon introduction of an alanine [73].

Gly83 and Gly94 [74, 75]. Helix B and its flanking loops comprises the so-called switch II region, which undergoes a large conformational change upon switching between the GDP and GTP-forms of EF-Tu. The rearrangement involves a shift in the direction of the helix by an angle of 42° and a shift of the position of the helix with respect to amino acid sequence by four positions. In this way, one helical turn unwinds at the N-terminus of the helix, while another is formed at the C-terminus upon switching from the GDP to the GTP-bound form [10, 11]. In the GDP-form, this helix is flanked by the glycines 83(84) and 94(95), which are conserved in all G-binding proteins. These glycines were supposed to act as pivots enabling the large conformational change to take place. This hypothesis was tested by mutation of these glycines to alanines either alone or in combination thereby reducing the flexibility of the main chain at these positions dramatically. Our results support this hypothesis and currently these mutants are being further characterised using pre-steady state kinetics to identify their specific roles.

The mutants' abilities to bind aa-tRNA was found to be impaired substantially. The most severe effects were observed for the mutants involving Gly83. The structure of the ternary complex shows that a number of residues located in helix B are engaged in the recognition of the major groove between the acceptor stem and the T-stem. Tyr87(88), **Lys89(90)** and **Asn90(91)** are all involved in the binding of the 5'-end and the conserved Asp86(87) interacts with the phosphates of G3 and A64 in a triangular pattern over the groove. Asp86 does not contribute directly to binding but serves as a structural regulator ensuring that the position of the acceptor helix is maintained. If a distortion of the acceptor helix takes place, the phosphates of the backbone may be brought too close to Asp86 causing an unfavourable repulsion. Surprisingly, the mutation of this residue to either asparagine or glutamic acid had no effect on the interaction with aa-tRNA [71]. Our *in vitro* characterisation of the glycine mutants indicates that the helix B's of these mutants are fixed in an alternative conformation thereby distorting an important part of the residues interacting with aa-tRNA. The conformational changes induced upon switching are largest in the N-terminus of helix B explaining why the Gly83Ala mutant is affected most severely.

Lys89 and Asn90 [69]. Helix B is a very interesting part of EF-Tu. It is part of the switch 2 region of G-binding proteins, which undergoes a large conformational change upon switching between the active and inactive forms of the protein (described above). The amino acids of helix B are evolutionary well-conserved with Lys89(90) and Asn90(91) being two of the most prominent examples. The side chains of these residues are solvent exposed in the GTP-form with the potential of interacting with ligands such as aa-tRNA. These two residues were therefore mutated to either alanine or glutamic acid.

A reduction in aa-tRNA affinity was observed upon mutation of either of these residues - in particular upon introduction of a glutamic acid. The structure of the ternary complex shows that Lys89 and Asn90 together with **Arg288**(300) form the docking site for the 5'-end of the tRNA. Arg288 interacts with the 5'-phosphate, Lys89 is positioned between the phosphates of G1 and G2 and Asn90 forms a hydrogen bond to the ribose of G1.

Arg288 [76]. The initial structure of the ternary complex suggested that Arg288(300) could form a salt bridge with the 5'-phosphate of G1. Apart from playing this rather specialised role in tRNA-binding, this residue was also suspected to be part of a signal transduction system transmitting the status of guanine-nucleotide binding to the rest of the molecule as the side chain of Arg288 was found to be hydrogen bonded to the switch 2 region at Asn90(91), which is in contact with the switch 1 region via a hydrogen bond to Ile62(63). Arg288 was therefore mutated to either alanine, isoleucine, lysine or glutamic acid.

The mutations had no effect on GDP/GTP binding or the GTPase activities. With respect to aa-tRNA binding the mutations of Arg288 to lysine had no effect, the mutation to isoleucine had only a slight effect, whereas the outcome of introducing an alanine was more pronounced and finally the introduction of a glutamic acid had the most severe effect. These results confirm the structure of the ternary complex: the introduction of a positive charge leads to a repulsion of the tRNA backbone and also the removal of charge decreases the affinity due to a loss of bonding energy. Apparently, also the size of the side chain is important as an isoleucine seems to cause a better packing of the 5'-binding pocket than an alanine.

Sprinzl and Graeser [77] studied the effect of removing the 5'-phosphate of tRNA. The modified tRNA could be aminoacylated and form a ternary complex. However, its rate in *in vitro* protein synthesis was slower than that of native tRNA. This could be due to an impaired interaction with the ribosome either due to the looser structural fit of the tRNA induced by EF-Tu or because the ribosome might interact directly with the 5'-phosphate of tRNA [78].

Also the Asp-tRNA synthetase belonging to the class II aa-tRNA synthetases possesses a positive cave for the binding of the 5'-phosphate of tRNA [79]. Similarly the binding of the CCA-end seems to be evolutionary conserved. Both EF-Tu and Asp-tRNA synthetase induce a helical curvature of the single-stranded 3'-end upon binding. Apart from these two common elements of recognition, EF-Tu and the synthetases seem to recognise different features of tRNA in accordance with their different roles.

When aa-tRNA binds to EF-Tu·GTP, the salt bridge formation pattern of Arg288 is changed from interacting with Asn90 to interacting with the 5'-phosphate of tRNA. Only the side chain moves during this shift. The switch 2 region is unaffected implying that Arg288 is not involved in regulating the allosteric mechanism of EF-Tu as confirmed by our experimental data. The salt bridge between Arg288 and the 5'-phosphate is important but not crucial. Then why is this arginine so well-conserved? The salt bridge could be of importance for the maintenance of an optimal structure of the ternary complex fitting perfectly to the ribosome thereby ensuring an optimal balance between speed and accuracy of protein synthesis.

Glu259 [80]. The structure of the ternary complex indicated that Glu259(271) was important for the binding of the 3'-CCA-Phe end of the aminoacylated tRNA. Its main chain CO group could form a hydrogen bond with the NH-part of the amino ester, and the side chain could form a hydrogen bond with the 2'-OH of the terminal ribose. Furthermore, the terminal adenine was stacked with Glu259 on one side with an antiparallel alignment of the dipole moments. Glu259 was mutated separately to either alanine, aspartic acid, glutamine or tyrosine in order to confirm these observations and to gain further insight into the role of this residue.

The Glu-to-Gln-substitution is an example of an isosteric substitution. There is no change in spatial requirements and the hydrogen and stacking potential is still present though slightly altered. As expected the mutant Glu259Gln behaves like the wild type. Introduction of an alanine at position 259 makes the fit of A76 into its pocket less perfect and causes an increased flexibility that may amplify to other parts of the aa-tRNA. The introduction of an aspartic acid, which is one CH_2-group shorter than glutamic acid, has a similar effect, most likely because of a loss of the hydrogen bond to the 2'-OH. Both Glu259Ala and Glu259Asp were observed to have pronounced defects in their ability of form a ternary complex. The introduction of a tyrosine at position 259 was expected to improve the stacking effect and create an EF-Tu mutant with an increased affinity for aa-tRNA. Furthermore, the terminal OH-group of the tyrosine should be able to participate in hydrogen-bond formation. However, the ability to bind aa-tRNA was completely lost, indicating that the tyrosine side chain is too bulky to replace a glutamic acid at this position.

The mutants show no significant deviations from the wild type with respect to binding of GDP/GTP and hydrolysis of GTP indicating that the mutations impact on aa-tRNA binding is not due to long-range structural rearrangements.

In summary, the strict conservation of Glu259 seems to be needed in order to maintain a very precise, tight packing around the 3'-end of aa-tRNA. Furthermore, it has been speculated that Glu259 through its interaction with the 2'-OH of the ribose is a central part of an isomerase activity carried out by EF-Tu. This regulatory activity ensures that all aa-tRNAs delivered to the ribosomal A-site have the amino acid attached to the 3' position of the ribose.

Lys2, Lys4, Arg7 and Lys9 [81, 82]. The function of the N-terminal region of EF-Tu (residues 1-10) remained obscure for many years. This part of the molecule seemed to be very flexible and therefore it was difficult to obtain any structural information about it. However, the presence of several well-conserved residues indicated that this region had to be of some importance. A basic residue is found at position 2 in more than 95% of all known EF-Tu species. Also the lysine at position 4 is well-conserved. However, in a few cases, including EF-Tu from *Thermus aquaticus*, an acidic residue is found at this position. Arg7 is almost 100% conserved among different EF-Tu species, whereas a lysine is found at the corresponding position of most species of EF-1α, the eukaryotic counterpart to EF-Tu. A 100% conserved lysine is found at position 9. We were particularly interested in these basic amino acids, since these were plausible candidates for making contacts to the backbone of tRNA upon formation of the ternary complex. Also chemical modification experiments had indicated that Lys2, Lys4(Glu4) and Arg7 are involved in formation of the ternary complex. When our work was initiated, the structure of the ternary complex was unknown, but inspection of the structure of EF-Tu·GDPNP showed that a cleft suitable for the accommodation of an RNA double helix was formed between domains 1 and 2 [83]. The N-terminal region was expected to be in close proximity of this cleft.

We decided to study the roles of Lys2, Lys4, Arg7 and Lys9 by mutational analysis and all four residues were changed separately to either alanine or glutamic acid. The mutants' abilities to form a ternary complex were affected in the following way: Lys4Ala behaved like the wild type, the Lys2 mutants were only moderately affected, Lys4Glu and Lys9Ala were affected significantly and to the same degree, whereas the abilities of the Arg7 mutants and Lys9Glu to form a ternary complex were completely lost.

The structure of the ternary complex, which was the first structure including the first ten amino acids, reveals that none of the residues are within contact distance from the aa-tRNA. Lys9 is part of a tight network of interactions. The side chains of Lys9 form a salt bridge with that of Glu70(71) and a hydrogen bond with the main chain oxygen of Thr71(72). Both of these interaction partners are situated centrally in domain 1 in the loop connecting β-strands b and c. The side chain of Thr71 is in contact with helix F via the side chain of Asp196(207), which is in contact with the side chain of

Tyr76(77). In the opposite direction, the main chain nitrogen of Thr71 forms a hydrogen bond with Tyr39(Asn39) at the beginning of the effector region.

The coiled structure of the N-terminal region protrudes from domain 1 and contacts domain 2. The major binding site in domain 2 is **Glu272**(Asp284), which forms a double salt bridge with Arg7. The acidity of this residue is conserved among different species of EF-Tu and EF-1α. In this way a connection between the effector region and domain 2 is established. Lys9 is not essential but important for the fine tuning of the intricate network of interactions allowing the perfect fit of the aminoacyl end of aa-tRNA into a well-defined cleft. Also Phe5 is likely to be part of the network connecting domains 1 and 2.

Lys 2 and 4 are solvent exposed and does not appear to form contacts with other parts of the protein or aa-tRNA. However, they seem to play a minor role in the binding of aa-tRNA. This could be by ensuring that residues 5–9 are positioned properly to allow the establishment of the connection between domains 1 and 2.

We believe that the formation of the salt bridges Arg7-Glu272 and Lys9-Glu70 are the last steps in a succession of movements leading to the switch from the GDP form to the GTP form, which is capable of binding aa-tRNA.

Glu272 [84]. Our results of the studies of point mutants located in the N-terminal region of EF-Tu (see above) indicated that a salt bridge formed between Arg7 of domain 1 and Glu272(Asp284) of domain 2 played an important role in maintaining the conformation capable of binding aa-tRNA.

This hypothesis was tested by mutating Glu272 to Arg either alone or in combination with the abovementioned Arg7Glu mutation. Our results show that the mutant Glu272Arg is as strongly impaired in binding aa-tRNA as the mutant Arg7Glu indicating that this residue is the second part of the salt bridge of interest. It should be noted that the mutants behave normally with respect to GDP/GTP-binding and stability excluding the possibility that any local structural effects caused by the mutations have propagated to other parts of the molecule.

Disappointingly, only part of the aa-tRNA binding capacity was restored when combining the two single-point mutations. However, the lack of full recovery of the broken salt bridge is not surprising, since a very precise geometry is required for salt bridge formation to take place and the introduction of mutations are apt to distort the backbone at the mutated position. Two slight distortions are certainly enough to weaken the potential for formation of a double salt bridge. However, we have successfully shown that also Glu272 is essential for forming a ternary complex in support of our hypothesis.

The domain 1/3 interface - Thr382 [85]. EF-Tu becomes phosphorylated at the fully conserved Thr382(394) located at the interface between domains 1 and 3. The role of this post-translational modification is unclear, and to study the effect of modifications at this position the four mutants T382S, T382A, T382D and T382Y were produced and characterised *in vitro*. The mutation to serine did not have any effect on the protein's ability to bind aa-tRNA whereas the other three mutations led to a severe loss of aa-tRNA binding capacity, which was particularly pronounced for the mutant T382Y.

In the GDP-form, Thr382 makes no direct contact with domain 1, whereas in the GTP-form it forms a hydrogen bond with the conserved Gln117(118) of helix C. In the ternary complex, Thr382 is not in direct contact with the tRNA but positioned next to the binding site of the T-stem. The introduction of an aspartate, which mimics the negative charge of the phosphate group, or a tyrosine, leads to an even more severe effect due to electrostatic repulsion and steric hindrance. Also the presence of a phosphate group at position 382 abolishes the formation of a ternary complex [86]. Two explanations are offered to account for the observed effects: (1) the mutations disturb the relative orientations of the three domains of EF-Tu thereby changing the relative orientation of the three areas of EF-Tu involved in aa-tRNA binding or (2) the mutations shift the dynamic equilibrium between the GDP and GTP forms in favour of the GDP-form.

The antibiotic kirromycin is a potent inhibitor of the translational activity of EF-Tu. It exerts its activity in complex with EF-Tu·GDP on the ribosome after delivery of aa-tRNA. This complex is apparently locked in a GTP-like conformation unable to dissociate from the ribosome thereby blocking any further protein synthesis [87]. A number of point mutants resistant to kirromycin have been selected by use of classical genetics [88-91]. The positions of these mutations map at the domain 1/3 interface of EF-Tu equivalent to Thr382. In general, these mutants seem to have a reduced affinity for aa-tRNA. Also in these cases, none of the residues are in contact with aa-tRNA. The Thr382 mutants were also found to have a reduced affinity for kirromycin in a pattern correlating perfectly with the pattern of reduced aa-tRNA affinity.

5. Summary and conclusion

We have constructed and characterised a large number of mutants using protein engineering techniques to shed light on the structure-function relations of elongation factor Tu from *E. coli*. The information obtained about the interaction of the elongation factor with aa-tRNA is particularly abundant. Basically, the effects of the mutations on EF-Tus ability to bind aa-tRNA are either direct i.e. the mutated residue are in contact with the aa-tRNA or indirect i.e. the mutated residue are involved in the maintenance of the structural integrity important for the binding of aa-tRNA.

Generally, the residues involved directly in the binding of aa-tRNA i.e. those that contribute to binding energy are important but not crucial. A lot of these contact points probably needs to be disrupted before the binding is completely abolished. The attached amino acid, the terminal adenine base and the 5'-end are docked into pockets, the exact packing of which are very critical for the binding as demonstrated by mutation of Glu259 and Arg288.

However, the mutation of the residues that are involved indirectly in formation of the ternary complex have a much more severe effect. Especially, the residues regulating interface contacts are involved in determining the relative positioning of the three contact points responsible for the binding of aa-tRNA. Even small displacements can be detrimental for the formation of the complex.

The fact that residues from all three domains of EF-Tu are involved in the binding of aa-tRNA minimises the degree of deviation tolerated from the perfect fit. Therefore the precise positioning of the three domains of EF-Tu relative to each other is of utmost importance for correct and tight binding of aa-tRNA. Especially, the domain 1/2 and 1/3 interfaces are prominent control points as shown by mutation of Arg7, Lys9 and Glu272 or Thr382, respectively. We still do not understand how the residues particularly at the domain 1/3 interface are involved in regulating the dynamic equilibrium between the active and inactive forms of the protein. This becomes particularly evident in the case of the His118-mutants, whose effects are difficult to understand by inspection of the available crystal structures. Also the domain 1/2/3 interface in the middle of the molecule serves as a control point regulating the docking of the 5'-end.

Helix B of the switch 2 region, which is involved in recognising the major groove between the acceptor stem and the T-stem needs to be in a very well-defined position as evident from the effects of mutating the flanking glycines 83 and 94. Also internal helical residues like Lys89 and Asn90 make important contributions to the binding. These interactions may be particularly important as they could be involved in the 10° bending between the acceptor stem and T-stem, which is induced upon binding to EF-Tu.

In some cases, the mutation of an invariant residue to a related residue with similar structure and function does not seem to have serious consequences for the interaction. It would, however, be very interesting to analyse in more detail how these mutant ternary complexes would be affected in their interaction with the ribosome, which is likely to have a platform for the interaction with aa-tRNA similar to that of EF-Tu. If this is the case, a ternary complex with even a very slight distortion from the perfect structure would have serious problems interacting properly with the ribosome - an interaction which controls both speed and accuracy of the synthesis of proteins.

Our mutagenesis studies of the interaction between EF-Tu and aa-tRNA are good examples of how structural and biochemical data supplement each other thereby increasing our knowledge of how macromolecules interact with each other in a recognition manner.

The idea of macromolecular mimicry has fueled the interest in understanding the translation process and a large body of information has started to accumulate. Most of the new knowledge is derived from structural studies and therefore our insight into the dynamic processes taking place on the ribosome is still limited. Biochemical studies are necessary to obtain the lacking details. Protein engineering will probably turn out to be a valuable technology. It would be very interesting to study the effect of domain swapping between translation factors. Especially, the swapping or mutation of tRNA-mimicry domains could enable the construction of an EF-G or IF1 responding to stop-codons. However, the target of future studies is even more likely to be the ribosomal components.

The resemblance between protein and nucleic acid as exemplified by the presence of tRNA-mimicry domains in EF-G, IF1/2 and the RFs may turn out to be a general feature of structural biology. Protein mimicry of DNA has been observed in the crystal structure of the uracil-DNA glycosylase-uracil glycosylase inhibitor protein complex [92, 93], and functional mimicry of a major autoantigenic epitope of the human insulin receptor by RNA has been presented [94].

Acknowledgements
We thankfully acknowledge the support by the Danish Biotechnology programme via the CISFEM and PERC centres. Figures 2 and 3 were kindly provided by Dr. Morten Kjeldgaard.

6. References

1. Bourne, H. R., Sanders, D. A. & McCormick, F. (1991) The GTPase superfamily: conserved structure and molecular mechanism, *Nature* **349,** 117-27.

2. Nyborg, J. & Kjeldgaard, M. (1996) Elongation in bacterial protein biosynthesis, *Curr. Opin. Biotechnol.* **7,** 369-75.
3. Louie, A. & Jurnak, F. (1985) Kinetic Studies of *Escherichia coli* Elongation Factor Tu-Guanosine 5'-Triphosphate-Aminoacyl-tRNA Complexes, *Biochem.* **24,** 6433-6439.
4. Faulhammer, H. G. a. J., R. L. (1987) Structural features in aminoacyl-tRNAs required for recognition by elongation factor Tu, *FEBS Lett.* **217,** 203-211.
5. Pingoud, A., Block, W., Wittinghofer, A., Wolf, H. and Fischer, E. (1982) The Elongation Factor Tu binds Aminoacyl-tRNA in the Presence of GDP, *J. Biol. Chem.* **257,** 11261-11267.
6. Rodnina, M. V., Pape, T., Fricke, R., Kuhn, L. & Wintermeyer, W. (1996) Initial binding of the elongation factor Tu.GTP.aminoacyl-tRNA complex preceding codon recognition on the ribosome, *J. Biol. Chem.* **271,** 646-52.
7. Kjeldgaard, M. & Nyborg, J. (1992) Refined structure of elongation factor EF-Tu from Escherichia coli, *J. Mol. Biol.* **223,** 721-42.
8. Polekhina, G., Thirup, S., Kjeldgaard, M., Nissen, P., Lippmann, C. & Nyborg, J. (1996) Helix unwinding in the effector region of elongation factor EF-Tu-GDP, *Structure* , 1141-1151.
9. Abel, K., Yoder, M. D., Hilgenfeld, R. & Jurnak, F. (1996) An α to β conformational switch in EF-Tu, *Structure* **4,** 1153-1159.
10. Berchtold, H., Reshetnikova, L., Reiser, C. O., Schirmer, N. K., Sprinzl, M. & Hilgenfeld, R. (1993) Crystal structure of active elongation factor Tu reveals major domain rearrangements [published erratum appears in Nature 1993 Sep 23;365(6444):368], *Nature* **365,** 126-32.
11. Kjeldgaard, M., Nissen, P., Thirup, S. & Nyborg, J. (1993) The crystal structure of elongation factor EF-Tu from Thermus aquaticus in the GTP conformation, *Structure* **1,** 35-50.
12. Nissen, P., Kjeldgaard, M., Thirup, S., Polekhina, G., Reshetnikova, L., Clark, B. F. & Nyborg, J. (1995) Crystal structure of the ternary complex of Phe-tRNAPhe, EF-Tu, and a GTP analog [see comments], *Science* **270,** 1464-72.
13. Kawashima, T., Berthet Colominas, C., Wulff, M., Cusack, S. & Leberman, R. (1996) The structure of the Escherichia coli EF-Tu.EF-Ts complex at 2.5 A resolution [see comments], *Nature* **379,** 511-8.
14. Wang, Y., Jiang, Y., Meyering-Voss, M., Sprinzl, M. and Sigler, P. B. (1997) Crystal structure of the EF-Tu:EF-Ts complex from *Thermus thermophilus*, *Nat. Struct. Biol.* **4,** 650-656.
15. Kjeldgaard, M., Nyborg, J. & Clark, B. F. (1996) The GTP binding motif: variations on a theme, *FASEB J.* **10,** 1347-68.
16. Robertus, J. D., Ladner, J. E., Finch, J. T., Rhodes, D., Brown, R. S., Clark, B. F. C. & Klug, A. (1974) Structure of yeast phenylalanine tRNA at 3 Å resolution., *Nature* **250,** 546-551.
17. Nissen, P., Kjeldgaard, M., Thirup, S., Clark, B. F. C. & Nyborg, J. (1996) The ternary complex of aminoacylated tRNA and EF-Tu-GTP. Recognition of a bond and a fold., *Biochimie* **78,** 921-933.
18. Bilgin, N., Ehrenberg, M., Ebel, C., Zaccai, G., Sayers, Z., Koch, M. H. J., Svergun, D. I., Barberato, C., Volkov, V., Nissen, P. and Nyborg, J. (1998) Solution structure of the ternary complex between aminoacyl-tRNA, elongation factor Tu and guanosine triphosphate, *Biochemistry* **37,** 8163-8172.
19. Nyborg, J. a. L., A. (1998) Protein biosynthesis: structural studies of the elongation cycle, *FEBS Lett.* **430,** 95-99.
20. Nyborg, J., Nissen, P., Kjeldgaard, M., Thirup, S., Polekhina, G. & Clark, B. F. C. (1996) Structure of the ternary complex of EF-Tu: macromolecular mimecry in translation., *Trends Biochem. Sci.* **21,** 81-82.
21. Ævarsson, A., Branzhikov, E., Garber, M., Zheltonosova, J., Chirgadze, Y., al-Karadaghi, S., Svensson, L. A. & Liljas, A. (1994) Three-dimensional structure of the ribosomal translocase: elongation factor G from Thermus termophilus., *EMBO J.* **13,** 3669-3677.
22. Czworkowski, J., Wang, J., Steitz, T. A. & Moore, P. B. (1994) The crystal structure of elongation factor G complexed with GDP, at 2.7 Å resolution., *EMBO J.* **13,** 3661-3668.
23. Langer, J. A. a. L., J. A. (1986) Elongation factor Tu localized on the exterior surface of the small ribosomal subunit, *J. Mol. Biol.* **187,** 617-621.
24. Girshovich, A. S., Bochkareva, E. S. and Vasiliev, V. D. (1986) Localization of elongation factor Tu on the ribosome, *FEBS Lett.* **197,** 192-198.
25. Girshovich, A. S., Kurtshalia, T. V., Ovchinnikov, Y. A. and Vasiliev, V. D. (1981) Localization of the elongation factor G on Escherichia coli ribosome, *FEBS Lett.* **130,** 54-59.
26. Moazed, D., Robertson, J. M., and Noller, H. F. (1988) Interaction of elongation factors EF-G and EF-Tu with a conserved loop in 23S RNA, *Nature* **334,** 362-364.
27. Hausner, T.-P., Atmadja, J. and Nierhaus, K. H. (1987) Evidence that the G2661 region of 23S rRNA is located at the ribosomal binding sites of both elongation factors, *Biochemie* **69,** 911-923.
28. Nyborg, J., Nissen, P., Kjeldgaard, M., Thirup, S., Polekhina, G. and Clark, B. F. C. (1996) Structure of the ternary complex of EF-Tu: macromolecular mimicry in translation, *Trends Biochem. Sci.* **21,** 81-82.
29. Liljas, A. (1996) Protein synthesis: Imprinting through molecular mimicry, *Current Biology* **6,** 247-249.
30. Heimark, R. L., Hershey, J. W. B. and Traut, R. R. (1976) Cross-linking of initiation factor IF2 to proteins L7/L12 in 70S ribosomes of Excherichia coli, *J. Biol. Chem.* **251,** 7779-7784.

31. Brock, S., Szkaradkiewicz, K. and Sprinzl, M. (1998) Initiation factors of protein biosynthesis in bacteria and their structural relationship to elongation and termination factors, *Mol. Microbiol.* **29,** 409-417.
32. Ito, K., Ebihara, K., Uno, M. & Nakamura, Y. (1996) Conserved motifs in prokaryotic and eukaryotic polypeptide release factors: tRNA-protein mimicry hypothesis, *Proc. Natl. Acad. Sci. USA* **93,** 5443-8.
33. Ito, K., Ebihara, K., and Nakamura, Y. (1998) The stretch of C-terminal acidic amino acids of translational release factor eRF1 is a primary binding site for eRF3 of fission yeast, *RNA* **4,** 958-972.
34. Nakamura, Y., Ito, K. & Isaksson, L. A. (1996) Emerging understanding of translation termination, *Cell* **87,** 147-150.
35. Agrawal, R. K., Penczek, P., Grassucci, R. A., Li, Y., Leith, A., Nierhaus, K. H. & Frank, J. (1996) Direct visualization of A-, P-, and E-site transfer RNAs in the Escherichia coli ribosome, *Science* **271,** 1000-2.
36. Stark, H., Orlova, E. V., Rinke-Appel, J., Junke, N., Mueller, F., Rodnina, M., Wintermeyer, W., Brimacombe, R. & van Heel, M. (1997) Arrangement of tRNAs in Pre- and Posttranslocational Ribosomes Revealed by Electron Cryomicroscopy, *Cell* **Vol. 88,** 19-28.
37. Malhotra, A., Penczek, P., Agrawal, R. K., Gabashvili, I. S., Grassucci, R. A., Jünemann, R., Burkhardt, N., Nierhaus, K. H. and Frank, J. (1998) Excherichia coli 70S ribosome at 15 Å resolution by cryo-electron microscopy: localization of fMet-$tRNA^{fMet}$ and fitting of L1 protein, *J. Mol. Biol.* **280,** 103-116.
38. Dube, P., Wieske, M., Stark, H., Schatz, M., Stahk, J., Zemlin, F., Lutsch, G. and van Heel, M. (1998) The 80S rat liver ribosome at 25 A resolution by electron cryomicroscopy and angular reconstitution., *Structure* **6,** 389-399.
39. Stark, H., Mueller, F., Orlova, E. V., Schatz, M., Dube, P., Erdemir, T., Zemlin, F., Brimacombe, R. & van Heel, M. (1995) The 70S Eschericia coli ribosome at 23 Å resolution: fitting the ribosomal RNA., *Structure* **3,** 815-821.
40. Ævarsson, A. (1995) Structure-Based Sequence Alignment of Elongation Factors Tu and G with Related GTPases Involved in Translation., *J. Mol. Evol.* **41,** 1096-1104.
41. Agrawal, R. K., Penczek, P., Grassucci, R. A. and Frank, J. (1998) Visualization of elongation factor G on the Escherichia coli 70S ribosome: The mechanism of translocation, *Proc. Natl. Acad. Sci. USA* **95,** 6134-6138.
42. Wilson, K. S. a. N., H. F. (1998) Mapping the position of translation elongation factor EF-G in the ribosome by directed hydroxyl radical probing, *Cell* **92,** 131-139.
43. Rodnina, M. V., Savelsbergh, A., Katunin, V. I. & Wintermeyer, W. (1997) Hydrolysis of GTP by elongation factor G drives tRNA movement on the ribosome, *Nature* **385,** 37-41.
44. Kimata, Y. a. K., K. (1994) Elongation factor 2 mutants dificient in diphthamide formation show temperature sensitive cell growth, *J. Biol. Chem.* **269,** 13497-13501.
45. Martemyanov, K. A., Yarunin, A. S., Liljas, A. and Gudkov, A. T. (1998) An intact conformation at the tip of elongation factor G domain IV is functionally important, *FEBS Lett.* **434,** 205-208.
46. Wilson, K. S. a. N., H. F. (1998) Molecular movement inside the translational engine, *Cell* **92,** 337-349.
47. Powers, T. & Noller, H. F. (1994) The 530 loop of 16S rRNA: a signal to EF-Tu?, *Trends. Genet.* **10,** 27-31.
48. Bowie, J. U., Reidhaar-Olson, J. F., Wendell, A. L. and Sauer, R. T. (1990) Deciphering the Message in Protein Sequences: Tolerance to Amino Acid Substitutions, *Science* **247,** 1306-1310.
49. Bordo, D. a. A., P. (1991) Suggestions for "Safe" Residue Substitutions in Site-directed Mutagenesis, *J. Mol. Biol.* **217,** 721-729.
50. Knowles, J. R. (1987) Tinkering with Enzymes: What Are We Learning?, *Science* **236,** 1252-1258.
51. Gibbs, C. S. a. Z., M. J. (1991) Identification of Functional Residues in Proteins by Charged-to-Alanine Scanning Mutagenesis, *Methods* **3,** 165-173.
52. Nilsson, J., Ståhl, S., Lundeberg, J., Uhlén, M. and Nygren, P.-Å. (1997) Affinity Fusion Strategies for Detection, Purification, and Immobilisation of Recombinant Proteins, *Protein Expr. Purif.* **11,** 1-16.
53. la Cour, T. F., Nyborg, J., Thirup, S. & Clark, B. F. (1985) Structural details of the binding of guanosine diphosphate to elongation factor Tu from E. coli as studied by X-ray crystallography, *EMBO-J* **4,** 2385-8.
54. Gümüsel, F., Cool, R. H., Weijland, A., Anborgh, P. A. and Parmeggiani, A. (1990) Mutagenesis of the NH_2-terminal domain of elongation factor Tu, *Biochim. Biophys. Acta* **1050,** 215-221.
55. Anborgh, P. H., Cool, R. H., Gümüsel, F., Harmark, K., Jacquet, E., Weijland, A., Mistou, M. Y. and Parmeggiani, A. (1991) Structure-function relationships of elongation factor Tu as studied by mutagenesis, *Biochimie* **73,** 1051-1059.
56. Parmeggiani, A., Swart, G. W. M., Mortensen, K. K., Jensen, M., Clark, B. F. C., Dente, L. and Cortese, R. (1987) Properties of a genetically engineered G domain of elongation factor Tu, *Proc. Natl. Acad. Sci. USA* **84,** 3141-3145.
57. Jensen, M., Cool, R. H., Mortensen, K. K., Clark, B. F. & Parmeggiani, A. (1989) Structure-function relationships of elongation factor Tu. Isolation and activity of the guanine-nucleotide-binding domain, *Eur-J-Biochem* **182,** 247-55.

58. Knudsen, C. R., Clark, B. F., Degn, B. & Wiborg, O. (1992) One-step purification of E. coli elongation factor Tu, *Biochem. Int.* **28,** 353-62.
59. Scarano, G., Krab, I. M., Bocchini, V., and Parmeggiani, A. (1995) Relevance of histidine-84 in the elongation factor Tu GTPase activity and in poly(Phe) synthesis: its substitution by glutamine and alanine., *FEBS Lett.* **29,** 214-218.
60. Boon, K., Vijgenboom, E., Madsen, L. V., Talens, A., Kraal, B. and Bosch, L. (1992) Isolation and functional analysis of histidine-tagged elongation factor Tu, *Eur. J. Biochem.* **210,** 177-183.
61. Ott, G., Faulhammer, H. G. and Sprinzl, M. (1989) Interaction of elongation factor Tu from Escherichia coli with aminoacyl-tRNA carrying a fluorescent reporter group on the 3' terminus, *Eur. J. Biochem.* **184,** 345-352.
62. Abrahamson, J. K., Laue, T. M., Miller, D. L. and Johnson, A. E. (1985) Direct determination of the association constant between elongation factor Tu:GTP adn aminoacyl-tRNA using fluorescence, *Biochemistry* **24,** 692-700.
63. Crechet, J. B. a. P., A. (1986) Characterization of the elongation factors from calf brain. 2. Functional properties of EF-1 alpha, the action of physiological ligands and kirromycin., *Eur. J. Biochem.* **161,** 647-653.
64. Hwang, Y.-W. a. M., D. L. (1987) A mutation that alters the nucleotide specificity of elongation factor Tu, a GTP regulatory protein, *J. Biol. Chem.* **262,** 13081-13085.
65. Abrahams, J. P., Kraal, B. and Bosch, L. (1988) Zone-interference gel electrophoresis: a new method for studying weak protein-nucleic acid complexes under native equilibrium conditions, *Nucl. Acids Res.* **16,** 10099-10108.
66. Tubulekas, I. a. H., D. (1993) A single amino acid substitution in elongation factor Tu disrupts interaction between the ternary complex and the ribosome, *J. Bacteriol.* **175,** 240-250.
67. Pingoud, A. & Urbanke, C. (1979) A Quantitave Assay for Ternary Complex Formation of Elongation Factor Tu, GTP, and Aminoacyl-tRNA, *Analytical Biochemistry* **92,** 123-127.
68. Andersen, C. & Wiborg, O. (1994) *Escherichia coli* elogation-factor-Tu mutants with decreased affinity for aminoacyl-tRNA, *Eur. J. Biochem.* **220,** 739-744.
69. Wiborg, O., Andersen, C., Knudsen, C. R., Clark, B. F. C. & Nyborg, J. (1996) Mapping Escherichia coli elongation factor Tu residues involved in binding of aminoacyl-tRNA, *J. Biol. Chem.* **271,** 20406-11.
70. Jonák, J., Anborgh, P. H. and Parmeggiani, A. (1994) Histidine 118 of elongation factor Tu: its role in aminoacyl-tRNA binding and regulation of the GTPase activity, *FEBS Lett.* **343,** 94-98.
71. Knudsen, C. R. & Clark, B. F. C. (1995) Site-directed mutagenesis of Arg58 and Asp86 of elogation factor Tu from *Escherichia coli*: effects on the GTPase reaction and aminoacyl-tRNA binding, *Prot. Eng.* **8,** 1267-1273.
72. Förster, C., Limmer, S., Zeidler, W. & Sprinzl, M. (1994) Effector region of the translation elongation factor EF-Tu.GTP complex stabilizes an orthoester acid intermediate structure of aminoacyl-tRNA in a ternary complex., *Proc. Natl. Acad. Sci. USA* **91,** 4254-4257.
73. Ahmadian, M. R., Kreutzer, R., Blechschmidt, B. and Sprinzl, M. (1995) Site-directed mutagenesis of Thermus thermophilus EF-Tu: the substitution of threonine-62 by serine or alanine, *FEBS Lett.* **377,** 253-257.
74. Knudsen, C. R., Kjaersgard, I. V., Wiborg, O. & Clark, B. F. (1995) Mutation of the conserved Gly94 and Gly126 in elongation factor Tu from Escherichia coli. Elucidation of their structural and functional roles, *Eur. J. Biochem.* **228,** 176-83.
75. Kjaersgard, I. V., Knudsen, C. R. & Wiborg, O. (1995) Mutation of the conserved Gly83 and Gly94 in Escherichia coli elongation factor Tu. Indication of structural pivots, *Eur. J. Biochem.* **228,** 184-90.
76. Rattenborg, T., Pedersen, G. N., Clark, B. F. C. and Knudsen, C. R. (1997) Contribution of Arg288 of Escherichia coli elongation factor Tu to translational functionality, *Eur. J. Biochem.* **249,** 408-414.
77. Sprinzl, M. a. G., E. (1980) Role of the 5'-terminal phosphate of tRNA for its function during protein biosynthesis elongation cycle, *Nucl. Acids Res.* **8,** 4737-4744.
78. Joseph, S. a. N., H. F. (1996) Mapping the tRNA neighborhood of the acceptor end of tRNA in the ribosome, *EMBO J.* **15,** 910-916.
79. Cavarelli, J., Rees, B., Ruff, M., Thierry, J. & Moras, D. (1993) Yeast $tRNA^{Asp}$ recognition by its cognate class II aminoacyl-tRNA synthetase, *Nature* **362,** 181-184.
80. Pedersen, G. N., Rattenborg, T., Knudsen, C. R. and Clark, B. F. C. (1998) The role of Glu259 in Escherichia coli elongation factor Tu in ternary complex formation, *Prot. Eng.* **11,** 101-108.
81. Mansilla, F., Knudsen, C. R., Laurberg, M. and Clark, B. F. C. (1997) Mutational analysis of Escherichia coli elongation factor Tu in search of a role for the N-terminal region, *Prot. Eng.* **10,** 927-934.
82. Laurberg, M., Mansilla, F., Clark, B. F. C. and Knudsen, C. R. (1998) Investigation of functional aspects of the N-terminal region of elongation factor Tu from Escherichia coli using a protein engineering approach, *J. Biol. Chem.* **273,** 4387-4391.
83. Sprinzl, M. (1994) Elongation factor Tu: a regulatory GTPase with an integrated effector, *Trends. Biochem. Sci.* **19,** 245-250.
84. Mansilla, F., Knudsen, C. R. and Clark, B. F. C. (1998) Mutational analysis of Glu272 in elongation factor 1A of E. coli, *FEBS Lett.* **429,** 417-420.

85. Plath, T., Knudsen, C. R., Bilgin, N., Erdmann, V., Clark, B. F. C. and Lippmann, C. (1998) Threonine 382 is essential for EF-Tu function, *Eur. J. Biochem.* (submitted).
86. Alexander, C., Bilgin, N., Lindschau, C., Mesters, J. R., Kraal, B., Hilgenfeld, R., Erdmann, V. A. & Lippmann, C. (1995) Phosphorylation of elongation factor Tu prevents ternary complex formation, *J. Biol. Chem.* **270,** 14541-7.
87. Parmeggiani, A. & Swart, G. W. (1985) Mechanism of action of kirromycin-like antibiotics, *Annu. Rev. Microbiol.* **39,** 557-77.
88. Kraal, B., Zeef, L.A., Mesters, J.R., Boon, K., Vorstenbosch, E.L., Bosch, L., Anborgh, P.H., Parmeggiani, A. and Hilgenfeld, R. (1995) Antibiotic resistance mechanisms of mutant EF-Tu species in Escherichia coli., *Biochem. Cell. Biol.* **73,** 1167-1177.
89. Mesters, J. R., Zeef, L.A., Hilgenfeld, R., de Graaf, J.M., Kraal, B. and Bosch, L. (1994) The structural and functional basis for the kirromycin resistance of mutant EF-Tu species in Escherichia coli., *EMBO J.* **13,** 4877-4885.
90. Abdulkarim, F., Liljas, L. and Hughes, D. (1994) Mutations to kirromycin resistance occur in the interface of domains I and III of EF-Tu:GTP, *FEBS Lett.* **352,** 118-122.
91. Abdulkarim, F., Ehrenberg, M. and Hughes, D. (1996) Mutants of EF-Tu defective in binding aminoacyl-tRNA, *FEBS Lett.* **382,** 297-303.
92. Savva, R. a. P., L.H. (1995) Nucleotide mimicry in the crystal structure of the uracil-DNA glycosylase-uracil glycosylase inhibitor protein complex., *Nat. Struct. Biol.* **2,** 752-757.
93. Mol, C. D., Arvai, A.S., Sanderson, R.J., Slupphaug, G., Kavli, B., Krokan, H.E., Mosbaugh, D.W. and Tainer, J.A. (1995) Crystal structure of human uracil-DNA glycosylase in complex with a protein inhibitor: protein mimicry of DNA., *Cell* **82,** 701-708.
94. Doudna, J. A., Cech, T.R. and Sullenger, B.A. (1995) Selection of an RNA molecule that mimics a major autoantigenic epitope of human insulin receptor., *Proc. Natl. Acad. Sci. USA* **92,** 2355-2359.

RNA STRUCTURE AND RNA-PROTEIN RECOGNITION DURING REGULATION OF EUKARYOTIC GENE EXPRESSION

GABRIELE VARANI, PETER BAYER, PAUL COLE, ANDRES RAMOS, LUCA VARANI

MRC Laboratory of Molecular Biology

Hills Road Cambridge CB2 2QH England

Gene expression in higher organisms is regulated after transcription through messenger RNA (mRNA) stability, transport and localization and through the excision of non-coding regions (pre-mRNA splicing). During these maturation events, mRNAs are complexed in ribonucleoprotein particles with proteins that recognize specific RNA sequences to affect these different regulatory steps. The assembly of large ribonucleoproteins performing RNA splicing (spliceosome) and translation (ribosome) also depends on recognition of spliceosomal snRNAs and ribosomal RNA, respectively, by constitutive and auxiliary protein factors. Understanding the molecular basis of RNA-protein recognition is therefore necessary to understand these regulatory events and to learn how to exogeneously regulate gene-expression.

1. Introduction

Biochemical and thermodynamic data on RNA-protein complexes reveal that intermolecular interaction occurs with different affinity in diverse systems to accommodate diverse regulatory requirements. For example, transfer RNA (tRNA) synthetase enzymes bind tRNA with modest affinity (the bimolecular dissociation constant $K_d \approx 10^{-6}$ M) to facilitate substrate release, whereas constitutive components of stable ribonucleoproteins

J. Barciszewski and B.F.C. Clark (eds.), RNA Biochemistry and Biotechnology, 195–216.

(such as the human U1A and U1 70K proteins) bind cognate RNAs much more tightly ($K_d < 10^{-9}$ M). Most RNA-binding proteins bind any RNA with low affinity, but biological function requires discrimination of cognate RNAs from non-cognate sequences present in large excess in the cell nucleus. Differences in binding energy between cognate and non-cognate RNAs define the specificity of the interaction. Affinity and specificity do not depend in a simple way on protein charge or size of intermolecular interface area. tRNA-synthetase complexes have very large interface areas compared to complexes involving RNP proteins, yet these other complexes have much tighter binding constants (see below). RNA structural diversity defines an enormous variety of specific interactions and is intimately related with the mechanisms by which molecular discrimination is achieved.

Numerous structures of DNA-protein complexes define an important paradigm in intermolecular recognition. Very often, DNA-protein recognition occurs by insertion of an α-helix into the major groove of double-stranded DNA [1]. Specific sequences are recognized by direct readout of hydrogen bond donors and acceptors on the DNA bases and by indirect readout of sequence-dependent DNA conformational features through van der Waals interactions and through electrostatic interactions with the negatively charged phosphodiester backbone. This paradigm cannot be applied to RNA recognition. First of all, the major groove of double helical RNA is too narrow and deep to allow insertion of protein secondary structural elements. In general, there is insufficient diversity of functionalities in the minor groove to allow effective sequence discrimination [2].

There is no known example of sequence specific recognition of RNA double helices. Sequence specific RNA-binding proteins recognize single-stranded regions, hairpin loops, internal loop or bulges, where the functional groups on the bases are accessible for recognition. Double-helical regions are recognized only when distortions in the double helix generated by internal loops or bulges or at the helix termini [3,4] allow access to base functionalities in the major groove. Another important difference arises from the fact that RNA elements often posses unique three-dimensional structures whose shape and charge distribution is recognized by cognate protein factors.

2. RNA-Binding Proteins Have Modular Structures and $\alpha\beta$ Protein Domains Constitute the Most Common Structural Unit in RNA Recognition

RNA-binding proteins have a modular structure comprising domains that achieve RNA recognition, and other domains that perform additional functions [5-7]; these often include binding other proteins. RNA-binding regions are composed of single or multiple copies of common RNA recognition units that sometimes constitute independent structural domains. Components of the RNA processing machinery are targeted to specific mRNAs through the recognition of specific RNA sequences and structures by these domains. Complexes functional in RNA processing or translational initiation are then formed by networks of RNA-dependent multiprotein complexes, targeted to specific mRNAs by interactions involving one or more of the components of the assembly. Thus, biological specificity (identification of a specific intracellular RNA target) and physical chemical specificity (RNA-protein molecular recognition) are closely related, but protein-protein interactions provide additional levels of regulation and the possibility of combinatorial control of gene expression.

The three most common RNA-binding modules, ribonucleoprotein (RNP), K-homology (KH) and double-stranded RNA binding (dsRBD) motifs, have compact, globular structures (Fig. 1) and constitute independent structural domains and RNA-binding motifs [8]. The dsRBD domain is a general, non-sequence specific double-stranded RNA-binding module [9-11]. Isolated domains bind the minor groove of double-stranded RNA without sequence specificity [12], but proteins containing multiple dsRBD domains specifically recognize certain RNA structures [12,13]. It is not clear whether this ability to bind specific RNA sequences is due to interdomain interactions, to cooperative interactions with additional protein factors or to the fact that the molecular target of dsRBD proteins in living organisms is not double stranded RNA, the accepted *in vitro* target of this class of proteins. KH domains appear to be non sequence specific single-stranded RNA-binding proteins associated with important metabolic functions. Depressed expression of the KH-protein FMR1 or a single amino acid substitution that unfolds the domain [14], lead to fragile-X syndrome [15], the most common cause of genetically inherited mental retardation in humans.

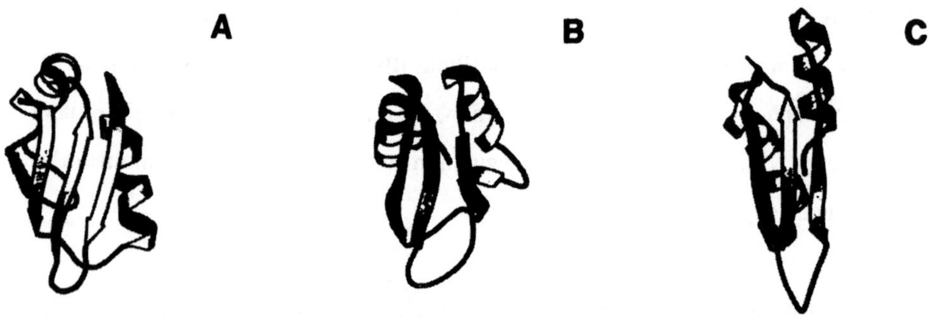

Figure 1. Three-dimensional structure of the three most common RNA-binding folds; RNP domain (a) [16,1726], KH domain (b) [14] and dsRBD domain (c) [10,11].

The RNP domain (also called RRM, RNA recognition motif) is the most common eukaryotic RNA-binding module. RNP proteins mediate RNA recognition in hundreds of proteins from the RNA processing machinery [5,18-20]. Animal, plant, fungal and prokaryotic cells contain RNP proteins in virtually all organelles, suggesting that this is an ancient protein fold associated with essential functions. Analysis of the sequence database indicates that this is one of the most common eukaryotic protein sequence motifs, providing an RNA-binding counterpart to zinc-finger or homeobox motifs for DNA. RNP domains are found in single or multiple copies in individual proteins (up to four, Fig. 2) and function primarily in targeting specific RNAs. Sometimes a single domain is sufficient to specify the RNA recognition ability of a given protein, but very often single

RNP domains do not function as independent RNA recognition units. Sequences immediately N- or C-terminal to the domain are often required for RNA recognition and in many cases of proteins containing multiple domains, two or more domains contribute to define the specificity of the protein. In addition to this primary ability to recognize RNA, RNP proteins also interact with other proteins. These interactions may either modulate or affect the specificity of the domain, and are critical in the assembly of multiprotein complexes that carry out various RNA processing events.

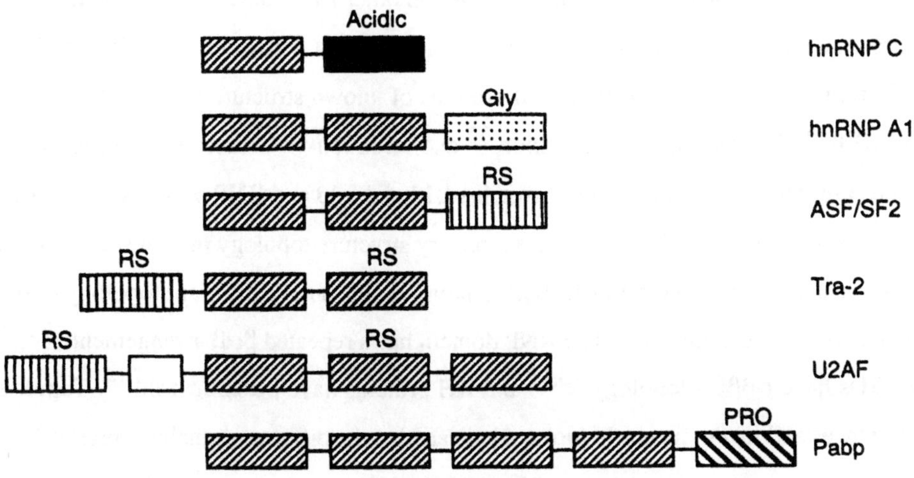

Figure 2. RNP proteins have modular structures and are composed of one or more RNP-type RNA-binding motifs (striped boxes) plus additional modules that mediate protein-protein interactions or other essential biological functions.

The RNP domain is identified by two highly conserved stretches of 6 and 8 amino acids separated by approximately 30 amino acids, named RNP-2 and RNP-1, respectively. The motif was first identified as a repeated sequence within the poly-Adenylate binding protein and in hnRNP A1 [21,22]. Sequence conservation outside these two repeats is low, but structure-based alignment of members of this family demonstrate the conservation of hydrophobic residues that stabilize the protein structure [7]. The two highly conserved amino acid sequences (RNP-1 and RNP-2) [23] are located in the central strands of a 4-stranded antiparallel β-sheet [17,24]. The β-sheet surface of RNP proteins is a generic RNA-binding platform to which specificity determinants are added in variable loops and in the terminal sequences of the domain [18]. Sequences related to RNP-1 and RNP-2 have been identified in prokaryotic RNA-binding proteins, including the bacteriophage T4 translational repressor RegA [25], initiation factor IF3 [26,27] and termination factor *Rho* [28,29].

RNP [17], KH [14] and dsRBD [10,11] are αβ proteins that present an antiparallel β-sheet on one face of the protein packed by an extensive hydrophobic core against an α–helical face (Fig. 1). The αβ structural theme is conserved in other RNA-binding proteins that do not share sequence homology with these three motifs. A split βαβ motif reminiscent of the RNP-fold is found in 6 of 9 ribosomal proteins of known structure (S6, L1, L6, L9, L12 and L30) [30], in translational elongation [30] and initiation [26,27] factors and in RegA [25]. In some cases (for example ribosomal proteins L12 and L30 and RNP proteins, or dsRBD and ribosomal protein S5) the similar secondary structure topology indicates a common evolutionary origin. However, αβ RNA-binding proteins have different topology of the secondary structure elements. The RNP domain has a repeated βαβ arrangement [17,24], dsRBDs have αβββα topology [10,11] and KH proteins have βααββα fold [14]. Topological differences and low sequence homology indicate that the different domains represent distinct, convergent evolutionary solutions towards a common structure for RNA recognition. This convergence could imply that the exposed β-sheet surface common to all such motives represent a convenient structural platform for RNA recognition. In support of this hypotheses, interactions between residues from the β-sheet surface and RNA bases are important determinants of specificity in RNP proteins. However, dsRBD proteins bind RNA through the face opposite the β-strand. Furthermore, there are now several examples of all helical RNA-binding proteins [31-33]. Thus, the common αβ structure may just reflect the high stability of this small protein folds and the ancient evolutionary origin of RNA-binding proteins. The preponderance of the αβ structural unit in RNA recognition, in contrast with the remarkable number of DNA-binding proteins with all α-helical structure, may nonetheless reveal a structural preference yet to be revealed.

3. New Methods of NMR Structural Investigation in Protein-RNA Recognition

Despite intensive efforts, relatively few structures of RNA-protein complexes have been determined either by NMR or X-ray crystallography. NMR studies of protein-RNA complexes have been severely limited by the relatively large molecular weight of such complexes, despite some notable success [34,35]. The traditional method to determine structure of intermolecular complexes by NMR has been based on recording filtered NOESY experiments on complexes containing isotopically labelled protein and unlabelled DNA or RNA [36,37]. The availability of isotopically labelled RNA provides the opportunity to do the reverse, i.e. mix labelled RNA with unlabelled protein, thereby facilitating extraction of distance constraints and the assignments of intermolecular NOE interactions to specific nucleic acid resonances [38]. NOE-based methods to determine structure of protein-nucleic acids complexes by NMR work well for systems characterized by tight binding ($K_d \approx 10^{-9}$ M), highly specific recognition and molecular weights of 40 kDa or less. However, these methods become increasingly less effective as the molecular weight increases and for systems (of great biochemical interest) characterized by weaker binding and poorer specificity. Several novel approaches have the potential to extend the range of RNA-protein structures that can be determined by NMR.

A first method uses paramagnetic spin labels to extract long-range intermolecular distance information in protein-RNA complexes [39]. This approach is based on electron-proton dipolar relaxation and is applicable also to DNA-protein complexes. We have inserted proxyl paramagnetic spin labels at specific sites on an RNA substrate. The presence of the unpaired electron on the nitroxide spin label increases the relaxation of NMR resonances in its vicinity. This effect can be quantitated by recording heteronuclear correlation spectra of a sample containing spin-labelled RNA and isotopically labelled protein, and is proportional to the inverse sixth power of the distance between the label and the reporter nucleus. This approach was applied to the complex between *Drosophila Staufen* protein and double stranded RNA. *Staufen* is an intensively studied protein containing multiple dsRBD domains, and its interaction with the *bicoid* mRNA 3'-untranslated region represents a paradigm in developmental biology [40]. Selected resonances in the protein are broadened by the spin label and map to specific regions of the protein. The sensitivity of the method is very high and this approach should remain effective at very high molecular weight, particularly using random fractional deuterated samples and/or line-narrowing techniques [41]. This technique can provide additional, longer-range information for systems that can be studied using NOE-based methods, but can also be applied to complex assemblies with molecular weight well in excess of the current limits for NMR structure determination.

A second approach we have employed is based on recording residual dipolar interactions. We have followed a newly introduced approach to induce partial orientation in biological samples in a liquid crystalline phase [42]. We have recorded coupled heteronuclear spectra of samples containing either labelled protein or labelled RNA, and measured residual dipolar splittings in dilute liquid crystalline solutions. In order to test whether the method would work with RNA-protein complexes, we have measured residual dipolar splitting for the 22 kDa complex of U1A protein, the structure of which had been previously determined [34,35,38]. The introduction of these additional constraints did not improve the structure of the RNA-bound protein significantly, but the structure was fully consistent with all recorded dipolar splittings. The greatest limitation of NMR-derived structures of RNA-protein complexes is the long-range order of the RNA in the structure. Locally, the RNA conformation is nearly as well defined as a protein structure, but the paucity of long-range NOE distance constraints in RNA (as compared to proteins) makes the overall structure much less precisely determined [43,44]. Residual dipolar splittings help reducing these limitations. Since there are relatively few NH resonances in an RNA molecule, we measured both NH and CH residual dipolar couplings for the RNA components of the complex. We were only able to record approximately 40 residual dipolar splittings due to the severe spectral overlap in the sugar region and to line broadening due to partial aggregation of the complex in the liquid crystalline solution. Refinement of the structure using these additional constraints leads to a significant improvement in the long-range definition of the RNA structure.

We have also used residual dipolar couplings to determine the relative orientation of protein and RNA components in the dsRBD-RNA complex described in a previous paragraph. Residual dipolar couplings are positive for NH residues located on β-sheet residues, and negative for residues from α-helical regions. NH and CH groups on the RNA bases are approximately perpendicular to the long-axis of the RNA double helix and the residual dipolar coupling constants are also positive. Therefore, the two α helices that determine the long axis of this elongated protein, must be roughly parallel to the long axis of the RNA double helix. In this way, we have been able to determine the relative orientation of protein and RNA in the complex.

4. Sequence-Independent Recognition of Double-Stranded RNA by Staufen Protein

Double stranded RNA-binding domain recognize RNA double helical regions of any sequence [9]. The structural analysis of the complex between the third double stranded RBD from Drosophila Staufen protein and double stranded RNA provides a rationale for the ability of dsRBD proteins to bind double stranded RNA. The protein is oriented with its long axis roughly parallel to the axis of the RNA double helix. There are no intimate contacts with the RNA bases, a distinctive difference with RNP proteins. On the contrary, the majority of contacts occur with the phosphodiester backbone and the RNA minor groove and several highly conserved basic amino acid side chains are in contact with the RNA phosphates or 2'-OH groups in the RNA minor groove. However, these interactions are poorly defined and probably very mobile. In addition, Staufen dsRBD3 interacts with the apical loop of the hairpin structure that was chosen as substrate. This was surprising, but raises the possibility that dsRBD proteins may interact *in vivo* with substrates other than purely double stranded RNAs. *In vitro*, Staufen dsRBD3 requires a contiguous stretch of at least 9-10 base pairs to bind double stranded RNA effectively, but there is no such site within its biological target, the 3'-untranslated region of *bicoid* mRNA [45]. Interactions with loop elements may overcome the requirements for very long stretches of purely double stranded RNA, that are extremely rare *in vivo*.

5. The RNP Domain Paradigm of RNA Recognition

Hundreds of proteins involved in RNA metabolism recognize substrates widely diverse in sequence and structure via RNP domains. Multiple RNP domains are required for tight binding to single-stranded RNA [46-48]. Single RNP domains can recognize 5-10 single-stranded RNA bases with high affinity and specificity only when the nucleotides are presented in a defined RNA structural context. For example, the human U1A protein recognizes 7 single stranded nucleotides in the context of a hairpin or internal loop with very high binding constant ($K_d \approx 10^{-11}$ M) [17,49]. Specificity is preserved when those 7 nucleotides are presented in the absence of RNA secondary structure [50], but the binding constant is reduced 100,000 fold, preventing effective discrimination [51]. The identity of RNA substrates of RNP proteins is often defined both by RNA structural features and by recognition of single-stranded nucleotides.

Crystallographic [52] and NMR [34] structures of the complexes of the human U1A protein with two distinct RNA substrates have revealed important aspects of the molecular basis of RNP-RNA recognition. The N-terminal RNP domain of U1A binds a hairpin loop during pre-mRNA splicing [53,54] and two related internal loops during autoregulation of polyadenylation [55-57]. Hairpin and internal loop substrates present similar single-stranded sequences in completely different structural contexts [51]. In both the crystal structure of the U1A-hairpin complex [52] and the solution structure of the internal loop complex [34], bases within the single-stranded loops are splayed out across the surface of the antiparallel β-sheet (Fig. 3). Nearly all hydrogen bond donors and acceptors from the single-stranded bases are recognized by an extensive hydrogen bonding network with protein residues from the β-sheet and from three loops connecting the secondary structural elements. An unusually large number of intermolecular contacts involve protein main chain hydrogen bond donors and acceptors, and all RNA bases are involved in intra- or intermolecular stacking interactions. Although there are few direct protein contacts to the phosphates, the RNA backbone follows a region of positive electrostatic potential in the protein to provide favorable electrostatic interactions [19,52].

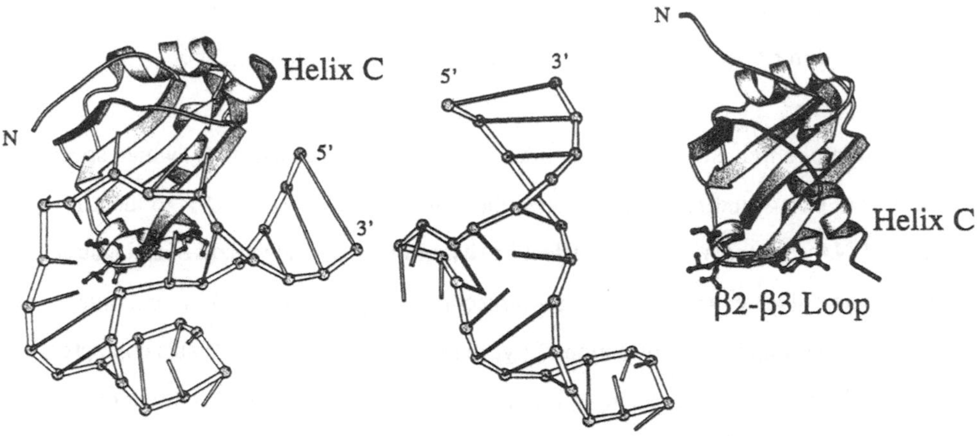

Figure 3. Structures of the free U1A protein (right), the free RNA internal loop (center) and the protein-RNA complex (left) [34]. Protein binding induces a dramatic change in the RNA conformation, and requires a sharp reorientation of helix C at the carboxy-end of U1A.

The comparison with the unbound RNA internal loop [44] and protein [16] structures (Fig. 3) reveals a complex recognition mechanism. A first set of intermolecular contacts involve rigid fit between the protein and the RNA double helical regions and three well ordered single-stranded nucleotides. The remaining five single stranded nucleotides are recognized instead by induced fit. The rigid interaction involves two loops within U1A (β1-helix A and β2-β3 loops) that have unique sequences in different RNP proteins, showing how the RNA structural context contributes to define substrate identity. Induced fit involve residues within RNP1 and RNP2 sequences which are important for sequence discrimination. RNA-binding reorients an α-helix within a region C-terminal to the domain that is essential for binding [51,53,58]. At the same time, protein binding orders five single-stranded nucleotides against the β-sheet surface through intermolecular stacking interactions with three very highly conserved aromatic amino acids. RNP1- and RNP2-like sequences with exposed aromatic residues are found in other RNA-binding [25-29] and single-stranded DNA-binding [59,60] proteins. Intermolecular stacking interactions with exposed aromatic side chains are common to $\alpha\beta$ proteins [34,52], tRNA synthetases [4,61] and viral coat proteins [62] and may be duplicated in ribosomal proteins [30] and translational regulators [25-29].

Recognition of single-stranded nucleotides is not the only mechanism by which specificity is achieved. RNA-binding proteins utilize the structural diversity of RNA to identify specific RNA structures to achieve effective discrimination. The variety of RNA structure defines many diverse, unique shapes for molecular recognition by proteins and other molecules that bind RNA. Formation of hydrogen bonds and van der Waals contacts with base functionality exposed on single stranded nucleotides allow additional fine tuning in the identification of the RNA target.

6. Conformational Flexibility in RNA-Protein Recognition

Structures of RNA-protein complexes have provided important insight into the mechanism of intermolecular recognition, but have also raised intriguing questions concerning the molecular origin of binding energy and sequence discrimination. Surprisingly, binding energy is not related to interface area, a measure of the van der Waals interaction energy. The very tight binding of the U1A protein ($K_d \approx 10^{-11}$ M) is remarkable, since binding requires an entropically costly disorder-order transition in the single-stranded RNA loops [34]. Changes in the distribution of molecular vibrations when exposed amino acid side chains become ordered at intermolecular interfaces contribute 15-25 kcal/mol to increase the free energy of protein-protein and protein-DNA interactions [63]. However, the NMR relaxation properties of arginine side chains in DNA-protein complexes indicate that motion is less restricted at the intermolecular interface than in the highly ordered hydrophobic core [64], suggesting that protein-nucleic acids interfaces may not be rigidly ordered. A recent study of the flexibility of protein side chains in an SH2 protein - phosphopeptide complex also showed that only part of the protein binding surface became ordered upon peptide binding [65]. The portion of the interface interacting with the peptide sequence C-terminal to a phosphorylated tyrosine was as mobile in the complex as in the free protein, rationalizing the relaxed specificity for this region of the phosphorylated peptide.

Several observations suggest that similar considerations apply to the U1A complexes. Firstly, the Arg 52 side chain in U1A is involved in 5 hydrogen bonding interactions in the crystal structure [52], but can be mutated to Lys without significant increase in the free energy of binding (< 0.5 kcal/Mol) [17,58]. The NMR data suggest a less ordered conformation, where each hydrogen bond is only present part of the time [34]. Secondly, a C->G mutation is an important determinant of the specificity of the U1A-related U2B" protein for its cognate RNA hairpin substrate [54]. In both NMR and X-ray structures [34,52], that cytosine is deeply buried at the intermolecular interface preventing fitting of a larger guanosine, and every cytosine functionality is recognized by the protein. Surprisingly, the C->G substitution only leads to a 10-fold decrease in binding constant [49]. An explanation consistent with several NMR observations is that the interface may be flexible enough to accommodate either base through energetically inexpensive local conformational adjustments. In the SH2-phosphopeptide complex [65], a number of residues with large amplitude of motion were deeply buried at the protein-peptide interface and involved in extensive inter- and intramolecular interactions. Studies of protein dynamics at intermolecular interfaces suggest that a fine balance between rigidity and flexibility may provide a compromise between complete specificity (at large entropic cost) and complete lack of selectivity.

In contrast to the situation observed with induced fit, it is more difficult to reorganize the intermolecular interface when the interaction involves highly ordered regions of the protein and RNA. Thus, it is costly to mutate a tightly packed Leu 49 in U1A at the junction between RNA helices and loops [34,66], while the highly conserved Arg 52 nearby can only be mutated to Lys, which has similar size and positive charge [17,58]. Measurement of side chain flexibility using deuterium NMR relaxation methods have shown that methyl-carrying side-chains from this region of the complex (Leu 49 and Met 51) become very rigid upon formation of the complex, but are relatively free to move in the free protein (T. Mittermaier, L. Kay, LV and GV, in preparation). Conformational flexibility in the free protein could be important for an exhaustive conformational search and the optimization of surface complementarity to provide large amounts of binding free energy though van der Waals interactions. Disruption of the RNA secondary structure reduces binding 100,000 fold [51,67]. The pre-formed RNA secondary structure provide a structural counterpart to the β-sheet of the protein in reducing the entropic costs of RNA folding [19] and provides large .mounts of binding free energy through electrostatic interactions.

7. RNA-Dependent Protein-Protein Interaction in the Regulation of Macromolecular Assemblies during Eukaryotic Gene Expression

A further level of biological selectivity, beyond RNA-binding, is provided by RNA-dependent protein-protein interactions. U1A autoregulates its own production by forming an RNA-dependent interaction with Poly(A) polymerase (PAP), the enzyme responsible for formation of the mature 3'-end of almost all eukaryotic mRNAs (Fig. 4) [55-57,68].

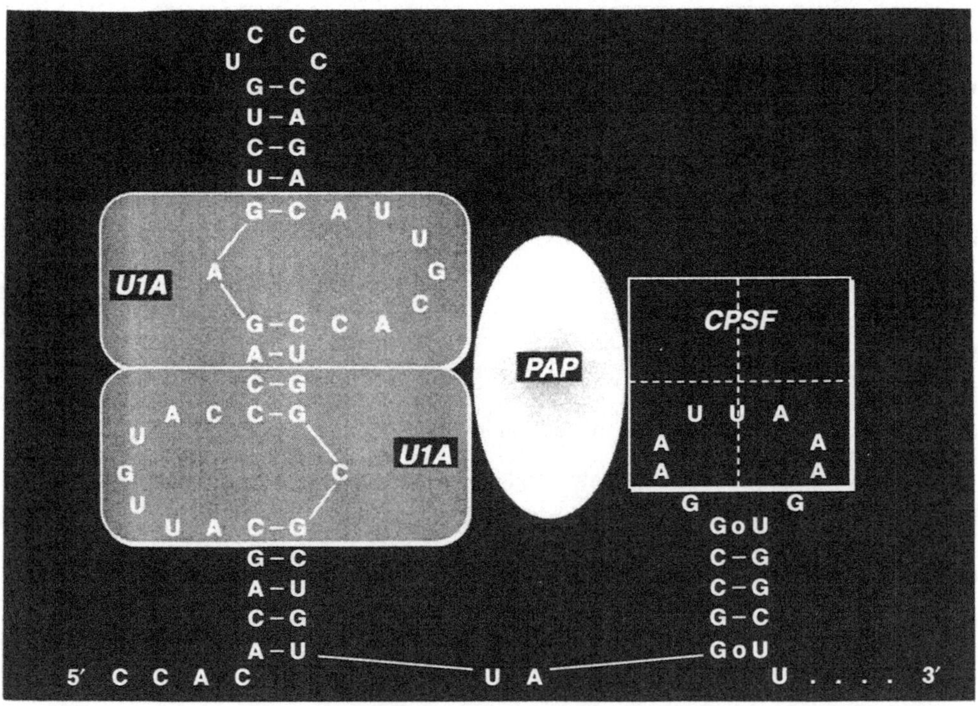

Figure 4. Regulation of polyadenylation of U1A pre-mRNA by human U1A protein. The polyadenylation inhibition element (PIE) RNA is recognized cooperatively by two U1A monomers [55-57,68]. This ternary complex then interact with poly(A) polymerase down regulating the activity of this enzyme.

PAP is an essential enzyme and its activity must not be indiscriminately affected by U1A. In the 40 kDa ternary complex between two U1A proteins and the U1A polyadenylation regulatory element RNA, interaction with RNA induces a conformational rearrangement in U1A (Fig. 3). The conformational shift allows protein-protein interactions to occur between U1A monomers to allow cooperative binding (Fig. 5), and productive interactions with the Poly(A)-polymerase enzyme. Protein and RNA conformational rearrangements are very common during molecular recognition and central to the formation of specific intermolecular complexes. The conformational changes occurring upon binding can activate RNA-binding proteins to regulate the formation of productive interactions with other proteins or RNAs.

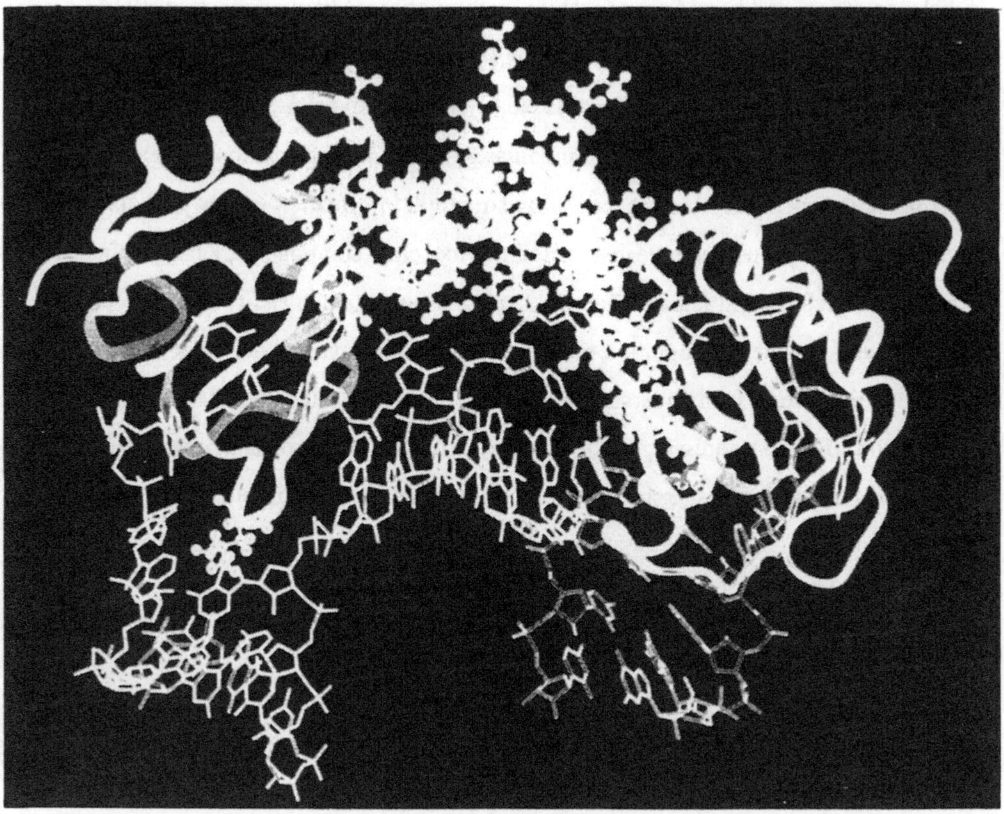

Figure 5. Structure of the ternary complex of two U1A proteins and the complete polyadenylation inhibition element RNA. Protein-protein interactions that mediate cooperativity in binding map to the C-terminal region of the protein domain, a region that immediately precedes the sequence responsible for interaction with Poly(A) polymerase.

8. Conclusions

Recent structures of RNA-protein complexes have revealed principles of specific intermolecular recognition by RNA-binding proteins. These structures demonstrate different ways in which protein β-sheets provide large surfaces for extensive interactions with RNA bases exposed in single-stranded regions. Exposed β-sheet surfaces are so common in RNA recognition to suggest a role as dominant as that of α-helices in DNA recognition. This dominance could originate from the stereochemical complementarity with RNA structure due to the natural right-handedness and concavity of antiparallel β-sheets [30,58,69].

A balance of induced fit and shape selectivity through rigid fit is common to tRNA-synthetase complexes and to the recognition of highly structured RNAs by RNP proteins. Formation of RNA-protein complexes is clearly a highly dynamic process: RNA structure directs protein binding, which in turn modulates the RNA conformation to create a unique intermolecular interface. Atomic resolution structural information is only available for a handful of RNA-protein complexes, excluding ribosomal components, KH domains and the arginine-rich protein family. The structural and functional diversity of RNA define a large variety of protein recognition mechanisms, and the next several years will continue to see exciting progress in understanding the thermodynamic and structural basis of recognition. Besides addressing fundamental questions in intermolecular recognition, these studies will underpin the search for compounds that interfere with critical RNA-protein recognition events to down-regulate gene expression and prevent replication of pathogenic viruses and bacteria [70].

9. References

(1) Steitz, T. A. (1990) Structural Studies of Protein-Nucleic Acid Interactions: the Sources of Sequence-Specific Binding, *Q. Rev. Biophys.* **23**, 205-280.

(2) Seeman, N. C.,Rosenberg, J. M., Rich, A. (1976) Sequence-Specific Recognition of Double-Helical Nucleic Acids by Proteins, *Proc. Natl. Acad. Sci. USA* **73**, 804-808.

(3) Rould, M. A.,Perona, J. J.,Söll, D., Steitz, T. A. (1989) Structure of E. coli Glutaminyl t-RNA Synthetase Complexes with $tRNA^{Gln}$ and ATP at 2.8 Å Resolution, *Science* **246**, 1135-1142.

(4) Rould, M. A.,Perona, J. J., Steitz, T. A. (1991) Structural Basis of Anticodon Loop Recognition by Glutaminyl-tRNA Synthetase, *Nature* **352**, 213-218.

(5) Varani, G., Nagai, K. (1998) RNA Recognition by RNP Proteins during RNA Processing and Maturation, *Ann. Rev. Biophys. Biomol. Struct.* **27**, 407-445.

(6) Biamonti, G., Riva, S. (1994) New Insights into the Auxiliary Domains of Eukaryotic RNA Binding Proteins, *FEBS Letters* **340**, 1-8.

(7) Birney, E.,Kumar, S., Krainer, A. R. (1993) Analysis of the RNA-Recognition Motif and RS and RGG Domains: Conservation in Metazoan pre-mRNA Splicing Factors, *Nucleic Acids Res.* **21**, 5803-5816.

(8) Nagai, K. (1996) RNA-Protein Complexes, *Curr. Op. Struct. Biol.* **6**, 53-61.

(9) St Johnston, D.,Brown, N. H.,Gall, J. G., Jantsch, M. (1992) A Conserved Double-Stranded RNA Binding Domain, *Proc. Natl. Acad. Sci. USA* **89**, 10979-10983.

(10) Bycroft, M.,Grünert, S.,Murzin, A. G.,Proctor, M., St Johnston, D. (1995) NMR Solution Structure of a dsRNA Binding Domain from *Drosophila* Staufen Protein Reveals Homology to the N-terminal Domain of Ribosomal Protein S5, *EMBO J.* **14**, 3563-3571.

(11) Kharrat, A.,Macias, M. J.,Gibson, T. J.,Nilges, M., Pastore, A. (1995) Structure of the dsRNA Binding Domain of *E.coli* RNAse III, *EMBO J.* **14**, 3572-3584.

(12) Clarke, P. A., Mathews, M. B. (1995) Interactions between the Double-Stranded RNA Binding Motif and RNA: Definition of the Binding Site for the Interferon-Induced Protein Kinase DAI (PKR) on Adenovirus VA RNA, *RNA* **1**, 7-20.

(13) Ferrandon, D.,Elphick, L.,Nüsslein-Volhard, C., St Johnston, D. (1994) Staufen Protein Associates with the 3'UTR of *bicoid* mRNA to form Particles that Move in a Microtubule-Dependent Manner, *Cell* **79**, 1221-1232.

(14) Musco, G.,Stier, G.,Joseph, C.,Castiglione Morelli, M. A.,Nilges, M.,Gibson, T. J., Pastore, A. (1996) Three-Dimensional Structure and Stability of the KH Domain: Molecular Insight into the Fragile X Syndrome, *Cell* **85**, 237-245.

(15) Siomi, H.,Siomi, M. C.,Nussbaum, R. L., Dreyfuss, G. (1993) The Protein Product of the Fragile X Gene, FMR1, Has Characteristic of an RNA-Binding Protein, *Cell* **74**, 291-298.

(16) Avis, J.,Allain, F. H.-T.,Howe, P. W. A.,Varani, G.,Neuhaus, D., Nagai, K. (1996) Solution Structure of the N-terminal RNP Domain of U1A Protein: The Role of C-terminal Residues in Structure Stability and RNA Binding, *J. Mol. Biol.* **257**, 398-411.

(17) Nagai, K.,Oubridge, C.,Jessen, T. H.,Li, J., Evans, P. R. (1990) Structure of the RNA-Binding Domain of the U1 Small Nuclear Ribonucleoprotein A, *Nature* **348**, 515-520.

(18) Burd, C. G., Dreyfuss, G. (1994) Conserved Structures and Diversity of Functions of RNA-Binding Proteins, *Science* **265**, 615-621.

(19) Nagai, K.,Oubridge, C.,Ito, N.,Avis, J., Evans, P. (1995) The RNP Domain: a Sequence-Specific RNA-Binding Domain Involved in Processing and Transport of RNA, *TIBS* **20**, 235-240.

(20) Mattaj, I. W. (1993) RNA Recognition: A Family Matter?, *Cell* **73**, 837-840.

(21) Adam, S. A.,Nakagawa, T.,Swanson, M. S.,Woodruff, T. K., Dreyfuss, G. (1986) mRNA Polyadenylate-Binding Protein: Gene Isolation and Sequencing and Identification of a Ribonucleoprotein Consensus Sequence, *Mol. Cell. Biol.* **6**, 2932-2943.

(22) Sachs, A. B.,Bond, M. W., Kornberg, R. D. (1986) A Single Gene from Yeast for both Nuclear and Cytoplasmic Polyadenylate Binding Proteins: Domain Structure and Expression, *Cell* **45**, 827-835.

(23) Query, C. C.,Bentley, R. C., Keene, J. D. (1989) A Common RNA Recognition Motif Identified within a Defined U1 RNA Binding Domain of the 70K U1 snRNP Protein, *Cell* **57**, 89-101.

(24) Hoffman, D. W.,Query, C. C.,Golden, B. L.,White, S. W., Keene, J. D. (1991) RNA-Binding Domain of the A Protein Component of the U1 Small Nuclear Ribonucleoprotein Analyzed by NMR Spectroscopy is Structurally Similar to Ribosomal Proteins, *Proc. Natl. Acad. Sci. USA* **83**, 2495-2499.

(25) Kang, C.-H.,Chan, R.,Berger, I.,Lockshin, C.,Green, L.,Gold, L., Rich, A. (1995) Crystal Structure of the T4 regA Translational Regulator Protein at 1.9 Å Resolution, *Science* **268**, 1170-1173.

(26) Garcia, C.,Fortier, P.-L.,Blanquet, S.,Lallemand, J.-Y., Dardel, F. (1995) Solution Structure of the Ribosome-Binding Domain of *E.coli* Translation Initiation Factor IF3. Homology with the U1A Protein of the Eukaryotic Spliceosome, *J. Mol. Biol.* **254**, 247-259.

(27) Biou, V.,Shu, F., Ramakrishnan, V. (1995) X-Ray Crystallography Shows that Translational Initiation Factor IF3 Consists of Two Compact α/β Domains Linked by an α-Helix, *EMBO J.* **14**, 4056-4064.

(28) Martinez, A.,Opperman, T., Richardson, J. P. (1996) Mutational Analysis and Secondary Structure Model of the RNP1-Like Sequence Motif of Transcription Termination Factor Rho, *J. Mol. Biol.* **257**, 895-908.

(29) Martinez, A.,Burns, C. M., Richardson, J. P. (1996) Residues in the RNP1-Like Sequence Motif of Rho Protein are Involved in RNA-Binding Affinity and Discrimination, *J. Mol. Biol.* **257**, 909-918.

(30) Liljas, A., Garber, M. (1995) Ribosomal Proteins and Elongation Factors, *Curr. Op. Struct. Biol.* **5**, 721-727.

(31) Huenges, M.,Rölz, C.,Gschwind, R.,Peteranderl, R.,Berglechner, F.,Richter, G.,Cacher, A.,Kessler, H., Gemmecker, G. (1998) Solution Structure of the Antitermination Protein NusB of *Escherichia coli*: a Novel All-Helical Fold for an RNA-Binding Protein, *EMBO J.* **17**, 4092-4100.

(32) Predki, P. F.,Nayak, L. M.,Gottlieb, M. B. C., Regan, L. (1995) Dissecting RNA-Protein Interactions: RNA-RNA Recognition by Rop, *Cell* **80**, 41-50.

(33) Hinck, A. P.,Markus, M. A.,Huang, S.,Grzesiek, S.,Kustonovich, I.,Draper, D. E., Torchia, D. A. (1997) The RNA-Binding Domain of Ribosomal Protein L11: Three-Dimensional Structure of the RNA-Bound Form of the Protein and its Interaction with 23S rRNA, *J. Mol. Biol.* **274**, 101-113.

(34) Allain, F.-H. T.,Gubser, C. C.,Howe, P. W. A.,Nagai, K.,Neuhaus, D., Varani, G. (1996) Specificity of Ribonucleoprotein Interaction Determined by RNA Folding during Complex Formation, *Nature* **380**, 646-650.

(35) Allain, F. H.-T.,Howe, P. W. A.,Neuhaus, D., Varani, G. (1997) Structural Basis of the RNA Binding Specificity of Human U1A Protein, *EMBO J.* **16**, 5764-5774.

(36) Otting, G., Wüthrich, K. (1990) Heteronuclear Filters in Two-Dimensional [1H-1H]-NMR Spectroscopy: Combined Use with Isotope Labeling for Studies of Macromolecular Conformation and Intermolecular Interactions, *Q. Rev. Biophys.* **23**, 39-96.

(37) Qian, Y. Q.,Otting, G.,Billeter, M.,Müller, M.,Gehring, W., Wüthrich, K. (1993) Nuclear Magnetic Resonance Spectroscopy of a DNA Complex with the Uniformly ^{13}C-Labeled

Antennapedia *Homeodomain* and Structure Determination of the DNA-Bound Homeodomain, *J. Mol. Biol.* **234**, 1070-1083.

(38) Howe, P. W. A.,Allain, F. H.-T.,Varani, G., Neuhaus, D. (1998) Determination of the NMR Structure of the Complex between U1A Protein and its RNA Polyadenylation Inhibition Element, *J. Biomol. NMR* **11**, 59-84.

(39) Ramos, A., Varani, G. (1998) A New Method to Detect Long-Range Protein-RNA Contacts: NMR Detection of Electron-Proton Relaxation Induced by Nitroxide Spin-Labeled RNA, *J. Am. Chem. Soc.*

(40) St Johnston, D. (1995) The Intracellular Localization of Messenger RNAs, *Cell* **81**, 161-170.

(41) Pervushin, K.,Riek, R.,Wider, G., Wüthrich, K. (1997) Attenuation T2 Relaxation by Mutual Cancellation by Dipole-Dipole Coupling and Chemical Shift Anisotropy Indicates an Avenue to NMR Structures of Very Large Biological Macromolecules in Solution, *Proc. Natl. Acad. Sci. USA* **94**, 12366-12371.

(42) Tjandra, N., Bax, A. (1997) Direct measurement of Distances and Angles in Biomolecules by NMR in a Dilute Liquid Crystalline Medium, *Science* **278**, 1111-1114.

(43) Varani, G.,Aboul-ela, F., Allain, F. H.-T. (1996) NMR Investigations of RNA Structure, *Prog. NMR Spectr.* **29**, 51-127.

(44) Gubser, C. C., Varani, G. (1996) Structure of the Polyadenylation Regulatory Element of the Human U1A pre-mRNA 3'-Untranslated Region and Interaction with the U1A Protein, *Biochemistry* **35**, 2253-2267.

(45) Ferrandon, D.,Koch, I.,Westhof, E., Nüsslein-Volhard, C. (1997) RNA-RNA Interaction is Required for the Formation of Specific *Bicoid* mRNA 3' UTR-STAUFEN Ribonucleoprotein Particles, *EMBO J.* **16**, 1751-1758.

(46) Tacke, R., Manley, J. L. (1995) The Human Splicing Factors ASF/SF2 and Sc35 Possess Distinct, Functionally Significant RNA Binding Specificities, *EMBO J.* **14**, 3540-3551.

(47) Kanaar, R.,Lee, A. L.,Rudner, D. Z.,Wemmer, D. E., Rio, D. C. (1995) Interaction of the Sex-Lethal RNA Binding Domains with RNA, *EMBO J.* **14**, 4530-4539.

(48) Shamoo, Y.,Abdul-Manam, N.,Patten, A. M.,Crawford, J. K.,Pellegrini, M. C., Williams, K. R. (1994) Both RNA-Binding Domains in Heterogeneous Nuclear Ribonucleoprotein A1 Contribute toward Single-Stranded-RNA Binding, *Biochemistry* **33**, 8272-8281.

(49) Hall, K. B., Stump, W. T. (1992) Interaction of N-Terminal Domain of U1A Protein with an RNA Stem-Loop, *Nucleic Acids Res.* **20**, 4283-4290.

(50) Harper, D. S.,Fresco, L. D., Keene, J. D. (1992) RNA Binding Specificity of a Drosophyla snRNP Protein that Shares Sequence Homology with Mammalian U1A and U2B" Proteins, *Nucleic Acids Res.* **20**, 3645-3650.

(51) Hall, K. B. (1994) Interaction of RNA Hairpins with the Human U1A N-Terminal RNA Binding Domain, *Biochemistry* **33**, 10076-10088.

(52) Oubridge, C.,Ito, N.,Evans, P. R.,Teo, C.-H., Nagai, K. (1994) Crystal Structure at 1.92 Å Resolution of the RNA-Binding Domain of the U1A Spliceosomal Protein Complexed with an RNA Hairpin, *Nature* **372**, 432-438.

(53) Scherly, D.,Kambach, C.,Boelens, W.,van Venrooij, W. J., Mattaj, I. W. (1991) Conserved Amino Acid Residues within and Outside of the N-Terminal Ribonucleoprotein Involved in U1 RNA Binding, *J. Mol. Biol.* **219**, 577-584.

(54) Scherly, D.,Boelens, W.,Dathan, N. A.,van Venrooij, W. J., Mattaj, I. (1990) Major Determinants of the Specificity of Interaction between Small Nuclear Ribonucleoproteins U1A and U2B" and their Cognate RNAs, *Nature* **345**, 502-506.

(55) van Gelder, C. W. G.,Gunderson, S. I.,Jansen, E. J. R.,Boelens, W. C.,Polycarpou-Schwartz, M.,Mattaj, I. W., van Venrooij, W. J. (1993) A Complex Secondary Structure in U1A pre-mRNA that Binds Two Molecules of U1A Protein is Required for Regulation of Polyadenylation, *EMBO J.* **12**, 5191-5200.

(56) Gunderson, S. I.,Beyer, K.,Martin, G.,Keller, W.,Boelens, W. C., Mattaj, I. W. (1994) The Human U1A snRNP Protein Regulates Polyadenylation via a Direct Interaction with Poly(A) Polymerase, *Cell* **76**, 531-541.

(57) Boelens, W. C.,Jansen, E. J. R.,van Venrooij, W. J.,Stripecke, R.,Mattaj, I. W., Gunderson, S. I. (1993) The Human U1 snRNP-Specific U1A Protein Inhibits Polyadenylation of its Own Pre-mRNA, *Cell* **72**, 881-892.

(58) Jessen, T. H.,Oubridge, C.,Teo, C. H.,Pritchard, C., Nagai, K. (1991) Identification of Molecular Contacts between the U1A Small Nuclear Ribonucleoprotein and U1 RNA, *EMBO J.* **10**, 3447-3456.

(59) Schindelin, H.,Marahiel, M. A., Heinemann, U. (1993) Universal Nucleic Acid Binding Domain Revealed by Crystal Structure of the *B.subtilis* Major Cold-Shock Protein, *Nature* **364**, 164-168.

(60) Schnuchel, A.,Wiltscheck, R.,Czisch, M.,Herrier, M.,Willimsky, G.,Graumann, P.,Marahiel, M. A., Holak, T. A. (1993) Structure in Solution of the Major Cold-Schock Protein from *Bacillus subtilis*, *Nature* **364**, 169-171.

(61) Cavarelli, J.,Rees, B.,Ruff, M.,Thierry, J.-C., Moras, D. (1993) Yeast $tRNA^{Asp}$ Recognition by Its Cognate Class II Aminoacyl-tRNA Synthetase, *Nature* **362**, 181-184.

(62) Valegård, K.,Murray, J. B.,Stockley, P. G.,Stonehouse, N. J., Liljas, L. (1994) Crystal Structure of an RNA Bacteriophage Coat Protein-Operator Complex, *Nature* **371**, 623-626.

(63) Spolar, R. S., Record, M. T. J. (1994) Coupling of Local Folding to Site-Specific Binding of Proteins to DNA, *Science* **263**, 777-784.

(64) Berglund, H.,Baumann, H.,Knapp, S.,Ladenstein, R., Härd, T. (1995) Flexibility of an Arginine Side Chain at a DNA-Protein Interface, *J. Am. Chem. Soc.* **117**, 12883-12884.

(65) Kay, L. E.,Muhandiram, D. R.,Farrow, N. A.,Aubin, Y., Forman-Kay, J. D. (1996) Correlation between Dynamics and High Affinity Binding in an Sh2 Domain Interaction, *Biochemistry* **35**, 361-368.

(66) Laird-Offinga, I. A., Belasco, J. G. (1995) Analysis of RNA-Binding Proteins by *in vitro* Genetic Selection: Identification of an Amino Acid Residue Important for Locking U1A Onto its RNA Target, *Proc. Natl. Acad. Sci. USA* **92**, 11859-11863.

(67) Tsai, D. E.,Harper, D. S., Keene, J. D. (1991) U1-snRNP-A Protein Selects a Ten Nucleotide Consensus Sequence from a Degenerate RNA Pool Presented in Various Structural Contexts, *Nucleic Acids Res.* **19**, 4931-4936.

(68) Gunderson, S. I.,Vagner, S.,Polycarpou-Schwarz, M., Mattaj, I. W. (1997) Involvement of the Carboxy Terminus of Vertebrate poly A Polymerase in U1A Autoregulation and in the Coupling of Splicing and Polyadenylation, *Genes Dev.* **11**, 761-773.

(69) Howe, P. W. A.,Nagai, K.,Neuhaus, D., Varani, G. (1994) NMR Studies of U1 snRNA Recognition by the N-Terminal RNP Domain of the Human U1A Protein, *EMBO J.* **13**, 3873-3881.

(70) Gait, M. J., Karn, J. (1995) Progress in Anti-HIV Structure-Based Drug Design, *TIBS* **13**, 430-438.

RNA-APTAMERS FOR STUDYING RNA PROTEIN INTERACTIONS

M. SPRINZL, H.-P. HOFFMANN, S. BROCK, M. NANNINGA AND V. HORNUNG
Laboratorium für Biochemie, Universität Bayreuth, Postfach 95440 Bayreuth, Germany

ABSTRACT

In vitro selection was used to isolate RNA molecules which specifically recognize translation factors of bacterial protein biosynthesis. Elongation factor Tu, elongation factor G and release factor 1 from *Thermus thermophilus* which carry a terminal histidine-tag were immobilized on Ni^{2+}-agarose and used for selection of RNA-aptamers from a library of random RNA sequences. RNA molecules which bind to particular translation factors were characterized by UV-melting, chemical modification and boundary analysis. These RNA-aptamers are expected to mimic the structure of domains of ribosomal RNA which interact with elongation factor Tu, elongation factor G and release factor 1 *in vivo*. Therefore, the *in vitro* selection of RNAs can be used as a tool to study protein RNA complexes which, due to their size and complexity, cannot be easily isolated and investigated by available chemical, physical or biochemical methods. Moreover, RNA-aptamers might be useful inhibitors of reactions which depend on RNA protein interactions. The high specificity and high affinity of RNA-aptamers for proteins may also permit their use as reporter molecules. Fluorescence labeling of RNA-aptamers is required for such applications.

1. Introduction

Three functionally distinct sites were identified on ribosomes [1]. These are the A-site, P-site and E-site where aminoacyl-tRNA, peptidyl-tRNA (in its function as donor of peptidyl residue) and tRNA (before it is released from the ribosome) are located, respectively. These sites are not invoked as rigid binding scaffolds but represent dynamic structures which specifically recognize translation factors, tRNAs or mRNA. Because of the dynamics of ribosomal sites it is often difficult to study ribosomal complexes leading to apparently conflicting results. In order to understand biochemical functions of ribosomal sites it is not sufficient to know the structure of ribosomes in just one functional state, since the alternative structures of ribosomes and especially of ribosomal RNA is one of the most important features of the protein synthesizing machinery.

The elongation cycle of protein biosynthesis starts with the binding of the aminoacyl-tRNA·EF-Tu·GTP ternary complex to the ribosomal A-site which results in codon-anticodon interaction between tRNA and mRNA. After GTP is cleaved, EF-Tu·GDP

J. Barciszewski and B.F.C. Clark (eds.), RNA Biochemistry and Biotechnology, 217–228.

leaves the ribosome. Thus it is possible for elongation factor G to interact with the A-site and translocate the newly formed peptidyl-tRNA to the P-site [2, 3].
Based on the X-ray structure analysis of elongation factor G [4, 5] and the Phe-$tRNA^{Phe}$·EF-Tu·GppNHp ternary complex [6] a molecular mimicry hypothesis was postulated which suggests that these translation factors and their complexes bound to ribosomal A-site have similar three-dimensional structures. The hypothesis was extended to complexes between release factor 1 (or release factor 2) and release factor 3 [7]. The complex between initiation factor 1, initiation factor 2 and fmet-$tRNA^{fmet}$ might also correspond to a structure which fills the ribosomal A- and P-sites [8] (Fig.1).

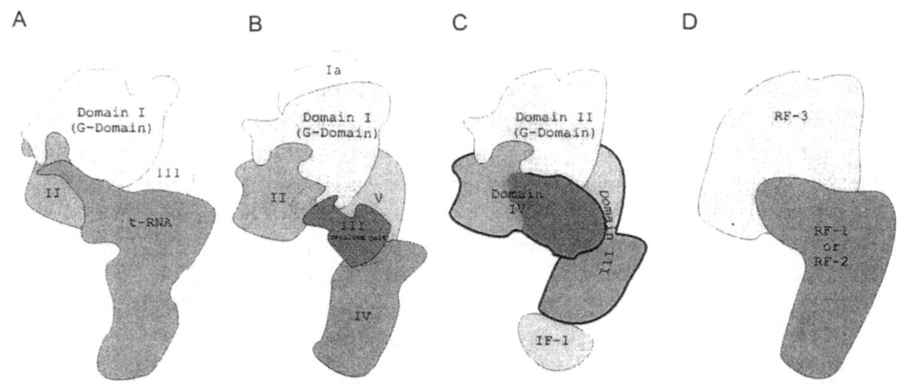

Fig. 1. Domain organization of (A) aminoacyl-tRNA•EF-Tu•GTP ternary complex, (B) EF-G and (C) a model of a IF-1 · IF2 complex. (D) represents the RF1/2•RF-3. (A) and (B) are derived from the three dimensional structures of aminoacyl-tRNA•EF-Tu•GTP ternary complex and EF-G•GDP, respectively. Only a part of domain III of EF-G is resolved in the X-ray structure. This part of domain III is indicated by dark color. Domain III may require a larger space, arbitrarily indicated by the broken line.

Relatively few details are known about the complex between EF-Tu, EF-G, IF-1, IF-2 and RF-1 with the ribosome. It was shown that the α-sarcin domain, a structural element around position 2670 of *T. thermophilus* 23S rRNA which consists of an RNA stem and a hairpin loop, is important for factor-related functions [9, 10, 11]. It has been suggested that the conformation of the α-sarcin loop changes during the elongation cycle. In anology to EF-Tu it is hardly possible to get hold of a complex between other translation factors and ribosomal components. Chemical modification of ribosomal RNA in the presence and absence of translation factors and tRNA suggested an overlap or at least topologically close binding sites for different A-site related components. More recently remarkable progress was achieved by electron microscopy of frozen ribosomes, which supports the results obtained by chemical mapping of ribosomal binding sites [12, 13].
The development of *in vitro* selection methods [14] offers a possibility to isolate RNA molecules reflecting sequence and structure of ribosomal RNA which contacts translation factors *in vivo.* Biochemical and structural characterization of such complexes may provide valuable information about the corresponding ribosomal complexes. It was especially tempting to test the mimicry hypothesis and ask whether

the translation factors which fit to the same site during different steps of translation interact with ribosomal RNA in the same way.
Conjugation of several histidines with a polypedtide chain is widely used for rapid isolation of overproduced proteins. We used the immobilization of His-tagged proteins on Ni^{2+}-agarose to develop a reproducible and highly adaptable procedure for *in vitro* selection of RNA (RNA-aptamers) which binds to translation factors. Several features of RNA-aptamers are biochemically important: i) the comparison of the structures of RNA-aptamers with know features of the corresponding ribosomal RNA may help to dissect the mechanism and regulation of the particular complexes ii) the usually high affinity of RNA-aptamers for their protein ligands may lead to specific inhibitors iii) aptamers which carry spectroscopic labels can serve as analytical tools to identify and characterize RNA binding proteins.

2. Results

2.1. *IN VITRO* SELECTION OF RNA-APTAMERS DIRECTED AGAINST IMMOBILIZED HISTIDINE-TAGGED PROTEINS

In vitro selection was used to identify RNA capable of binding to elongation factor Tu from *T. thermophilus*. The RNAs were selected from a pool of approximately 10^{13} RNA molecules consisting of a 50 nucleotide random region and two constant flanking regions. The RNA was prepared by transcription with T7 RNA polymerase from a DNA library which was created by synthesizing degenerate deoxyribooligonucleotides (Fig. 2). After each selection step the RNA bound to immobilized protein was isolated and reverse transcribed into cDNA.

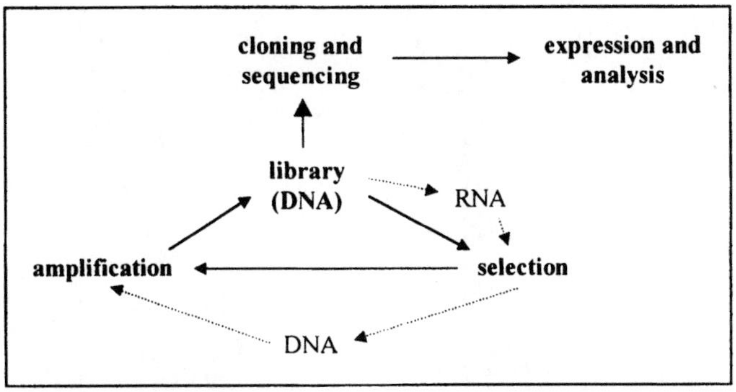

Fig. 2. Scheme for *in vitro* selection and amplification of RNA-aptamers. Starting from a DNA library which is transcribed into RNA, the RNA is selected for binding. After reverse transcription into DNA, the DNA is amplified and forms a new, selected library for a new round of selection. After enrichment of binding RNAs, the corresponding DNA is cloned and sequenced. RNA can then be transcribed from isolated clones and used for further analysis

The cDNA was amplified and the RNA obtained by transcription was submitted to the next selection cycle. After several rounds of selection (typically 5-8) the amount of RNA molecules that bound to the protein was increased up to 20% of total RNA. The

corresponding DNA was cloned and sequenced. RNA prepared by transcription of the respective clones was then used for further functional and structural analysis.
Proteins with terminal histidine residues interact with Ni^{2+} and can be immobilized on Ni^{2+}-NTA-agarose. The immobilized proteins can also be used for the isolation of functional RNAs in *in vitro* selection experiments. In order to prevent the selection of Ni^{2+}-binding RNAs [15], it is essential to perform a preselection of RNA on free Ni^{2+}-NTA-agarose prior to each selection cycle to remove the Ni^{2+}-binding RNA. Moreover, for the selection of protein binding RNAs the Ni^{2+}-NTA-agarose has to be fully saturated with protein. To select RNA-aptamers against His-tagged proteins it proved advantageous to use a batch procedure instead of affinity chromatography. It has to be kept in mind that the slow dissociation of the histidine-tagged protein from the Ni^{2+}-NTA-agarose during chromatography provides free Ni^{2+}-binding sites and increases the probability to isolate Ni^{2+}-binding RNA [15].

2.2. SEQUENCE AND STRUCTURE OF EF-Tu BINDING RNA-APTAMERS

The RNA-aptamers obtained by *in vitro* selection using His-tagged EF-Tu·GTP share a short consensus sequence, 5'-ACCG-3', which is extended to 5'-ACCGAAG-3' in some of the RNA-aptamers [16]. Figure 3 compares the consensus sequence of the RNA-aptamers with the corresponding sequence of the α-sarcin domain of 23S rRNA of the large ribosomal submit which is believed to be part of the ribosomal binding site for both EF-Tu and EF-G [10].

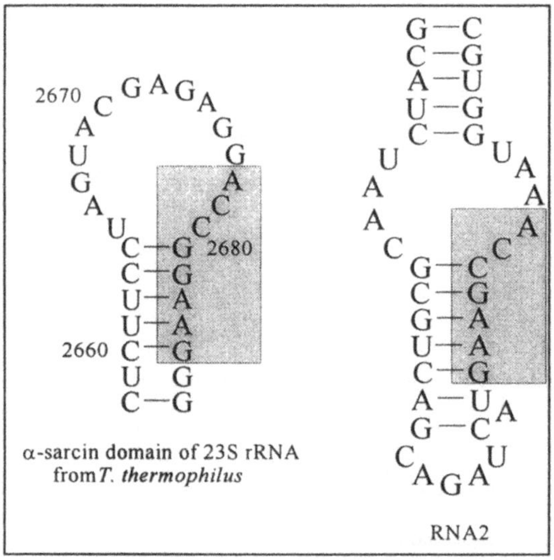

Fig. 3: Comparison of sequence and secondary structure of the selected RNA2 and the α-sarcin domain of 23S rRNA. The consensus sequences of the RNAs are shaded.

So far, the main interest focused on the highly conserved loop region of the α-sarcin domain [10, 17], especially on the GAGA sequence element which forms a tetraloop in the isolated α-sarcin-RNA [18, 19]. Recent evidence suggests that not only the loop region is important for elongation but that the stem region of this domain might play an equally crucial role in the recognition of EF-Tu [20]. The results of the selection sustain that the short 5′-ACCG-3′ sequence at the junction of loop and stem of the open structure of the α-sarcin domain (Fig. 4) contacts EF-Tu in the ribosomal A-site. The predicted secondary structures which could be confirmed by extensive enzymatic and chemical structure probing have the potential to form relatively stable hairpins with a flexible central part containing the consensus sequence. The structural similarity of the selected RNAs to the secondary structures of the α-sarcin domain reveals that the RNAs which bind to EF-Tu present the consensus sequence in a partly single stranded, flexible structural context. Thus, EF-Tu specific RNA-aptamers resemble the open conformation of the α-sarcin domain (Fig. 4).

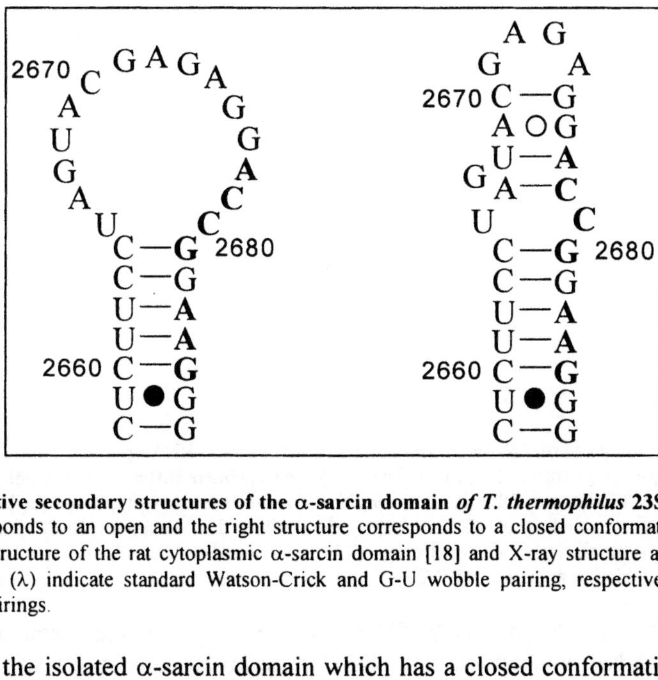

Fig. 4: Alternative secondary structures of the α-sarcin domain *of T. thermophilus* 23S rRNA. The left structure corresponds to an open and the right structure corresponds to a closed conformation based on the NMR-derived structure of the rat cytoplasmic α-sarcin domain [18] and X-ray structure analysis [19]. The symbols (-) and (λ) indicate standard Watson-Crick and G-U wobble pairing, respectively; (ϒ) indicates noncanonical pairings.

Remarkably, the isolated α-sarcin domain which has a closed conformation [18] is not able to bind to EF-Tu and is reported to be thermodynamically unstable [21]. The analysis of RNA-aptamers binding to EF-Tu suggests that for binding EF-Tu the α-sarcin domain adopts an open conformation in the ribosome which cannot be detected on the isolated rRNA domain. As proposed, a switch between an open and closed conformation of the α-sarcin domain might be important for the regulation of EF-Tu binding [11, 22]. With regard to this possible regulation of factor binding, the consensus sequence is located in an interesting region of the α-sarcin domain. A conformational change at the junction of stem and loop of the RNA can be easily recognised by a protein contacting this region.

2.3. INTERACTION OF EF-Tu, RIBOSOMES AND tRNA

As a mimic of ribosomal RNA the RNA-aptamers offer a possibility to identify the binding regions of rRNA on the surface of EF-Tu. For this purpose binding experiments were performed with the isolated domains of EF-Tu and EF-Tu in complex with aminoacyl-tRNA or elongation factor Ts. There is no competition between the RNA-aptamers and aminoacyl-tRNA for binding to EF-Tu and the EF-Tu·GTP·aa-tRNA·RNA-aptamer quaternary complex is readily formed [16]. In the ternary complex of EF-Tu·GppNHp·Phe-tRNA^{Phe} the aminoacyladenosine is located on domain II of EF-Tu close to the interface between domain I and domain II. The 5'-phosphate and part of the TψC stem of aminoacyl-tRNA are in contact with domain III (Fig. 5).

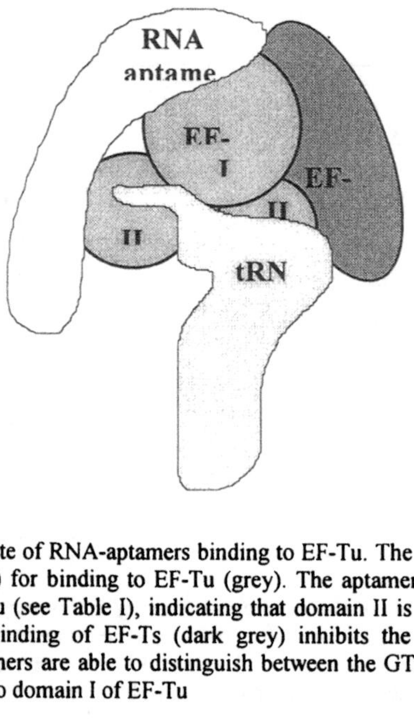

Fig. 5: Possible binding site of RNA-aptamers binding to EF-Tu. The RNA-aptamers (white) do not compete with aa-tRNA (light grey) for binding to EF-Tu (grey). The aptamers differ in their binding affinity to the isolated domains of EF-Tu (see Table I), indicating that domain II is most important for binding the RNA-aptamer. However, the binding of EF-Ts (dark grey) inhibits the binding of RNA-aptamers to EF-Tu. Moreover, the RNA-aptamers are able to distinguish between the GTP- and GDP-form of EF-Tu, indicating that they might also bind to domain I of EF-Tu

Despite the presence of aminoacyl-tRNA large parts of domain I and domain II are accessable for RNA-aptamer binding. Band retardation experiments with the isolated domains of EF-Tu show that domain I/II and domain II/III are sufficient for binding the RNA-aptamers, whereas domain I and domain III alone do not exhibit detectable binding capability (Table I).

Table I: Binding of RNA-aptamers to the isolated domains of elongation factor Tu as determined by band retardation assays. Domains resulting in band retardation are indicated by (+).

Domain	I	I / II	II / III	III
Interaction	-	+	+	-

Obviously domain II is critical for binding the aptamers. However, since the RNA-aptamers distinguish between the GDP- and GTP- form of EF-Tu, the nucleotide binding domain I may also be involved in the interaction. The interaction of the aptamers with domain I was also confirmed by the observation that EF-Ts and the RNA-aptamers compete for binding to EF-Tu. A schematic presentation of the possible topology of different EF-Tu complexes based on present and previous observations is depicted in Figure 5.

2.4. RNA-APTAMERS AS INHIBITORS OF EF-Tu RELATED FUNCTIONS

RNA-aptamers offer the possibility to inhibit or modify specific functions of RNA binding proteins. Blocking a functional site of EF-Tu with an RNA-aptamer should accordingly influence EF-Tu functions that depend on the formation of the particular complex. An important feature of the EF-Tu ribosome contact is the stimulation of the GTPase activity of EF-Tu by ribosomes. The hydrolysis of EF-Tu bound GTP occurs after codon-specific binding of aa-tRNA·EF-Tu·GTP complex to the ribosome [2, 3]. The formation of the quaternary complex between the EF-Tu, GTP, aa-tRNA and RNA-aptamer may block the surface on EF-Tu which is required for this interaction. EF-Tu binding RNA-aptamers reduce the GTPase activity of EF-Tu (Fig. 6).

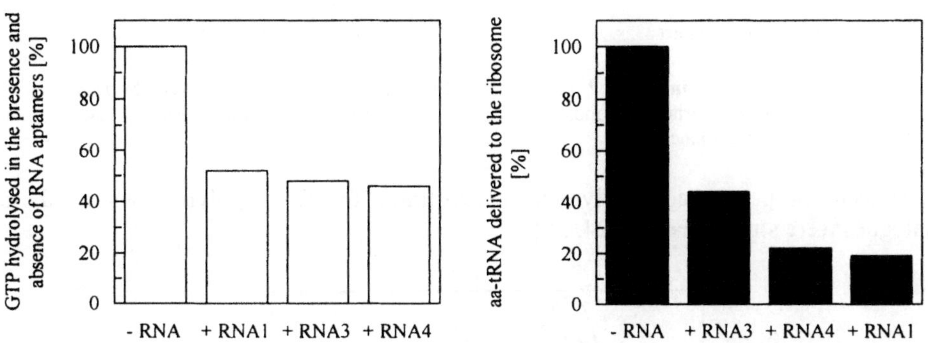

Fig. 6: Inhibition of factor dependent functions. (A) inhibition of GTPase activity of EF-Tu in the presence of RNA-aptamers. **(B)** inhibition of EF-Tu dependent delivery of aa-tRNA to the ribosome in the presence of RNA-aptamers.

It has to be kept in mind that the binding of aminoacyl-tRNA·EF-Tu·GTP ternary complex to the ribosomal A-site is a complex process consisting of several steps [2, 3, 23]. Only partial inhibition of the mentioned functions by RNA-aptamers directed against EF-Tu might reflect this complexity.

2.5. RNA-APTAMERS AS FLUORESCENT REPORTER MOLECULES

Due to the high specificity and high affinity the RNA-aptamers may be valuable tools to assay proteins. A prerequisite for such an application is the introduction of spectroscopic reporter groups into the RNA-aptamer. The easiest way to introduce for example a fluorescent group into RNA is the direct incorporation of fluorescent nucleotide analogues by transcription with T7 RNA polymerase. The nucleotide analogues suitable for labelling RNA-aptamers have to fulfil a number of criteria. First, they must be substrates for T7 RNA polymerase. Second, after incorporation into the RNA the quantum yield of the fluorescent analogue has to be sufficiently high and should provide a specific change in fluorescence upon RNA protein interaction. Third, the incorporation of the analogue must not disturb the structure and protein binding properties of the RNA-aptamer. Figure 7 shows the structures of two fluorescent adenosine analogues, formycine and 2-aminopurine.

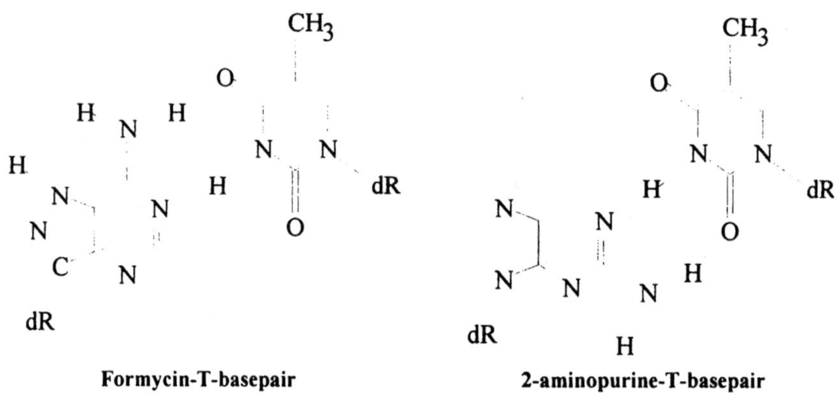

Fig. 7: Formation of formycin-T and 2-aminopurine-T-basepairs. Replacement of ATP by formycin-5'-triphosphate or 2-aminopurine-5'-triphosphate during T7-transcription of the aptamers leads to the incorporation of fluorescent labels.

Fluorescent properties and activities as substrates for RNA polymerases of these analogues were studied earlier [24, 25].

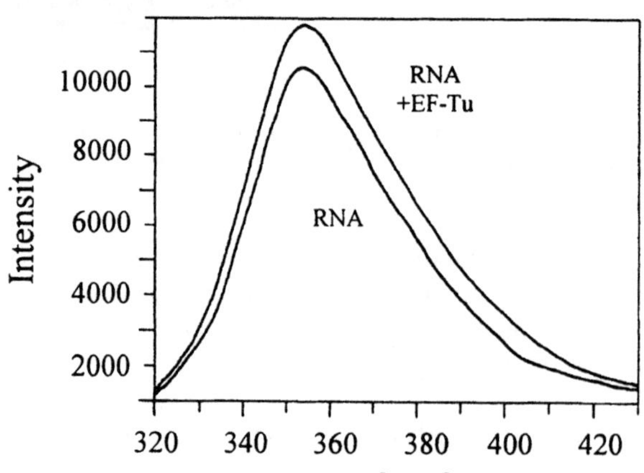

Fig. 8: Fluorescence spectrum of 2-aminopurine-RNA in the absence and presence of EF-Tu·GDP. 30% of the total amount of ATP has been replaced by 2-aminopurine during T7-transkription of an RNA-aptamer. The addition of EF-Tu results in an increase in fluorescence intensity.

The incorporation of both formycin and 2-aminopurine by T7 RNA polymerase results in fluorescent RNA-aptamers which are still capable of binding to EF-Tu.
The quantum yield of formycin is lower than that of 2-aminopurine. On the other hand, formycin disturbs the RNA structure to a lesser extend than 2-aminopurine. As Figure 8 demonstrates, the addition of EF-Tu·GDP to the RNA-aptamer in which adenosines were partly substituted by 2-aminopurines leads to an increase of fluorescence.

2.6. SELECTION OF RNA-APTAMERS WHICH BIND TO OTHER TRANSLATION FACTORS

RNA-aptamers offer a simple and efficient way to identify properties of different proteins with respect to RNA binding. This is especially useful for proteins like translation factors as their interaction with rRNA is difficult to analyze due to size and complexity of ribosomes. The strategy which was used for EF-Tu is also applicable to other translation factors like EF-G or RF-1. The *in vitro* selection resulted in RNAs with a defined consensus sequence (Fig. 9). Aptamers which bind EF-G have a consensus sequence which can be found in the thiostrepon domain of 23S rRNA of the large ribosomal subunit.

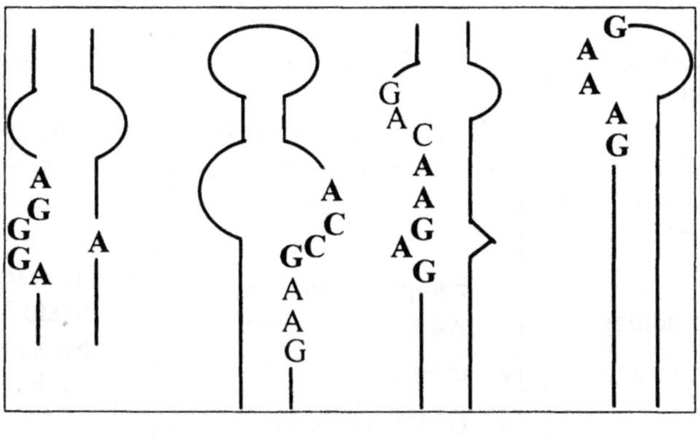

Fig. 9: Secondary structures and consensus sequences of RNA-aptamers binding to translation factors and to nickel cations. Bold letters depict the consensus sequences, plain letters indicate the extended consensus sequences which can be found in most but not all of the selected aptamers.

This domain is the place of major interaction with the antibiotic thiostrepton and the ribosomal protein L11 [26, 27]. There is also evidence that the thiostrepton region of 23S rRNA is involved in EF-G-dependent functions [9]. Aptamers which bind to release factor 1 should also be related to rRNA sequences present in the ribosomal A-site. A consensus sequence was identified in the RNA-aptamers binding to RF-1 which is different from the consensus sequences of the EF-Tu and EF-G binding RNA-aptamers. Presently it is not known how this consensus sequence is related to the structure of rRNA.

The comparison of secondary structures and consensus sequences in the RNA-aptamers shown in figure 9 reveal the selectivity of the *in vitro* selection method to study the interaction of complicated and structurally dynamic complexes. The selection of RNA-aptamers binding to translation factors resulted in molecules binding with high affinity (K_d in nM range). In the case of EF-Tu and EF-G we could provide evidence that the consensus sequences of the RNA-aptamers are related to the structure and function of the ribosomal A-site. However, the consensus sequences of the respective translation factor-binding aptamers are different and thus reflect the structural and functional dynamics of the ribosomal A-site. Further investigation of the complexes between RNA-aptamers and translation factors might help to understand the mechanism of factor-ribosome interaction at a molecular level.

3. Summary and outlook

RNA-aptamers selected for protein binding can be easily isolated using immobilized histidine-tagged proteins and used in different biochemical fields (Fig. 10). RNA binding proteins can be replaced by relatively simple RNA-aptamers which mimic the structure and function of RNA domains binding to the respective proteins *in vivo*.

Mechanism of translation	**Inhibitors**	**Reporters**
Does the RNA-protein complex mimic an *in vivo* situation?	Does the selected RNA inhibit the function of the protein?	Can the aptamers be used for the detection of EF-Tu?
• sequence homology • structure homology • mechanism	• complex formation with the ribosome • translation • GTP hydrolysis	• fluorescent aptamers to study interactions between proteins and nucleic acids

Fig. 10: Different ways to use aptamers binding to translation factors.

However, the structures of RNA-aptamer protein complexes at molecular resolution will be necessary for detailed analysis of such complexes. The structures can be determined only by high resolution nuclear magnetic resonance spectroscopy (NMR) or X-ray analysis of the corresponding complexes. The size of many available RNA-aptamers is still too large for such investigations. Therefore, new ways including chemical synthesis of appropriate RNA-aptamers with minimized motives have to be evolved. A combination of chemical and enzymatic RNA synthesis may be necessary to develop specifically labeled RNA-aptamers from the available lead substances derived from *in vitro* selection. A similar scenario is probably valid for the development of RNA based inhibitors. *In vitro* selection methods can provide inhibitors or useful catalytic RNA molecules. RNA with its ability to form tertiary structures is probably more suitable for selection of structural variants than DNA. On the other hand, RNA is labile and the isolated RNA-aptamers have to be modified to increase their stability in biological fluids. Structures of complexes between RNA-aptamers and proteins may be very useful as leads to develop nucleic acid based inhibitors of protein biosynthesis and other cellular processes.

REFERENCES

1. Gnirke, A., Geigenmüller, U., Rheinberger, H.-J., and Nierhaus, K.H. (1989) The allosteic three-site model for the ribosomal elongation cycle. *J.Biol.Chem.* **264**, 7291-7301.
2. Rodnina, M.V., Fricke, R., and Wintermeyer, W. (1994) Transient conformational states of aminoacyl-tRNA during ribosome binding catalyzed by elongation factor Tu. *Biochem.* **33**, 12267-12275.
3. Rodnina, M.V., Fricke, R., Kuhn, L., and Wintermeyer, W. (1995) Codon-dependent conformational change of elongation factor Tu preceding GTP hydrolysis on the ribosome. *EMBO J.* **14**, 2613-2619.
4. Czworkowski, J. and Moore, P.B. (1997) The Conformational Properties of Elongation Factor G and the Mechanism of Translocation. *Biochem.* **36**, 10327-10334.
5. AEvarsson, A., Brazhnikov, E., Garber, M., Zheltonosova, J., Chirgadze, Y., Al-Karadaghi, S., Svensson, L.A., and Liljas, A. (1994) Three-dimensional structure of the ribosomal translocase: elongation factor G from *Thermus thermophilus*. *EMBO J.* **13**, 3669-3677.
6. Nissen, P., Kjeldgaard, M., Thirup, S., Polekhina, G., Reshetnikova, L., Clark, B.F.C., and Nyborg, J. (1995) Crystal structure of the ternary complex of Phe-tRNA^{Phe}, EF-Tu, and a GTP analog. *Science* **270**, 1464-1472.
7. Ito, K., Ebihara, K., Uno, M., and Nakamura, Y. (1996) Conserved motifs in prokaryotic and eukaryotic polypeptide release factors: tRNA-protein mimicry hypothesis. *Proc.Natl.Acad.Sci.USA* **93**, 5443-5448.
8. Brock, S., Szkaradiewicz, K., and Sprinzl, M. (1998) Initiation factors of protein biosynthesis in bacteria and their structural relationship to elongation and termination factors. *Mol.Microbiol.* **29**, 409-417.
9. Moazed, D., Robertson, J.M., and Noller, H.F. (1988) Interaction of elongation factors EF-G and EF-Tu with a conserved loop in 23S RNA. *Nature* **334**, 362-364.
10. Hausner, T.-P., Atmadja, J., and Nierhaus, K.H. (1987) Evidence that the G^{2661} region of 23S rRNA is located at the ribosomal binding sites of both elongation factors. *Biochimie* **69**, 911-923.
11. Wool, I.G., Glück, A., and Endo, Y. (1992) Ribotoxin recognition of ribosomal RNA and a proposal for the mechanism of translocation. *Trends Biochem.Sci.* **17**, 266-269.
12. Stark, H., Rodnina, M.V., Rinke-Apel, J., Brimacombe, R., Wintermeyer, W., and van Heel, M. (1997) Visualization of elongation factor Tu on the *Escherichia coli* ribosome. *Nature* **389**, 403-406.
13. Agrawal, R.K., Penczek, P., Grassucci, R.A., and Frank, J. (1998) Visualization of elongation factor G on the Escherichia coli 70S ribosome: the mechanism of translocation [see comments]. *Proc.Natl.Acad.Sci.U.S.A.* **95**, 6134-6138.
14. Tuerk, C. and Gold, L. (1990) Systematic evolution of ligands by exponential enrichment: RNA ligands to bacteriophage T4 DNA polymerase. *Science* **249**, 505-510.
15. Hofmann, H.-P., Limmer, S., Hornung, V., and Sprinzl, M. (1997) Ni2+-binding RNA motifs with an asymetric purine-rich internal loop and a G-A base pair. *RNA* **3**, 1289-1300.
16. Hornung, V., Hofmann, H.-P., and Sprinzl, M. (1998) In vitro selected RNA molecules that bind to elongation factor Tu. *Biochem.* **37**, 7260-7267.
17. Endo, Y. and Wool, I.G. (1982) The Site of Action of α-Sarcin on Eukaryotic Ribosomes. *The Journal of Biological Chemistry* **257**, 9054-9060.
18. Szewczak, A.A., Moore, P.B., Chan, Y.-L., and Wool, I.G. (1993) The conformation of the sarcin/ricin loop from 28S ribosomal RNA. *Proc.Natl.Acad.Sci.USA* **90**, 9581-9585.
19. Correll, C.C., Munishkin, A., Chan, Y.-L., Ren, Z., Wool, I.G., and Steitz, T.A. (1998) Crystal structure of the ribosomal RNA domain essential for binding elongation factors. *Proc.Natl.Acad.Sci.USA* **95**, 13436-13441.
20. O'Connor, M. and Dahlberg, A.E. (1996) The influence of base identity and base pairing on the function of the α-sarcin loop of 23S rRNA. *Nucleic Acids Res.* **24**, 2701-2705.
21. Szewczak, A.A. and Moore, P.B. (1995) The sarcin/ricin loop, a modular RNA. *J.Mol.Biol.* **247**, 81-98.
22. Mesters, J.R., Potapov, A.P., de Graaf, J.M., and Kraal, B. (1994) Synergism between the GTPase Activities of EF-Tu·GTP and EF-G·GTP on empty ribosomes. *J.Mol.Biol* **242**, 644-654.

23. Rodnina, M.V., Pape, T., Fricke, R., Kuhn, L., and Wintermeyer, W. (1996) Initial binding of the elongation factor Tu ·GTP·aminoacyl-tRNA complex preceding codon recognition on the ribosomes. *J.Biol.Chem.* **271**, 646-652.
24. Ward, D.C. and Reich, E. (1969) Fluorescence studies of nucleotides and polynucleotides. *The Journal of Biological Chemistry* **244**, 1228-1237.
25. Piccirilli, J.A., Moroney, S.E., and Benner, S.A. (1991) A C-nucleotide base pair: methylpseudouridine-directed incorporation of formycin triphosphate into RNA catalyzed by T7 RNA polymerase. *Biochem.* **30**, 10350-10356.
26. Rogers, M.J., Bukhman, Y.V., McCutchan, T.F., and Draper, D.E. (1997) Interaction of thiostrepton with an RNA fragment derived from the plastid-encoded ribosomal RNA of the malaria parasite. *RNA* **3**, 815-820.
27. Xing, Y. and Draper, D.E. (1995) Stabilization of a ribosomal RNA tertiary structure by ribosomal protein L11. *J.Mol.Biol* **249**, 319-331.

PROBING OF RIBONUCLEOPROTEIN COMPLEXES WITH SITE-SPECIFICALLY DERIVATIZED RNAs

MARIA M. KONARSKA, PAVOL KOIS[1], MA SHA[2], NAÏMA ISMAÏLI, E. HILARY GUSTAFSON, and JEFFREY McCLOSKEY

The Rockefeller University
1230 York Avenue
New York, NY 10021

1 - present address: Comenius University, Organic Chemistry Department, SK-84215 Bratislava, Slovakia
2 – present address: Ciphergen Biosystems, Inc., Palo Alto, CA 94306

ABSTRACT

Splicing of nuclear pre-mRNA takes place in a complex structure known as the spliceosome. Elucidation of the molecular mechanism of such a complex process requires the development of a suitably simplified in vitro system. A system presented here is based on the use of two separate, trans-acting RNA molecules encompassing the two splice sites, which are together assembled into spliceosome complexes and undergo both steps of splicing. The small size of the 5' splice site (5'SS) RNA substrate allows for its use as a convenient probe to study the spliceosome structure and function. This chapter describes a simple chemical modification-derivatization strategy used to introduce selected chemical groups at specific internal positions within the RNA ribose backbone. The strategy is based upon the coupling of a haloacetyl adduct to a thiol residue in the phosphodiester bond. This method is applied to derivatize the 5'SS RNA with heterobifunctional photo-crosslinking reagents and to probe the spliceosomal components located in proximity of the active site of the complex. Several crosslinked species can be detected under splicing conditions. These include U1 and U6 snRNAs, and a number of protein products; hPrp8p, p114, p70, p54, and p27. While the 5'SS RNA crosslinks to hPrp8p and p70 are formed through the 5' exon segment (positions –4 and –3), the remaining contacts detected in this system represent adducts formed through the intron sequence (positions +4 to +8). Interestingly, introduction of large photoreactive groups near the 5' splice junction (positions –2 to +3) interferes with spliceosome formation and thus no crosslinks can be detected at these positions. The information concerning the structural distribution of components in the spliceosome will aid in the future biochemical and genetic analyses of the spliceosome.

J. Barciszewski and B.F.C. Clark (eds.), RNA Biochemistry and Biotechnology, 229–240.

1. The use of in vitro simplified systems to study complex biochemical reactions

A number of processes involved in eukaryotic gene expression are carried out by large, multicomponent complexes. For example, assembly of complexes that carry out transcription, splicing, and translation, require a large number of components and multiple dynamic changes involved in their proper positioning. In particular, both spliceosomes and ribosomes consist of multiple proteins and RNA molecules that, when properly assembled, form together catalytically active enzymes. Due to the remarkable complexity of these complexes, development of simplified systems is needed to allow a more detailed biochemical analysis of the processes. In these approaches, the classic and well-tested concept of "divide and conquer" is frequently applied. First, identification of components of the given machinery is achieved by purification of complexes and analysis of its individual elements. Ideally, this phase is followed by reconstitution of the active complex from the purified components. Such an approach has been very fruitful in the analysis of ribosomes, transcription factors and identification of snRNP proteins involved in splicing. However, it does not solve the problem posed by the high complexity of the studied systems. An alternative course is to develop a simplified system, which does not reproduce the full complexity of the original process, but faithfully mimics its selected features. We have used such a scheme to study the mechanism of pre-mRNA splicing.

A more detailed description of this system is presented in the previously published reports [1, 2]. The most important feature of the simplified in vitro splicing system used in the described studies is that the two splice sites of pre-mRNA are provided in trans on two separate RNA molecules. While the 3' splice site (3'SS), polypyrimidine tract and the branch site elements are present on one, longer RNA transcript, the 5'SS is provided as a short, synthetic RNA oligonucleotide that represents a 5'SS consensus sequence (AAG/GUAAGUAT, where "/" corresponds to the exon/intron border). This trans-splicing system uncouples two important stages in the spliceosome assembly; binding of the 5'SS by U1 snRNP and its subsequent interaction with U4/U5/U6 and U2 snRNPs, leading to splicing complex B formation [2]. This feature allows for an efficient preparation of a homogenous complex B suitable for biochemical studies. In addition, since the 5'SS consensus is presented in the absence of any flanking sequences normally present in pre-mRNA, interactions of the 5'SS with regulatory RNA-binding proteins are limited. Finally, because of the small size of the 5'SS RNA used in these experiments (11-18 nt), a number of chemical and molecular biology manipulations can be applied to generate a flexible, sensitive, and specific probe to biochemically analyze a highly complex and dynamic spliceosomal particle [3, 4].

The detailed analysis of the spliceosome is largely limited by the difficulty in monitoring RNA:protein interactions within the particle. Introduction of photoreactive groups at specified positions within the RNA generates a sensitive and flexible tool to probe the distribution of components of ribonucleoprotein complexes. We have applied such a technique to analyze the structure of splicing complex B in proximity to the 5'SS RNA [4]. The obtained information concerning the distribution of RNA and protein components of the spliceosome is important in defining the candidate molecules that are suspected of building the catalytic site of the complex. Since the 5'SS represents one of the two substrates for the first step of splicing (together with the branch site upstream of the 3'SS), it must be, by definition, located at the catalytic site within the active spliceosome complex. While the precise positioning of the 5'SS within the complex may change during the spliceosome assembly, the protein and RNA molecules identified in direct proximity of the 5'SS within splicing complex B are likely to participate in formation of the active site.

2. Protein composition of the spliceosome

Perhaps the most systematic approach toward the biochemical characterization of mammalian spliceosomal proteins is based on the affinity purification of snRNPs and splicing complexes followed by two-dimensional gel electrophoresis [5-9]. The identified polypeptides include hnRNP, snRNP, and non-snRNP splicing factors. In some cases, changes in protein composition can be correlated with various stages of complex assembly [6] consistent with the dynamic nature of the interactions involved. In parallel, a combination of genetic and biochemical approaches in yeast revealed a large number of proteins (so called PRP gene products) involved in various steps of splicing. With the ongoing description of new splicing factors both in yeast and mammalian systems, a growing number of yeast-human orthologs have been identified [9, 10]. A number of excellent, comprehensive reviews were recently published that extensively discuss the complexity of spliceosomal factors [9-16]. Here we present only a brief characterization of a selected subset of these factors that are expected to control formation of catalytically active spliceosomes.

Many of the spliceosome assembly steps correspond to changes in various RNA:RNA interactions occurring through a series of mutually exclusive base-pairing contacts [11-15]. It is expected that a class of ATP-ases facilitate at least some of these RNA rearrangements. This class of proteins, characterized by the presence of conserved motifs, termed DEAD and DEAH boxes, is thought to participate in unwinding of RNA:RNA segments [17]. In addition, proofreading functions affecting splicing fidelity have also been proposed as a possible role for these proteins [18]. The known DEAD/DEAH ATP-ases include Prp5 and UAP56 - involved in U2 snRNP addition, Prp2, Brr2 (and its human ortholog U5-200 kDa), Prp28 (and its human ortholog U5-100 kDa) – required before the first step of splicing, Prp16, Prp22 and Prp43 – required for the second step and mRNA and intron release from the spliceosome [9, 13]. Another spliceosomal protein, Snu114/U5-116 kDa, has been identified as a member of the GTP-ase family [19]. This component of U5 snRNP shares extensive sequence similarity with EF-2, required for the translocation step during translation [20].

A number of spliceosomal proteins are involved in direct interactions with pre-mRNA. One of them, the U5 snRNP-specific yeast Prp8 protein and its mammalian ortholog hPrp8/p220 [21, 22], can be crosslinked to precursor RNA at the 5'SS, branch site, polypyrimidine tract, and the 3' SS region [3, 6, 23-26]. The Prp8p forms crosslinks with the intron lariat intermediate, indicating that this protein remains in close proximity to the splice sites at later stages of the reaction [23-25]. A number of additional mammalian factors have been identified that directly contact pre-mRNA during different stages of splicing [26]. By correlating genetic and biochemical data it has been possible to identify a selected group of proteins present in close proximity to, and perhaps at the active site of the spliceosome. Significantly, our analysis of hPrp8 interaction with the 5'SS suggests that this protein may be directly responsible for recognition of the GU dinucleotide at the 5'SS [3, 27], again implicating this factor in close proximity of the catalytic center.

Finally, the recently identified SR family of regulatory proteins represents an important group of mammalian splicing factors (reviewed in [28]). This large family contains splicing factors such as ASF/SF2, and SC35, and a number of additional proteins that contain characteristic serine (S) -arginine (R) -rich domains. Similar domains have been found in a growing number of other proteins, including U1 snRNP 70kD protein, U2AF65, Drosophila proteins su(wa), tra, and tra-2 [28-30]. Typically, in addition to the RS domain, the SR proteins contain also an RNA-recognition motif (RRM) [28]. Both these domains are thought to be involved in the SR protein-mediated RNA annealing, selection of splice sites, and splicing enhancer function [31-37].

3. UV-crosslinking analysis of RNA:protein interactions

Perhaps one of the most frequently used crosslinking approaches used in studies of ribonucleoprotein complexes relies on the natural photoreactivity of the RNA bases at 254 nm [3, 6, 25, 38-41]. This method allows for detection of crosslinked polypeptides located in the direct proximity ("zero Å" distance) of the RNA. Alternatively, the photoreactive base analog (e.g. 5-BrU, 5-IU, 4-thioU, 2-azidoA, 8-azidoA) can be placed at a selected position within the RNA, thus defining the site of the formed crosslink [3, 24, 42-47]. Another variation on this theme is to introduce into the RNA probe base analogs that can be subsequently derivatized with various photoreactive groups coupled through a longer linker. An example of this method is the use of the convertible nucleoside O6-(4-chlorophenyl)-inosine, which after treatment with cysteamine disulfide forms an adenine analog containing an active thiol group at position N6 [26]. A variety of thiol-specific crosslinking groups can then be coupled to such modified RNA, generating a site-specifically derivatized probe. A similar strategy can be applied by introducing base analogs containing a primary amino group that can be subsequently derivatized with amino-specific crosslinkers [48].

4. Site-specific derivatization of RNA

Photoreactive groups can also be attached through the phosphodiester backbone of the RNA. We have developed a simple two-step chemical modification strategy to couple selected adducts at specific internal positions within the RNA substrate and applied it to study the spliceosomal components in proximity of the 5'SS RNA. The strategy involves a post-synthetic derivatization of RNA by coupling the haloacetyl adduct to the phosphodiester backbone through a phosphorothioate residue. A similar

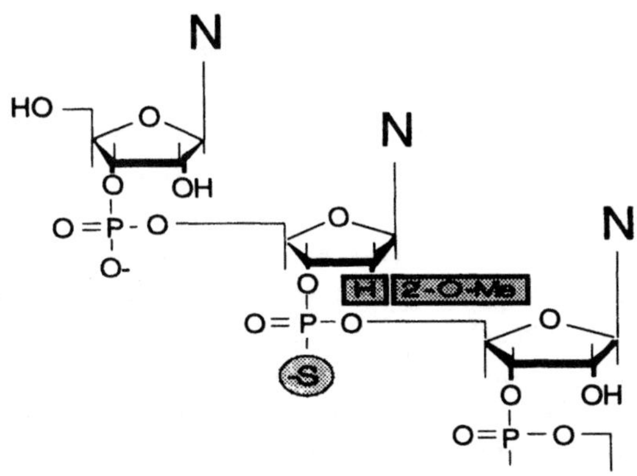

Figure 1. Schematic representation of a fragment of RNA modified for site-specific derivatization. The 2' hydroxyl group of the ribose adjacent to the phosphorothioate is replaced with either a 2' deoxy (H) or a 2'-O-methyl (2'OMe) group.

derivatization of DNA through phosphorothioate groups has been previously used in a number of systems [49-51]. In addition, similar techniques have been used in studies of RNA:protein and RNA:RNA interactions [52-56]. However, in the unmodified RNA the triester formed upon derivatization with haloacetyl reagents is unstable in the presence of the adjacent 2' hydroxyl group and becomes hydrolyzed [57]. Thus, this technique was limited to derivatization of the 5'-terminal phosphorothioate in RNA [53-56]. This restriction can be bypassed by removing the reactive 2' hydroxyl group adjacent to the derivatized phosphorothioate and replacing it with a single deoxyribose [52] or a 2'-O-methyl group [4] (see Fig. 1). Direct comparison of the effects of these two groups does not show detectable differences in the efficiency of the reaction or the stability of products, allowing for the selection of deoxy or 2'-O-Me modification based on the particular requirements of the experimental system.

	Exon	Intron
5'SS RNA:	**A A G**	**/G U A A G U A dT**
	-3 -2 -1	+1 +2 +3 +4 +5 +6 +7 +8

5'SS RNA	5' **AAGGUAAGT**AT 3'
template DNA	3' NNNTTCCATTCATA 5'

Figure 2. The RNA:DNA oligonucleotide substrate for extension of the 5'SS RNA with phosphorothioate deoxynucleotides. The 5'SS RNA oligonucleotide (in bold letters) containing a single deoxynucleotide (dT) at the 3' end is annealed to the DNA "splint" oligonucleotide. The protruding 5' nucleotides serve as a template for DNA polymerase that extends the 5'SS RNA with deoxynucleotides (in outlined letters). Addition of thio-dATP and dTTP yields the 5'SS RNA with thiol group at position +6, while extension in the presence of thio-dTTP and dATP produces the +7-thio 5'SS RNA.

The most obvious and general approach in preparing site-specifically modified RNA substrates involves chemical synthesis of RNA oligonucleotides containing a single phosphorothioate and adjacent deoxy or 2'-O-methyl groups. Such oligonucleotides can be used directly as the 5'SS substrates, as in the case of trans-splicing reaction (see below), or can be ligated to specific RNA transcripts to generate longer cis-splicing RNA substrates. An alternative strategy involves enzymatic extension of an RNA oligonucleotide with two deoxynucleotides, one of which contains a thiol group. This method, using Klenow polymerase, allows for the simultaneous introduction of a phosphorothioate and labeling of the substrate.

Since 5'SS RNAs containing just 3 ribonucleotides spanning the 5' splice junction in the context of an otherwise DNA sequence undergo both steps of splicing, the resulting chimeric RNA/DNA molecule serves as a functional splicing substrate [27]. The enzymatic extension procedure requires a bridging DNA oligonucleotide that is complementary to the 3'-terminal portion of the RNA substrate and contains an unpaired overhang at its 5' end that serves as a template for primer extension of the RNA molecule [58].

Subsequently, coupling of the phosphorothioate group with the haloacetyl photoreactive reagent [59-61] is carried out by nucleophilic displacement of halogen (Fig. 4). Derivatized RNAs are then used to form RNP complexes, reactions are UV-irradiated, and crosslinked products are analyzed.

5. Thiol-specific photocrosslinkable reagents

To prepare site-specifically derivatized substrates, two photoreactive crosslinking reagents, azidophenacyl bromide (APB) or benzophenone 4-iodoacetamide (BPI), are covalently attached through the thiol groups introduced at internal positions in the 5'SS RNA (Fig. 3). The azidophenacyl (APA) group positions the photoreactive nitrene ~9Å from the phosphorothioate, the distance comparable in length to approximately three nucleotides in the RNA chain, while the benzophenone (BP) group provides a longer, ~15Å linker. In addition to the size difference, these two groups differ also in chemical properties and the mechanism of crosslinking. The azido group in APA is photolyzed to a highly reactive nitrene that indiscriminately forms covalent bonds to neighboring molecules by electrophilic addition [60, 61]. In the BP reactions, UV light irradiation induces the excited Π^* triplet state, which preferentially reacts with C-H bonds of nearby molecules [59]. This preference for C-H bonds generates highly efficient, often remarkably site-specific crosslinking, since in the absence of the suitable C-H bond the excited probe relaxes to its initial state rather than reacting with water. Because of the chemical properties of the reagent, BP-derivatized probes are particularly effective at hydrophobic sites. Benzophenone probes have several advantages over aryl azides (e.g. APA). BP is chemically more stable and can be more easily manipulated at ambient light (it is activated at 350-360 nm), thus avoiding potentially damaging short UV wavelengths. However, APA groups are smaller, and react randomly with surrounding molecules, generating a less biased pattern of crosslinking.

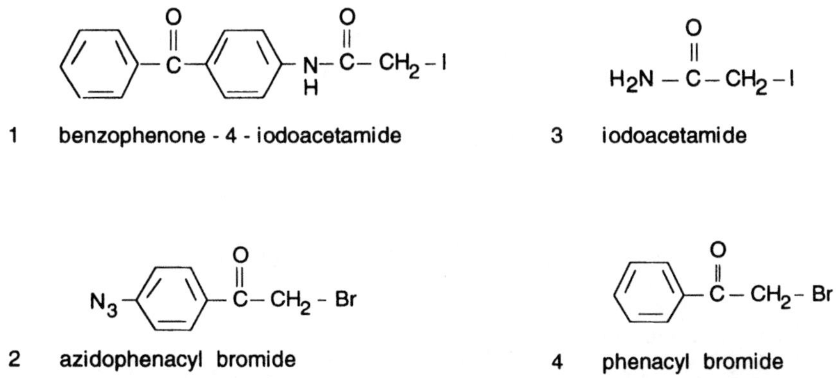

Figure 3. Schematic structures of chemical reagents used for derivatization of thiol-modified RNAs. 1. Benzophenone-4-iodoacetamide, 2. Azidophenacyl bromide, 3. Iodoacetamide, 4. Phenacyl bromide.

In addition to photo-reactive APA and BP, other groups can also be conjugated with RNA using this technique (Fig.3). For example, phenacyl (PA) and acetamide (AA) groups can be used as site-specific modification-interference probes [4]. Similar photocrosslinking approaches were used in studies of a large number of nucleic acid:protein and RNA:RNA interactions, including tRNA:cognate synthetase complex [52], RNase P:tRNA complex[53, 54], HDV ribozyme[55], TaqI restriction endonuclease:DNA complex [49], RNA polymerase II initiation complex [50], the ribosome [42, 47], and the spliceosome [4].

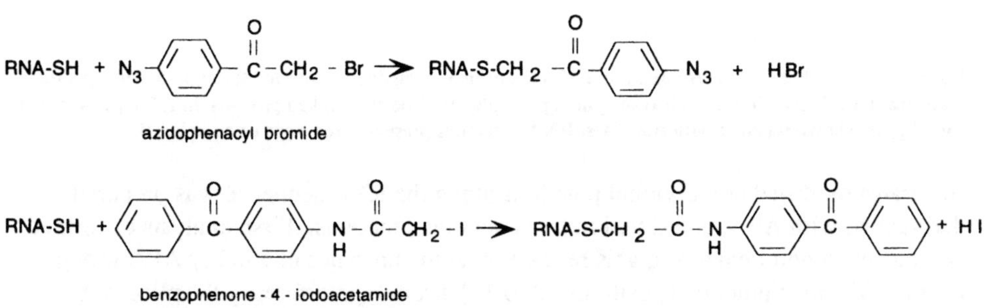

Figure 4. Schematic representation of conjugation of thiol-modified RNA (RNA-SH) with haloacetyl reagents, azidophenacyl bromide (top) and benzophenone-4-iodoacetamide (bottom).

6. Analysis of the spliceosomal components crosslinked to the 5'SS RNA

We have used the above technique in splicing reactions containing the 5'SS RNA oligonucleotides derivatized with APA or BP to analyze the spliceosomal components present in close proximity of the 5' splice site [4]. A series of the 5'SS RNA

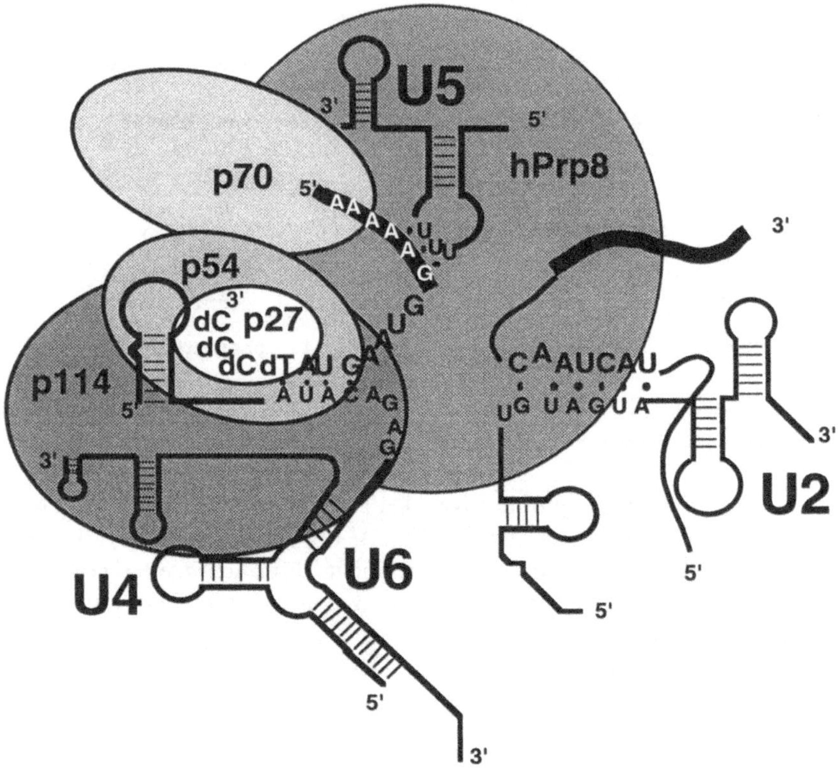

Figure 5. Schematic representation of spliceosomal components detected through the site-specifically derivatized 5'SS RNA in the context of splicing complex B. The crosslinked proteins; hPrp8p, p114, p70, p54, and p27, are shown together with snRNA-snRNA structures present at the stage of complex B.

substrates modified at individual positions along the RNA sequence was prepared, derivatized with APA, and tested in trans-splicing reactions. First, analysis of formed complexes in non-denaturing gels revealed that the presence of a bulky APA group near the 5' splice junction (positions –2 to +3) strongly interferes with spliceosome assembly. Similar results were also obtained with the BP-derivatized 5'SS RNA substrates. This interference appears to represent steric hindrance caused by a large (9Å) APA group, since analogous 5'SS RNA substrates derivatized with smaller groups, phenacyl (PA, 6Å), and acetamide (AA, 3Å), exhibited progressively lower levels of inhibition. This result is reminiscent of our earlier observation that a methyl (2Å) or an iodo (2.15Å) group at position 5 of uridine at the conserved GU dinucleotide at the 5'SS (pos. U+2 in the intron) also strongly inhibits spliceosome formation and splicing. The GU dinucleotide can be crosslinked to hPrp8 in splicing complex B [3, 27]. Thus, the detected interference of the spliceosome assembly results most likely from a misalignment in the precise interaction between the 5'SS and Prp8, suggesting a close, specific contact between these molecules. Consistent with this notion, the 5'SS RNAs derivatized with either APA or BP within the 5' exon segment (exon pos. –3, -4) and subjected to UV irradiation with 302 nm light, produce crosslinks to hPrp8 and one other protein of thus far unknown identity, p70.

In contrast, photoreactive groups placed within the intron segment detect U6 snRNA (pos. +3 to +8) and several proteins, p114, p54, and p27. There are some obvious differences between crosslinking profiles obtained with APA and BP groups. While both p114 and p27 can be detected with APA, the efficiency of crosslink formation is significantly higher with the BP-derivatized RNA. On the other hand, p54 crosslink can be detected only with APA-derivatized 5'SS RNA. As expected, the BP-mediated 5'SS RNA:p114 crosslink appears to be highly specific and homogenous, based on profiles of proteolytic digestions with various proteases and chemical reagents. Finally, the efficiency of the 5'SS RNA crosslinking to U6 snRNA is greater with the APA- than with BP-derivatized substrate. In both cases, the photoreactive groups are placed within the intron sequence that is thought to pair with U6 snRNA, and the decrease in crosslinking efficiency of BP-5'SS RNA probably reflects the hydrophobic nature of BP group. Finally, the background crosslinking signal can be assessed by comparing the profile of reactions that contain underivatized 5'SS RNA. Only trace amounts of U6 snRNA crosslink can be detected with 302 nm UV, while 254 nm UV light generates in addition a typical crosslink to hPrp8 [3, 27].

The differential 5'- and 3'-end labeling of the 5'SS RNA substrates containing a single phosphorothioate at pos. +6 or +7 distinguishes between crosslinks formed within the exon and intron segments. In these substrates, a single riboU residue is present at pos. +2, allowing for separation of upstream and downstream segments by RNase A digestion. The hPrp8 and p70 crosslinks are formed through thiol-nonspecific conjugation of APA and BP groups, since these crosslinks are lost upon RNase A digestion. This is consistent with results of site-specific derivatizations placing photoreactive groups witin the exon segment, at pos. –3 and –4. In contrast, U6 snRNA, p114, p54, and p27 crosslinks correspond to thiol-specific derivatization products within the intron segment (pos. +4 to +8).

7. Concluding remarks

The described method adds to the growing list of possible modification and derivatization techniques that can be used in molecular biology of RNA. While this method has its own limitations, it also offers several advantages over other frequently used crosslinking strategies. The capacity to derivatize the ribose backbone may be important in cases where high sequence conservation does not allow for base modifications. Most importantly, this technique allows for selection of the site of modification with a single nucleotide resolution, without any restrictions imposed by the sequence of the target RNA. While the photoreactive probes used here (APA ~9Å, BP ~ 15Å in length) are large enough to cause steric interference when present at a site of tight RNA:protein interaction, their relatively small size makes them a useful tool for probing of a variety of complex assemblies containing nucleic acids. In addition, the same strategy can be used for a precise modification-interference analysis, by analogy to the previously employed techniques [4, 62, 63].

The resulting RNA:protein crosslink products are stable and thus it is possible to subject them to further analysis, including polyacrylamide/SDS gel electrophoresis, immunoprecipitation, etc. Importantly, if the crosslink represents a homogenous population of molecules attached at a single site within a protein, it is possible to map the crosslink location using a series of specific enzymatic and chemical reactions. This is less typical of APA-derived crosslinks, because of the indiscriminate mode of crosslinking. However, this lack of specificity in crosslink formation results in a less biased profile of detected crosslinks.

8. References

1. Konforti, B.B., M.J. Koziolkiewicz, and M.M. Konarska. (1993) Disruption of base pairing between the 5' splice site and the 5' end of U1 snRNA is required for spliceosome assembly., *Cell* **75**, 863 - 873.
2. Konforti, B.B. and M.M. Konarska. (1995) A short 5' splice site RNA oligo can participate in both steps of splicing in mammalian extracts, *RNA* **1**, 815-827.
3. Reyes, J.L., *et al.* (1996) The canonical GU dinucleotide at the 5' splice site is recognized by p220 of the U5 snRNP within the spliceosome, *RNA* **2**, 213-225.
4. Sha, M., *et al.* (1998) Probing of the spliceosome with site-specifically derivatized 5' splice site RNA oligonucleotides, *RNA* **4**, 1069-1082.
5. Bennet, M., *et al.* (1992) Protein components specifically associated with prespliceosome and spliceosome complexes., *Genes & Dev* **7**, 1986-2000.
6. Chiara, M.D., *et al.* (1996) Identification of proteins that interact with exon sequences, splice sites, and the branchpoint sequence during each stage of spliceosome assembly, *Mol. Cell. Biol.* **16**, 3317-3326.
7. Laggerbauer, B., J. Lauber, and R. Luhrmann. (1996) Identification of an RNA-dependent ATPase activity in mammalian U5 snRNPs, *Nucleic Acids Res* **24**, 868-75.
8. Neubauer, G., *et al.* (1998) Mass spectrometry and EST-database searching allows characterization of the multi-protein spliceosome complex, *Nature Genetics* **20**, 46-50.
9. Will, C.L. and R. Lührmann. (1997) Protein functions in pre-mRNA splicing, *Curr. Opin. Cell Biol.* **9**, 320-328.
10. Krämer, A. (1996) The structure and function of proteins involved in mammalian pre-mRNA splicing, *Annu. Rev. Biochem.* **65**, 367-409.
11. Nilsen, T.W. (1994) RNA-RNA interactions in the spliceosome: Unraveling the ties that bind, *Cell* **78**, 1-4.
12. Nilsen, T.W., *RNA-RNA interactions in nuclear pre-mRNA splicing*, in *RNA Structure and Function*, R.W. Simons and M. Grunberg-Manago, Editors. 1998, Cold Spring Harbor Laboratory Press: New York. p. 279-307.
13. Staley, J.P. and C. Guthrie. (1998) Mechanical devices of the spliceosome: Motors, clocks, springs, and things, *Cell* **92**, 315-326.
14. Hertel, K.J., K.W. Lynch, and T. Maniatis. (1997) Common themes in the function of transcription and splicing enhancers, *Curr. Opin. Cell Biol.* **9**, 350-357.
15. Newman, A. (1997) RNA splicing: out of the loop, *Curr. Biol.* **7**, 418-420.
16. Wang, J. and J.L. Manley. (1997) Regulation of pre-mRNA splicing in metazoa, *Curr. Opin. Genet. Dev.* **7**, 205-211.
17. Fuller-Pace, F.V. (1994) RNA helicases: modulators of RNA structure, *Trends Cell Biol.* **4**, 271-274.
18. Burgess, S.M. and C. Guthrie. (1993) Beat the clock: paradigms for NTPases in the maintenance of biological fidelity, *Trends Biochem. Sci* **18**, 381-384.
19. Fabrizio, P., *et al.* (1997) An evolutionarily conserved U5 snRNP-specific protein is a GTP-binding factor closely related to the ribosomal translocase EF-2, *EMBO J.* **16**, 4092-106.
20. Abel, K. and F. Jurnak. (1996) A complex profile of protein elongation: translating chemical energy into molecular movement, *Structure* **4**, 229-238.
21. Anderson, K. and M.M. Moore. (1997) Bimolecular exon ligation by the human spliceosome, *Science* **276**, 1712-1716.
22. Beggs, J.D., S. Teigelkamp, and A.J. Newman. (1995) The role of PRP8 protein in nuclear pre-mRNA splicing in yeast, *J Cell Sci Suppl* **19**, 101-5.

23. Teigelkamp, S., A.J. Newman, and J.D. Beggs. (1995) Extensive interactions of PRP8 protein with the 5' and 3' splice sites during splicing suggest a role in stabilization of exon alignment by U5 snRNA, *Embo J* **14**, 2602-12.
24. Wyatt, J.R., E.J. Sontheimer, and J.A. Steitz. (1992) Site-specific cross-linking of mammalian U5 snRNP to the 5' splice site before the first step of pre-mRNA splicing, *Genes & Dev.* **6**, 2542-2553.
25. Umen, J.G. and C. Guthrie. (1995) A novel role for a U5 snRNP protein in 3' splice site selection, *Genes Dev* **9**, 855-68.
26. MacMillan, A.M., *et al.* (1994) Dynamic association of proteins with the pre-mRNA branch region, *Genes Dev.* **8**, 3008-3020.
27. Reyes, J.L., *Biochemical and genetic analysis of the interaction of Prp8 with the 5' splice site during pre-mRNA splicing*, in *Laboratory of Molecular Biology and Biochemistry*. 1998, The Rockefeller University: New York.
28. Fu, X.-D. (1995) The superfamily of arginine/serine-rich splicing factors, *RNA* **1**, 663-680.
29. Manley, J.L. and R. Tacke. (1996) SR proteins and splicing control, *Genes Dev* **10**, 1569-79.
30. Valcárcel, J. and M.R. Green. (1996) The SR protein family: Pleiotropic functions in pre-mRNA splicing, *Trends Biochem. Sci.* **21**, 296-301.
31. Crispino, J.D., B.J. Blencowe, and P.A. Sharp. (1994) Complementation by SR proteins of pre-mRNA splicing reactions depleted of U1 snRNP, *Science* **265**, 1866-1869.
32. Fu, X.-D. and T. Maniatis. (1992) The 35-kDa mammalian splicing factor SC35 mediates specific interactions between U1 and U2 small nuclear ribonucleoprotein particles at the 3' splice site, *Proc. Natl. Acad. Sci. USA* **89**, 1725-1729.
33. Fu, X.-D., *et al.* (1992) General splicing factores SF2 and SC35 have equivalent activities in vitro, both affect alternative 5' and 3' splice site selection, *Proc. Natl. Acad. Sci.* **89**, 11224-11228.
34. Kohtz, J.D., *et al.* (1994) Protein-protein interactions and 5'-splice-site recognition in mammalian mRNA precursors, *Nature* **368**, 119-124.
35. Tarn, W.Y. and J.A. Steitz. (1995) Modulation of 5' splice site choice in pre-messenger RNA by two distinct steps, *Proc Natl Acad Sci U S A* **92**, 2504-8.
36. Zahler, A.M., *et al.* (1993) Distinct functions of SR proteins in alternative pre-mRNA splicing, *Science* **260**, 219-222.
37. Zuo, P. and J.L. Manley. (1994) The human splicing factor ASF/SF2 can specifically recognize pre-mRNA 5' splice sites., *Proc.Natl.Acad.Sci.USA.* **91**, 3363 - 3367.
38. Gaur, R.K., J. Valcarcel, and M.R. Green. (1995) Sequential recognition of the pre-mRNA branch point by U2AF65 and a novel spliceosome-associated 28-kDa protein, *RNA* **1**, 407-417.
39. Sawa, H. and Y. Shimura. (1992) Association of U6 snRNA with the 5'-splice site region of pre-mRNA in the spliceosome, *Genes and Dev.* **6**, 244-254.
40. Sawa, H. and J. Abelson. (1992) Evidence for a base-pairing interaction between U6 small nuclear RNA and 5' splice site during the splicing reaction in yeast, *Proc Natl Acad Sci U S A* **89**, 11269-73.
41. Wu, S. and M.R. Green. (1997) Identification of a human protein that recognizes the 3' splice site during the second step of pre-mRNA splicing, *Embo J* **16**, 4421-32.
42. Sylvers, L.A. and J. Wower. (1993) Nucleic acid-incorporated azidonucleotides: Probes for studying the interaction of RNA and DNA with proteins and other nucleic acids, *Bioconjugate Chem.* **4**, 411-8.
43. Query, C.C., S.A. Strobel, and P.A. Sharp. (1996) Three recognition events at the branch-site adenine, *Embo J* **15**, 1392-402.
44. Sontheimer, E.J. and J.A. Steitz. (1993) The U5 and U6 snRNAs as active site components of the spliceosome, *Science* **262**, 1989-1996.
45. Gott, J.M., *et al.* (1991) A specific, UV-induced RNA-protein cross-link using 5-bromouridine-substituted RNA, *Biochemistry* **30**, 6290-6295.
46. Willis, M.C., *et al.* (1993) Photocrosslinking of 5-iodouracil-substituted RNA and DNA to proteins, *Science* **262**, 1255-7.
47. Wower, J., *et al.* (1994) Synthesis of 2,6-diazido-9-(beta-D-ribofuranosyl)purine 3',5'-bisphosphate: incorporation into transfer RNA and photochemical labeling of Escherichia coli ribosomes, *Bioconjug Chem* **5**, 158-61.
48. Langer, P.R., A.A. Waldrop, and D.C. Ward. (1981) Enzymatic synthesis of biotin-labeled polynucleotides: novel nucleic acids affinity probes, *Proc. Natl. Acad. Sci. U.S.A.* **78**, 6633-6637.
49. Mayer, A.N. and F. Barany. (1995) Photoaffinity cross-linking of TaqI restriction endonuclease using an aryl azide linked to the phosphate backbone, *Gene* **153**, 1-8.

50. Lagrange, T., *et al.* (1996) High-resolution mapping of nucleoprotein complexes by site-specific protein-DNA photocrosslinking: organization of the human TBP-TFIIA-TFIIB-DNA quaternary complex, *Proc Natl Acad Sci U S A* **93**, 10620-5.
51. Cassetti, M.C. and B. Moss. (1996) Interaction of the 82-kDa subunit of the vaccinia virus early transcription factor heterodimer with the promoter core sequence directs downstream DNA binding of the 70-kDa subunit, *Proc. Natl. Acad. Sci. U.S.A.* **93**, 7540-7545.
52. Musier-Forsyth, K. and P. Schimmel. (1994) Acceptor helix interactions in a class II tRNA synthetase: Photoaffinity cross-linking of an RNA miniduplex substrate, *Biochemistry* **33**, 773-779.
53. Harris, M.E., *et al.* (1994) Use of photoaffinity crosslinking and molecular modeling to analyze the global architecture of ribonuclease P RNA, *Embo J* **13**, 3953-63.
54. Burgin, A.B. and N.R. Pace. (1990) Mapping the active site of ribonuclease P RNA using a substrate containing a photoaffinity agent, *EMBO J.* **9**, 4111-4118.
55. Rosenstein, S.P. and M.D. Been. (1996) Hepatitis Delta Virus ribozymes fold to generate a solvent-inaccessible core with essential nucleotides near the cleavage site phosphate, *Biochemistry* **35**, 11403-11413.
56. Wang, J.F., W.D. Downs, and T.R. Cech. (1993) Movement of the guide sequence during RNA catalysis by a group I ribozyme, *Science* **260**, 504-8.
57. Gish, G. and F. Eckstein. (1988) DNA and RNA sequence determination based on phosphorothioate chemistry, *Science* **240**, 1520-1522.
58. Hausner, T.P., L.M. Giglio, and A.M. Weiner. (1990) Evidence for base-pairing between mammalian U2 and U6 small nuclear ribonucleoprotein particles, *Genes Dev* **4**, 2146-56.
59. Dormán, G. and G.D. Prestwich. (1994) Benzophenone photoprobes in biochemistry, *Biochemistry* **33**, 5661-5673.
60. Hixson, S.H. and S.S. Hixson. (1975) P-Azidophenacyl bromide, a versatile photolabile bifunctional reagent. Reaction with glyceraldehyde-3-phosphate dehydrogenase, *Biochemistry* **14**, 4251-4.
61. Hixson, S.H., *et al.* (1980) Bifunctional aryl azides as probes of the active sites of enzymes, *Ann N Y Acad Sci* **346**, 104-14.
62. Conway, L. and M. Wickens. (1989) Modification interference analysis of reactions using RNA substrates, *Methods Enzymol.* **180**, 369-379.
63. Rymond, B.C. and M. Rosbash. (1988) A chemical modification/interference study of yeast pre-mRNA spliceosome assembly and splicing, *Genes Dev* **2**, 428-39.

THE IRE MODEL FOR FAMILIES OF RNA STRUCTURES

Selective Recognition by Binding Proteins (IRPs), NMR Spectroscopy and Probing with Metal Coordination Complexes

E.C. THEIL, Y. KE, Z. GDANIEC* and H. SIERZPUTOWSKA-GRACZ
Children's Hospital Oakland Research Institute
747 52nd Street
Oakland, CA 94609
U.S.A.

and

Department of Biochemistry
North Carolina State University
Raleigh, NC 27695-7622
U.S.A.

**Institute of Bioorganic Chemistry, Polish Academy of Sciences*
Noskowskiego Str. 12/14
Poznan 61-704
Poland

1. Introduction

Regulation of mRNA is likely an ancient form of genetic control, with the current rapid expansion of knowledge driven by advances in biotechnology. The regulation of ferritin mRNA by iron in animals, an older example of mRNA regulation, has been extended to other mRNAs of iron and oxidative metabolism. The regulatory element (IRE) is conserved in animal mRNAs encoding the transferrin receptor (TfR), m-aconitase, erythroid aminolevulinate synthase (eALAS) and Nramp2 (an ion transport protein), as well as ferritin. [1-4]. [The sequence identity of an IRE is >95% in the mRNAs of different animals, but is only 36-60% among different IRE containing mRNAs in the same animal]. An IRE can regulate either mRNA ribosome binding (translation) or degradation (stability/turnover). Control of ribosome binding is the mechanism of IRE-dependent regulation for ferritin, eALAS synthesis and m-aconitase, while control of nuclease binding/activity appears to be the mechanism of IRE-dependent regulation for TfR and Nramp2 synthesis. A common secondary structure is shared among the iso-IREs which includes a hairpin loop (CAGUGX) and a base paired stem with an interruption or "hinge" in mid-stem. Stem base pair sequence is mRNA-specific as is the structure of the "hinge". Among iso-IREs, the greatest variation occurs in the lower stem [1].

J. Barciszewski and B.F.C. Clark (eds.), RNA Biochemistry and Biotechnology, 241–247.

2. Selective Recognition by Binding Proteins (IRPs)

MRNAs containing the IRE sequence(s) are differentially regulated either through control of ribosome and initiation factor binding or nuclease binding. The only IRE-specific binding proteins currently identified are IRP1 and IRP2, which are structural homologues of m-aconitase [5-9]. IRP 1 can actually acquire aconitase activity in the cytoplasm when an FeS cluster forms in the protein that also impedes RNA binding [10]. Apo-m-aconitase itself does not appear to bind mRNAs [11], but the m-aconitase mRNA contains an IRE and is regulated by IRPs, 11-14] suggesting that the evolution of IRE-RNAs and IRP-proteins may have involve shared progenitors.

The depression of ferritin mRNA by iron is greater than the depression of other IRE-containing mRNAs *in vivo*. Differential binding of IRP1 and IRP2 could explain such observations, if, for example, a larger fraction of the ferritin-IRE were complexed with IRP than the other mRNAs. When IRP1 binding to IRE sequences of comparable length (28 - 30 nucleotides) was compared *in vitro* [15] , using a 15-fold molar excess of recombinant IRP, the binding of IRP1 was found to be comparable for the iso-IREs from ferritin, eALAS, TfR and m-aconitase mRNAs. In contrast, IRP2 binding under the same conditions, showed large differences among the iso-IREs (Figure 1). The ferritin-IRE bound IRP2 > 10-fold better than the other iso-IREs; similar results were obtained using natural IRPs in cell extracts [15]. Thus *in vivo*, a larger fraction of ferritin mRNA is likely to be repressed by IRP binding than other IRE-containing mRNAs, which would create a larger effect on translation observed [14], when IRP binding was blocked by excess cytoplasmic iron.

Structural differences among the iso-IREs include different stem sequences and a different hinge structure in the middle of the stem; all IREs have the same hairpin loop structure of CAGUGX. The hinge in the ferritin-IRE is a conserved sequence of four nucleotides-UGC/C, in an internal loop/bulge, IL/B, whereas other iso-IREs only have a conserved C in a bulge. Deletion of a single nucleotide, U6, in the conserved UGC/C of the ferritin-IRE changes IRP2 binding (Figure 1).

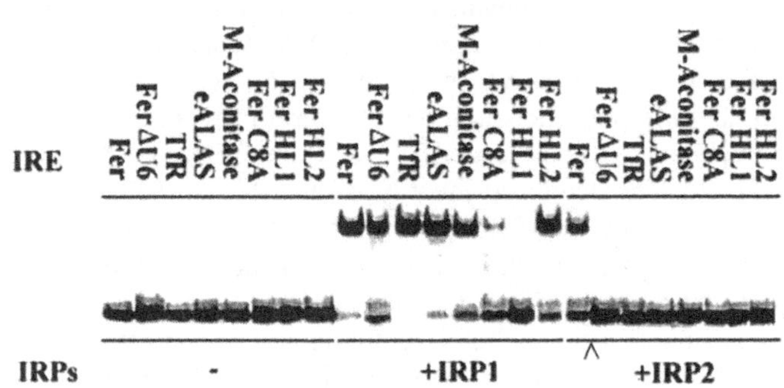

Figure 1. Sensitivity of IRP1 and IRP2 binding to the structure of the IRE(RNA) internal loop/bulge. 5'- ^{32}P-IREs were incubated with or without purified, recombinant IRP1 or IRP2 at a ratio of 1:15, RNA:protein. Fer-ferritin-IRE, TfR-transferrin receptor IRE, Fer ΔU6-IRE -IRE(delete U and converts IL/B to a bulge), eALAS .(^) Ferritin IRE binds IRP2. Figure taken from[15].

Deletion of U6, converts the IL/B to the C-bulge of the other iso-IREs [15]. The results contrast with IRP2 binding to the wild type ferritin IRE, but are comparable to IRP2 interactions with natural iso-IREs with a C-bulge IREs (Figure 1). Thus, the ferritin-IRE specific IL/B contributes significantly to the specificity of IRP2 interactions and to the differential regulation of ferritin mRNA and m-aconitase mRNA, e.g., observed in rat liver [14]. Since cells regulate both the relative amounts of the different IREs and the relative amounts of the two IRPs, the potential range of mRNA responses to a single signal is enormous. The wide range of responses possible for RNA/protein interactions in the IRE/IRP family is analogous to those for DNA/protein interactions in the steroid receptor/gene family.

3. NMR Spectroscopy

Examination of the three dimensional structure, by NMR spectroscopy and molecular modeling, of the ferritin–IRE and a consensus C-bulge IRE showed that the hairpin has a dynamic terminal loop and hinge region, [16-17]. The hinge region contains a conserved C residue, required for IRP binding, which is disordered in both the ferritin-IRE and the C-bulge consensus IRE [16-17], in the absence of protein. The internal loop/bulge in the ferritin IRE with the C residue (Figure 2), forms a pocket that binds

Effect of pH on the imino/imino region of the NMR spectrum of a G-U bp (*), that closes the IL/B of the ferritin IRE (2D-NOESY)

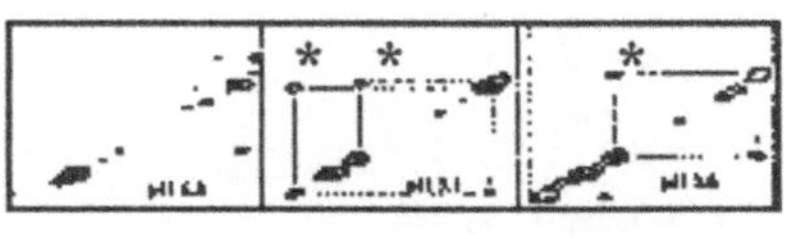

pH6.8 pH5.1 pH3.5
zero G-U 2 G-U 1 G-U
* * *

[G26 has 2 crosspeaks (G-U5 and G-U6?) depending on pH]

Superposition of NMR-assisted models for the ferritin-IRE in the region of the dynamic (pH sensitive)

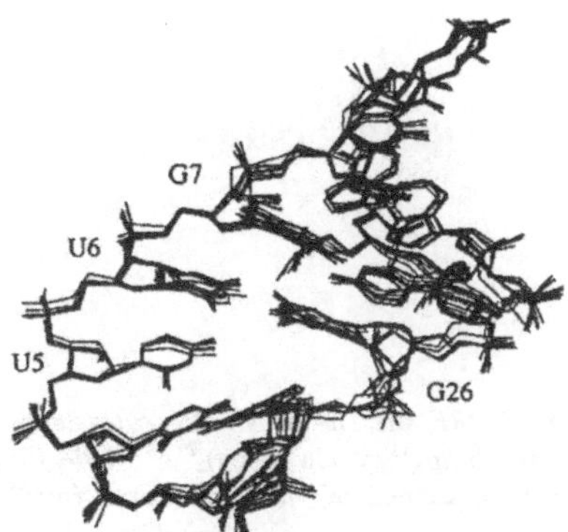

Figure2. The dynamic internal loop/bulge region of the (frog H) ferritin-IRE The internal loop/bulge (C8, G7, U6, U5, G26) region of the animal ferritin-IRE structure. [Note that IREs are >95% conserved among species but only 36-60% conserved among is iso-IREs in the same species Left: The effect of pH on the IL/B structure of the ferritin IRE (the 2D-NOESY from the NMR analysis of the ferritin-IRE MC-SYM). Right. Superposition was carried out with nucleotides U3-U10 and A23-A28, bracketing the internal loop/bulge region. Data taken from [16].

Co(III) hexammine [16]and Mg (Sierzputowska-Gracz and Theil, unpublished observations), based on perturbations of the 1H-NMR spectrum. Inhibition of binding of transition metal complexes (TMCs) also indicate Mg and Mn binding in the IL/B (see section 4). The conformation of the IL/B is pH dependent (Figure 2) and accounts for the 10-fold enhancement of IRP2 binding, shown in Figure 1, compared to other iso-IREs.

4. Probing with Metal Coordination Complexes

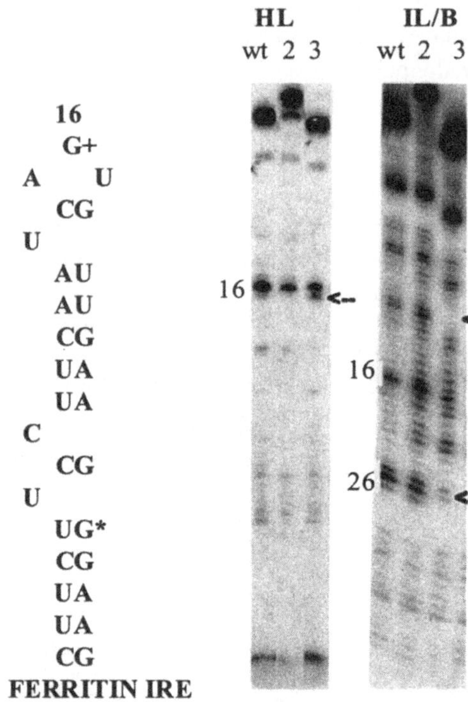

Ru(tpy)bpy Cu(phen)/CuP
(Arrows show TMC sites altered by IREmutation)

Figure 3. IRE sites in mRNA are cut selectively by transition metal complexes (TMCs): Ru(tpy)bpy, Cu $(phen)_2^{2+}$. Wild type or mutated frog ferritin IREs in full-length, capped *in vitro* transcripts (n=1000 nt) were analyzed by primer extension after RNA cutting with TMCs. ^{32}P-cDNAs, synthesized from the RNA fragments by reverse transcription, were separated on a polyacrylamide gel calibrated with a reverse transcription reaction run with ddXTPs. The IRE is numbered 1-30: Left. IRE structure: + Ru (tpy)bpy site selective for HL;* Cu $(phen)_2^{2+}$ selective for IL/B. Right. TMC cutting: **Ru(tpy)bpy** lanes: wt,2,3: wild type, HL-1 (G18A), IL-1(ΔU6, C8); **Cu(phen)** $_2^{2+}$ lanes wt ,2,3: wild type, HL-1 (G18A), IL-1 (ΔU6,C8).] Note that Cu $(phen)_2^{2+}$ also cannot bind to C-bulge in 30-mer oligoribonucleotides of Fer- ΔU6 or TfR-IREc (Ke and Theil, to be published)]. Data taken from [20].

The richness of RNA tertiary structure, as subtle as protein tertiary structure, has been difficult to exploit in the development of chemical probes because knowledge and visualization of RNA 3D structure has been relatively limited. However, RNA structures can recognize small molecules with great selectivity, distinguishing between such similar structures as caffeine and theophylline [18]. Such compounds have the potential for design matched to the folds and distortions in RNA (and DNA). To a first approximation, "sizes" of an RNA helix bend, for example, could be detected with transition metal complexes (TMCs), if the metal were redox active and cleaved the RNA at or near the TMC binding site. In addition, TMCs only require ng amounts of RNA for study (Figure 3).

The ferritin IRE was used to test TMC probes. Using classical, well characterized "off the shelf" ligands, TMCs were found which targeted different subdomains of the IRE. $Cu(phen)_2^{2+}$, for example, mainly targeted G26/27 in the IL/B region of the ferritin-IRE [19] and $Ru(tpy)bpy^{2+}$ recognized G16 in the hairpin loop [20]. Conversion of the IL/B to a C-bulge by mutation of the ferritin IRE eliminated cut sites in the IL/B (Figure 3); similar results were obtained when natural IREs, which varied in the hinge region structure, were probed with Cu $(phen)_2^{2+}$ (Ke and Theil, to be published). Mutation of the HL also changed RNA recognition by Ru $(tpy)bpy^{2+}$. However, mutation in the HL itself (G18A, lane 2, Figure 3) had less effect on structure, probed with Ru $(tpy)bpy^{2+}$, than did mutation in the IL/B (HL, lane 3, Figure 3).

Recent studies show that when the IRE base sequence is synthesized with deoxysugars and U/T replacement, the largest effect on recognition by Ru $(tpy)bpy^{2+}$ is also in the IL/B [21]. Mn inhibits Ru $(tpy)bpy^{2+}$ binding suggesting that the IL/B also accommodates Mn. Results of TMC studies of the ferritin IRE emphasize the existence of interactions between the HL and the IL/B regions of the RNA that, to date, have been difficult to show by any other means. The use of TMCs to study other RNA structures is likely to reveal features that have also escaped detection by other experimental approaches.

5. Summary

The IRE family has constant hairpin loops combined with variable internal loop/bulge (IL/B) or C-bulges and stem base pairs. Transition metal complexes (TMCs) and NMR spectroscopy can detect IRE-specific variations in tertiary structure. Chemical probing of RNA structure by TMCs requires only very small amounts (ng) of RNA, in contrast to physical probing of RNA structure by NMR spectroscopy, which indicates a role for TMCs in the analysis of tertiary structures in other RNAs.

Differences in cellular iron response, exemplified by ferritin and m-aconitase induction in rat liver, coincide with IRE variations in structure. IRE-specific differences in the sequence and tertiary structure of the IRE hinge region in the middle of the base-paired stem, an IL/B or C-bulge, are crucial for IRE-specific, differential binding of IRP2, but not IRP1. The differential binding of IRP 2, which reflects differences in RNA tertiary structure, explains the differential iron effect on m-aconitase and ferritin induction *in vivo*. Multiple IRE/IRP interactions are analogous to multiple DNA/steroid receptor interactions in providing specific variations in signal response and may represent a very ancient form of regulation from the RNA world.

6. Acknowledgements

The work of the senior author has been supported in part by the Extramural Hematology Program of the NIH-NIDDK (DK 20251).

7. References

1. Theil, E.C., (1998) The Iron Responsive Element (IRE) Family of mRNA Regulators, *Regulation of Iron Transport and Uptake Compared in Animals, Plants, and Microorganisms*, IN: Metal Ions in Biological Systems, A. Sigel and Sigel, H. (Eds.), Marcel Dekker, Inc., New York , Vol. **35,** pp. 403-434
2. Rouault, T.A., and Klausner, R.D., (1996), Post-transcriptional regulation of genes of iron metabolism in mammalian cells, *J. Biol. Inorg. Chem.* **1,** 494-499.
3. Hentze, M.W., and Kuhn, L.C., (1996), Molecular control of vertebrate iron metabolism: mRNA-based regulatory circuits operated by iron, nitric oxide, and oxidative stress, *Proc. Natl. Acad. Sci. USA* **93,** 8175-8182.
4. Andrews, N.C., and Levy, J., (1998), Iron is hot: An update on the pathophysiology of hemochromatosis, *BLOOD* **92,** 1845-1851.
5. Rouault, T.A., Stout, C.D., Kaptain, S., Harford, J.B., and Klausner, R.D., (1991), Structural Relationship between an Iron-Regulated RNA-Binding Protein (IRE-BP) and Aconitase: Functional Implications, *Cell* **64,** 881-883 .
6. Samaniego, F. Chin, J., Iwai, K., Rouault, T.A., and Klausner, R.D., (1994), Molecular Characterization of a Second Iron-responsive Element Binding Protein, Iron Regulatory Protein 2: Structure, Function, and Post-translational Regulation, *Biological Chemistry* **269,** 30904-30910.
7. Guo, B., Brown, F.M., Phillips, J.D., Yu,Y., and Leibold, E.A., (1995), Characterization and Expression of Iron Regulatory Protein 2 (IRP2). Presence of Multiple IRP2 Transcripts Regulated by Intracellular Iron Levels, *J. Biol. Chem.* **270,** 16529-16535.
8. Pantopoulos, K., Gray, N.K., and Hentze, M.W., (1995), Differential regulation of two related RNA-binding proteins, iron regulatory protein IRPA and IRPB, *RNA* **1,** 155-163.
9. Iwai, K., Klausner, R.D., and Rouault, T.A., (1995), Requirements for iron-regulated degradation of the RNA binding protein, iron regulatory protein 2, *EMBO J* **14,** 5350-5357.
10. Haile, D.J., Rouault, T.A., Tang, C.K., Chin, J., Harford, J.B., and Klausner, R.D., (1992), Reciprocal control of RNA-binding and aconitase activity in the regulation of the iron-responsive element binding protein: Role of the iron-sulfur cluster, *Proc. Natl. Acad. Sci. USA* **89,** 7536-7540.
11. Zheng, L., Kennedy, M.C., Blondin, G.A., Beinert, H., and Zalkin, H., (1992), Binding of Cytosolic Aconitase to the Iron Responsive Element of Porcine Mitochondrial Aconitase mRNA, *Arch. of Biochem. and Biophys.* **299,** 356-360.
12. Kim, H.-Y., LaVaute, T., Iwai, K., Klausner, R.D., and Rouault, T.A., (1996), Identification of a Conserved and Functional Iron-responsive Element in the 5'-Untranslated Region of Mammalian Mitochondrial Aconitase, *J. Biol. Chem.* **271,** 24226-24230.
13. Gray, N.K., Pantopoulos, K., Dandekar, T., Ackrell, B.A.C., and Hentze, M.W., (1996), Translational regulation of mammalian and *Drosophila* citric acid cycle enzymes via iron-responsive elements, *Proc. Natl. Acad. Sci. USA* **93,** 4925-4930.
14. O.S., Chen, Schalinske, K.L., and Eisenstein, R.S., (1997), Dietary Iron Intake Modulates the Activity of Iron Regulatory Proteins and the Abundance of Ferritin and Mitochondrial Aconitase in Rat Liver, *J. Nutrition* 238-248.
15. Ke, Y., Wu, J., Leibold, E.A., Walden, W.E., and Theil, E.C., (1998), Loops and Bulge/Loops in Iron-responsive Element Isoforms Influence Iron Regulatory Protein Binding *Journal of Biological Chemistry* **273,** 23637-23640.

16. Gdaniec, Z., Sierzputowska-Gracz, H., and Theil, E.C., (1998), Iron Regulatory Element and Internal Loop/Bulge Structure for Ferritin mRNA Studied by Cobalt(III) Hexammine Binding, Molecular Modeling, and NMR Spectroscopy, *Biochemistry* **37,** 1505-1512.
17. Addess, K.J., Basilion, J.P., Klausner, R.D., Rouault, T.A., and Pardi, A., (1997), Structure and Dynamics of the Iron Responsive Element RNA: Implications for Binding of the RNA by Iron Regulatory Binding Proteins, *J. Mol. Biol.* **274,** 72-83.
18. Jenison, R.D., Gill, S.C., Pardi, A., and Polisky, B., (1994), High-Resolution Molecular Discrimination by RNA, *Science* **263,** 1425-1429.
19. Wang, Y.-H., Sczekan, S.R., Theil, E.C., (1990), Structure of the 5' untranslated regulatory region of ferritin mRNA studied in solution, *Nuc. Acids Res.* **18,** 4463-4468.
20. Thorp, H.H., McKenzie, R.A., Lin, P.-N., Walden, W.E., and Theil, E.C., (1996), Cleavage of functionally relevant sites in ferritin mRNA by oxidizing metal complexes, *Inorganic Chemistry* **35,** 2773-2779.
21. Cliftan, S.A., Theil, E C. , and Thorp, H. H., (1998), Oxidation of Guanines in the Iron Responsive Element RNA: Similar Structures from Chemical Modification and Recent NMR Studies, *Chemistry and Biology* **in press**.

FUNCTIONAL ANALYSIS OF RNA SIGNALS IN THE HIV-1 GENOME BY FORCED EVOLUTION

BEN BERKHOUT and ATZE T. DAS
Department of Human Retrovirology
Academic Medical Center
University of Amsterdam
Meibergdreef 15
1105 AZ Amsterdam
The Netherlands

1. Abstract

The human immunodeficiency virus type 1 (HIV-1) is well-known for its genetic variability, which leads for instance to the development of drug-resistant virus variants. The molecular basis for the genetic flexibility of this virus is the error-prone Reverse Transcriptase enzyme. We employ the genetic flexibility of HIV-1 to study the function of regulatory viral RNA motifs in tissue culture infections by a method termed forced evolution. Mutant viruses with specific alterations in RNA motifs that cause a severe replication defect were cultured for a prolonged period to select for revertant viruses with improved replication characteristics. This method turned out to be very productive for the molecular analysis of structured RNA motifs that control different steps of the viral replication cycle. We will review the results obtained in the analysis of two RNA stem-loop structures that are encoded within the repeat region (R element) of the HIV-1 genome: the TAR and polyA hairpins. The different repair strategies observed for mutant forms of these hairpin structures will be described in detail, followed by a discussion of the functional insight gained by these studies. Because the R region forms both the extreme 5' and 3' end of the retroviral RNA genome, the two structured RNA motifs could in theory have distinct replicative functions at either end of the genome. The 5' copy of the TAR hairpin forms the binding site for the viral Tat *trans*-activator protein and is important for transcriptional up-regulation of viral gene expression. The polyA hairpin motif occludes part of the AAUAAA polyadenylation signal and appears to be critical for down-regulation of 5' polyadenylation. The same sequence is used efficiently as a polyadenylation signal at the 3' end of HIV-1 RNA, which is due to the presence of upstream enhancer elements. Putative additional functions of these RNA hairpins, e.g. in packaging of HIV-1 RNA into virions and the process of reverse transcription, will be discussed.

J. Barciszewski and B.F.C. Clark (eds.), RNA Biochemistry and Biotechnology, 249–275.

2. Introduction

2.1. STRUCTURED RNA SIGNALS IN THE REPEAT REGION OF THE HIV-1 GENOME

Retroviral RNA genomes are terminally redundant, and the 97-nucleotide repeat (R) region of HIV-1 RNA encodes both the TAR and polyA hairpin motifs (Figure 1). The TAR RNA hairpin structure is important for optimal transcription from the viral promoter in the long terminal repeat (LTR). In particular, the upper part of the TAR structure has been shown to be important for binding of the viral Tat *trans*-activator protein that triggers high level expression through interaction with the cellular transcription machinery [1-5]. The R region also encodes sequences that are important for polyadenylation of the viral transcripts [6-8]. The AAUAAA polyadenylation signal is embedded within a stem-loop structure that we termed the polyA hairpin [9-11]. Thus, the R region encodes a tandem hairpin motif that is present at both ends of HIV-1 transcripts. It is generally assumed that TAR is functional as the Tat-binding site in the 5'R context, but integrated proviruses may trigger the transcription of downstream cellular genes through activation of the 3' LTR promoter and Tat binding to the 3' TAR element [12]. In contrast, the viral polyadenylation signal should be functional exclusively in the 3'R context. Several mechanistic models have been proposed to explain this differential polyadenylation site usage within the HIV-1 genome [7,13-20]. Recent evidence indicates that the presence of the AAUAAA signal within the hairpin structure is critical for this regulation of polyadenylation [21].

A translational component of Tat/TAR-mediated activation of HIV-1 gene expression has been reported initially [22]. The 5' TAR structure interferes with mRNA translation in Xenopus oocytes [23,24] and in cell-free assays [25-27], and this repression could be overcome by addition of the Tat protein. Two mechanistic explanations have been proposed for TAR-mediated repression of translation. First, the 5' terminal TAR hairpin may inhibit translation in *cis* by interfering with the binding of translation initiation factors or ribosomes to the mRNA cap structure [25]. Second, TAR may activate the double-stranded RNA-dependent kinase PKR [26,28,29]. The activated form of this kinase phosphorylates and thereby inactivates the translation initiation factor eIF-2, causing inhibition of translation in *trans*.

Retroviruses also use the terminal repeat sequence in the process of reverse transcription. Reverse transcription is initiated near the 5' end of the genome at the primer-binding site (PBS, see Figure 1), and a cDNA of the 5'R region is synthesized. Upon removal of the RNA template strand by RNaseH action of the Reverse Transcriptase enzyme (RT), this cDNA anneals to the 3'R region and reverse transcription is resumed. There is some preliminary evidence that specific sequence and/or structure motifs within R are required for this strand-transfer reaction in murine leukemia virus [30].

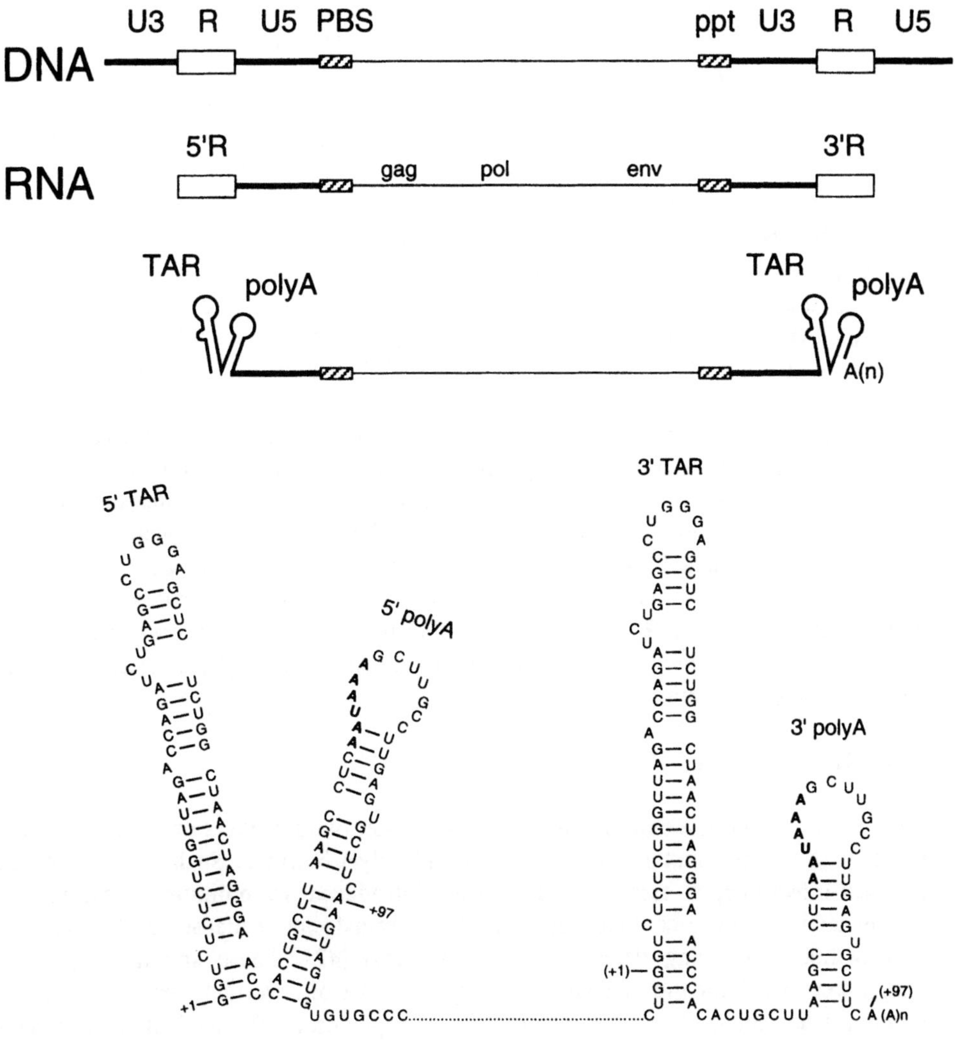

Figure 1. RNA structures within the R repeat region of the HIV-1 genome. The upper line represents the structure of the HIV-1 proviral DNA (not drawn to scale, e.g. the central protein-encoding part of the genome is shown in truncated form). Highlighted are the two long terminal repeats, of which the subdomains are indicated (U3, R, U5). The unique elements flanking the 5' LTR (PBS, primer-binding site) and the 3' LTR (ppt, polypurine track) are also shown. The viral RNA (second line) contains a 5' and a 3' R element that encode a tandem hairpin motif (third line). The upstream hairpin is termed TAR, the downstream hairpin is the polyA hairpin. Details of the RNA secondary structure of the hairpins in the 5'R and 3'R region of the mature, polyadenylated HIV-1 RNA are shown below. Polyadenylation is blocked at 5'R, but occurs efficiently 19 nucleotides downstream of the AAUAAA signal (indicated in bold) in the 3'R at position 97. Nucleotide positions in both R regions are relative to the transcription initiation site (+1) in the 5'R. The two hairpins are connected without a single unpaired nucleotide in the 5'R region, raising the possibility of coaxial stacking [31].

In vitro studies indicated that TAR is involved in dimerization of HIV-2 RNA [32], and results from an electron microscopy study are consistent with the involvement of 5'R sequences in formation of the HIV-1 RNA dimer [33]. Furthermore, a role for both the TAR and polyA hairpin in RNA packaging was suggested [34-36]. Although the exact function of these two R region hairpins in packaging of the viral genome remains to be determined, it is possible that these structures are part of the packaging signal that is recognized by the viral Gag protein during virion assembly. Alternatively, the effect may be more indirect, as the complete leader RNA may be required for correct folding and presentation of the actual packaging signal. The packaging function of TAR was shown to be independent of the Tat protein [34]. Both the TAR RNA element and the viral Tat protein have been suggested to stimulate the process of reverse transcription [37-39], although no mechanistic details have been reported.

Thus, a pleiotropy of functions have been attributed to the two hairpin structures of the 5'R region of HIV-1 RNA, and it remains possible that the 3'R motifs will also have multiple functions. For example, both hairpins may confer protection against cellular exonucleases. To date, these results have been obtained in a variety of experimental systems. The biologically most relevant assay system is that of the replicating virus, and replication studies with mutant viruses support the idea that the structured 5'R and 3'R motifs play multiple roles in the viral life cycle [37,40,41]. We will now discuss attempts to address some of these issues with the method of forced evolution.

2.2. FORCED EVOLUTION

Tissue culture evolution experiments were performed with replication-impaired virus mutants with specific mutations in the TAR and polyA hairpins. Such HIV-1 variants are *quasi*-infectious, a term that has been introduced in poliovirus research to describe an RNA genome that can replicate in transfected cells at greatly reduced levels without producing detectable progeny virus [42]. These smouldering virus mutants were used for the genesis of an extensive set of revertant viruses, of which the genome has changed to allow more efficient replication. The analysis of revertant viruses is a classical virological method, but we use this forced evolution approach in a systematic manner for the study of RNA signals that regulate HIV-1 replication.

The method of forced evolution allows us to study the repair of mutated RNA signals on a laboratory time scale, but the underlying principles are very similar to the mechanism of positive Darwinian evolution as proposed for real time biology. Reversion is a process that consists of two, independent steps. First, mutations will be introduced during low-level virus replication in the course of transcription or reverse transcription of the retroviral genome. Because the mutations will be random in nature, evolution is a chance process, and independent experiments will likely yield different types of revertants. Second, virus variants with improved replication properties will outgrow the original mutant. In other words, selection of the most fit virus variant is the driving force behind this *in vitro* evolution approach. The forced evolution approach has proven to be particularly valuable in the analysis of regulatory RNA motifs of prokaryotic viruses [43-46], eukaryotic viruses [32,47-50] and viroids [51]. We will describe genetic reversion experiments that were performed to increase our understanding of the potentially multi-functional TAR and polyA hairpin motifs. The possibilities of the forced evolution method and some of the putative problems that one may encounter will be discussed.

3. Results

3.1. FORCED EVOLUTION OF THE TAR HAIRPIN

We introduced a 14-nucleotide substitution in the lower left part of the TAR stem, resulting in partial opening of the TAR hairpin (Figure 2). This Xho+10 mutation was introduced in both the 5' and 3' LTRs of the full-length, infectious pLAI plasmid. The virus replication system does not allow one to individually study the contribution of the 5' or 3' TAR motifs to virus replication. Even if the mutation is introduced solely in the 5'R element, the mutant sequence is inherited in a dominant manner in both R regions [52]. The mutant virus was unable to establish a productive infection in the CD4-positive SupT1 cell line, but this replication-impaired HIV-1 mutant was used as starting material for long-term cultures to allow the generation of faster replicating revertant viruses. Small syncytia were first detected around day 45, and increased virus replication was measured over the next 5 months. Proviral DNA samples were PCR-amplified at several times and the LTR-TAR region was sequenced. The reversion pathway is shown in the upper part of Figure 2. As happens frequently upon disruption of an existing RNA structure, the lower part of TAR was able to fold an alternative structure (not shown), which was abolished by the first G51A mutation. The subsequent two mutations (A3U and G4U) allow refolding of a TAR-like stem with increasing stability. A single A8G mutation was selected at this point, which triggers the formation of four additional basepairs and closure of the large internal loop. Thus, starting with a destabilized TAR mutant (ΔG = -14.7 kcal/mole), a revertant was selected after 154 days of culture with a stability (ΔG = -21.9) that is similar to that of the wild-type TAR element (ΔG = -24.8). We demonstrated that these additional LTR-TAR changes did cause the reversion phenotype by subcloning of these revertant sequences in the original Xho+10 mutant, thereby restoring replication capacity [32].

Because reversion is a chance process, it was expected that a different repair pathway would be observed in an independent evolution experiment. Indeed, a completely different route towards restoration of the TAR function is described in the lower part of Figure 2. Initially, the virus seemed to stabilize the alternative folding of the Xho+10 mutant by the G52A change (not shown), but this structure did maintain the extreme 5' end of the HIV-1 transcript unpaired. Next, we observed a deletion of five nucleotides (Δ -8CTGTA-4) just upstream of position +1, which

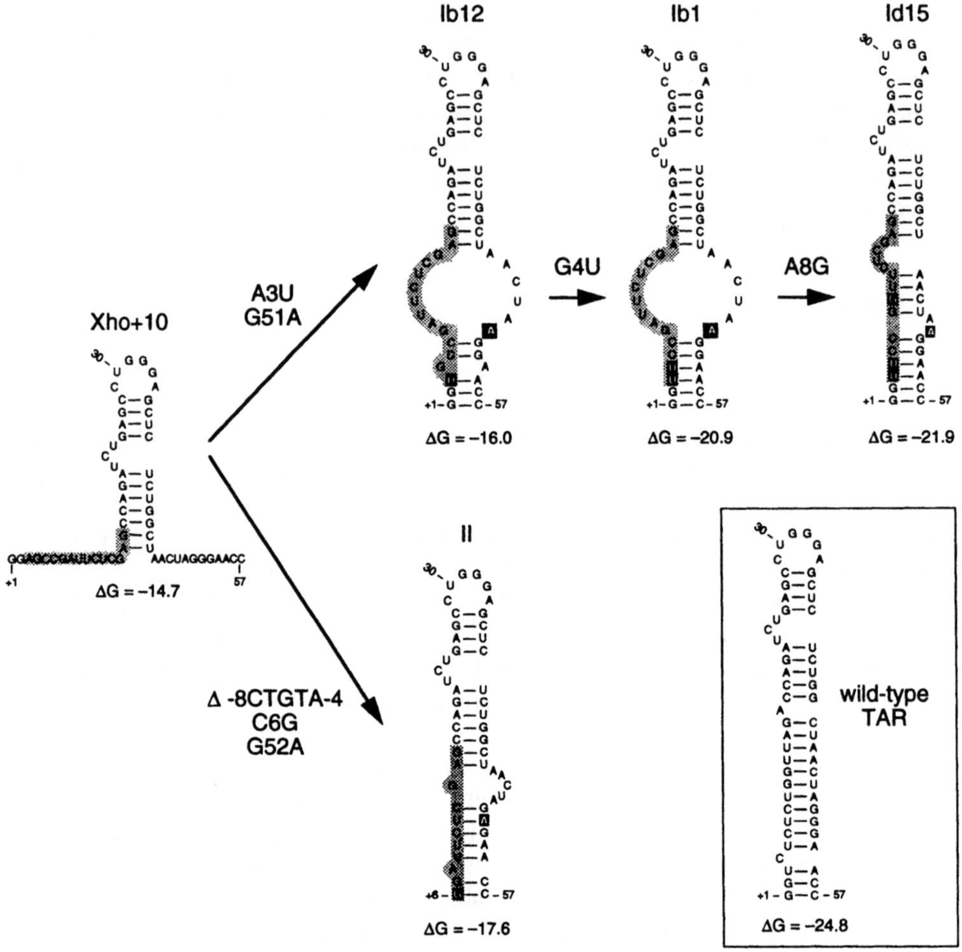

Figure 2. Repair of an opened TAR hairpin. The RNA secondary structure of the mutant Xho+10 is shown on the left, and the evolution in two independent cultures is indicated. The mutated region of TAR is shaded, and additional mutations in the revertants are marked by black boxes. The upper pathway uses 4 consecutive nucleotide substitutions to gradually restore basepairing in the lower TAR stem region. The initial G51A substitution was observed on day 32 of the evolution experiment, the A3U change was detected first at day 53. The subsequent G4U and A8G changes were seen at day 82 and 154, respectively. The bottom panel shows another TAR revertant observed after 79 days in an independent culture. This reversion pathway uses another strategy to close the TAR stem, see the text for details. The mutations include deletion of upstream promoter sequences (-8CTGTA-4) to shift the transcription initiation site to TAR position +6. The wild-type TAR structure is included for comparison. The free energy of the hairpins (including terminal stacks) was calculated with the Zuker algorithm [53]. All ΔG values are indicated in kilocalories per mole.

caused a shift in the transcription start site to a further downstream position. Diffuse start sites were demonstrated in the +6 to +11 region, but a more precise usage of the +6 start site was observed after mutation of this new transcription start site (C6G). Thus, the last two steps in this repair pathway did effectively remove the 5' dangling end, but did not restore the thermodynamic stability of TAR. At the same time, the unpaired nucleotides immediately downstream of TAR end up in the new stem, and the excessive nucleotides are put in a large internal loop. It is apparently essential for the virus to avoid unpaired nucleotides flanking the 5' TAR stem. In fact, elimination of the unpaired nucleotides may be as important as restoration of the TAR helix stability. It is possible that an unpaired 5' end on HIV-1 RNA triggers exonucleolytic attack or negatively affect virus replication in another way. The selective pressure to avoid unpaired nucleotides immediately downstream of TAR is currently unknown, but these nucleotides may interfere with coaxial stacking of the TAR and polyA hairpins [31].

3.2. FORCED EVOLUTION OF THE POLYA HAIRPIN

We constructed two HIV-1 mutants with either a stabilized or destabilized polyA hairpin (Figure 3). The polyA hairpin (ΔG = -15.3) was stabilized in mutant A by deletion of two bulging nucleotides on the right-hand side of the stem and by modification of one G-U basepair into G-C (new ΔG = -25.7). Mutant B contains four nucleotide substitutions on the left hand-side of the hairpin that were designed to open the central and lower stem segments (ΔG = -6.7). It was realized afterwards that these substitutions will induce a slight rearrangement in the basepairing scheme, resulting in the formation of a more stable hairpin (ΔG = -11.4). To minimize effects due to changes in the nucleotide sequence, the mutations did mimic natural sequence variation seen in other virus isolates. For instance, three of the substituted nucleotides in mutant B are also present in the sequence of the ANT-70 isolate [54].

The B mutant virus demonstrated a severe delay in replication, and the A mutant was replication-impaired. Fast-replicating revertants were obtained upon prolonged culturing of both mutants, indicating that mutant A represents a *quasi*-infectious virus. Second-site mutations became fixated in the polyA hairpin of the progeny. Mutant A with the stabilized hairpin (ΔG = -25.7) acquired additional mutations that destabilize the hairpin (Figure 3). At day 18, a C-G basepair in the top of the stem is replaced by a weaker U-G basepair in clone A18-2 (ΔG = -22.6). Subsequently, two alternative repair steps were observed at day 57. Clone A57-2 replaced a C-G basepair by a C A mismatch, thereby reducing the free energy to -16.0. Clone A57-1 used a similar strategy to attack one of the bottom basepairs, which results in a loss of the last three basepairs, thereby reducing the free energy to ΔG = -17.1. Only the latter sequence was recovered upon prolonged culturing, indicating a replication-advantage for this revertant.

Mutant B with a destabilized polyA stem acquired additional mutations that stabilized the hairpin (Figure 3). The stem was first stabilized by an A-to-G substitution in the central stem region (clone B57-2, ΔG = -17.1), and subsequently

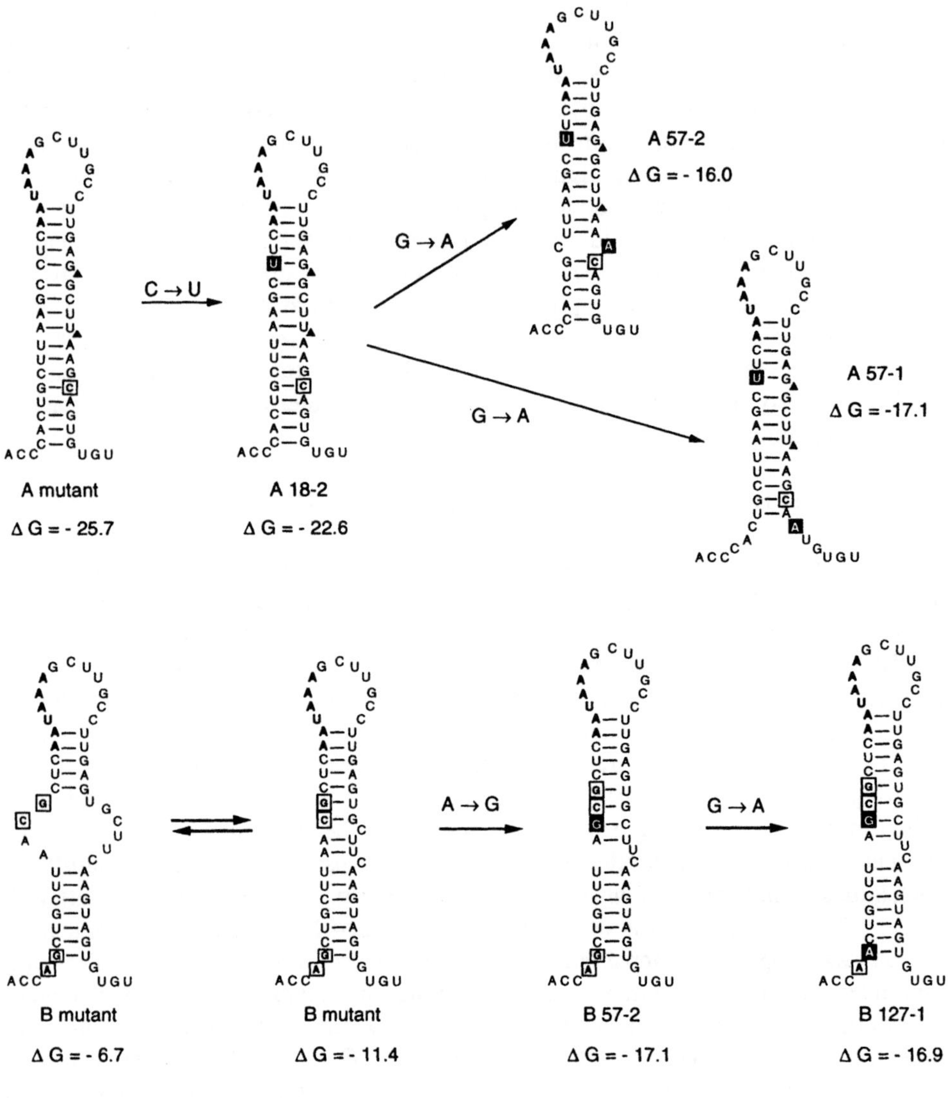

Figure 3. Repair of a stabilized/destabilized polyA hairpin. The evolution of the stabilized mutant A is shown on top, and that of the destabilized mutant B at the bottom. The substitutions present in the mutant hairpins are indicated in open boxes, and the position of two bulged nucleotides that were deleted in mutant A are indicated by ▲. Additional mutations that were acquired during the evolution are in black boxes. The free energies of the structures are indicated in kilocalories per mole. The C-to-U change in mutant A was observed at day 18, followed by two alternative G-to-A changes that were present in individual clones analyzed at day 57. At later timepoints up to day 200, the A57-1 revertant dominated the culture. The B mutant acquired two mutations, the A-to-G change was observed first at day 57, and the G-to-A change was present at day 127. The ΔG of the wild-type polyA hairpin is -15.3 kcal/mol.

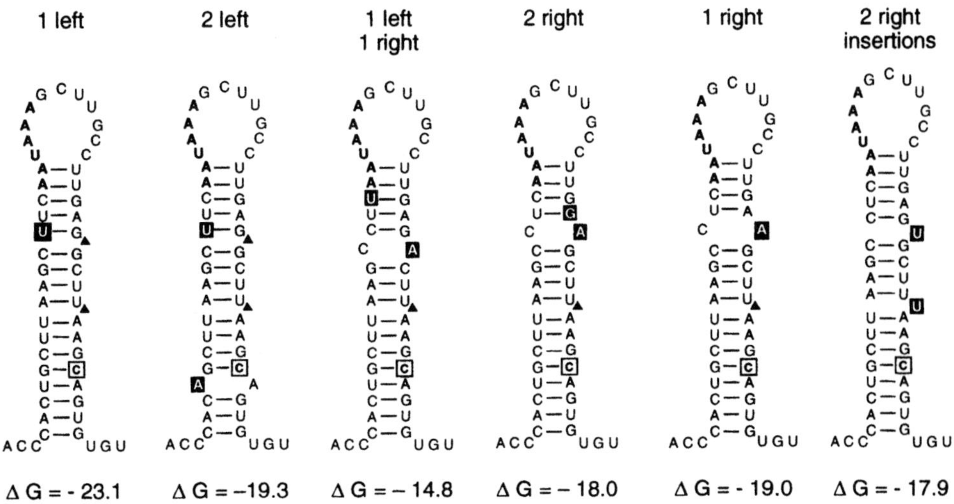

Figure 4. Several routes for repair of the stabilized polyA hairpin. The mutant A (see Figure 3) was allowed to regain replication capacity in 17 independent cultures. A wide variety of revertants were obtained, and a representative of each group is shown. All revertants acquired at least one additional nucleotide change within the hairpin (marked in black box). Indicated is the number and position (e.g. 1 left) of the acquired mutation(s). The predicted structure with the calculated helix stability are shown. A unique revertant acquired two nucleotide insertions at the right hand side of the stem (see the text for further details). A similar, but less extensive spectrum of revertants has been described previously for mutant B with the opened hairpin [55].

a G-to-A mutation was observed in the lower part of the stem (clone B127-1, ΔG = -16.9). The hairpin stability was only marginally affected by the latter mutation. In fact, this change may have been selected because of sequence requirements, as the wild-type sequence is restored. The role of the acquired mutations in the phenotypic reversion of the polyA mutants was verified by introduction of these sequence changes in the context of the original mutants [37].

To generate a large number of revertant viruses, we performed long-term cultures of mutants A and B in the format of a 24-well tissue culture plate. Two alternative strategies were observed for the repair of the destabilized B hairpin [55]. One group acquired one or multiple hits that led to rearrangement and optimization of the central stem domain. Another group choose to improve the bottom part of the helix, either by a mutation on the left or right hand side of the stem. At least 13 different escape routes were observed for the stabilized mutant A, and we grouped the different solutions by the number and position of the acquired mutations. A representative of each group is shown in Figure 4. It appears that there are many more mutational ways to disrupt an existing hairpin than to build a new hairpin structure, which may be a more general principle. Contrasting with this enormous genetic flexibility is the invariable drift towards RNA structures with a wild-type-like

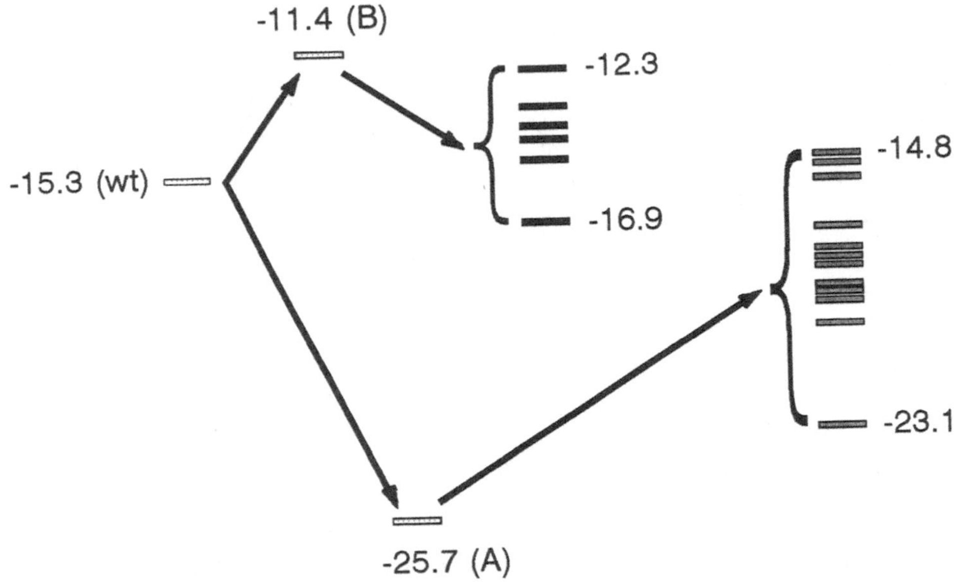

Figure 5. Restoration of the thermodynamic stability of the polyA hairpin. Plotted is the thermodynamic stability of the polyA hairpin in wild-type HIV-1, the A and B mutants and the revertants thereof. The free energy (ΔG in kcal/mole) was calculated according to the Zuker algorithm [53].

thermodynamic stability (Figure 5). The stable hairpin A was consistently remodelled into a hairpin with less basepairing potential (ΔG = -14.8/-23.1) and the stability of the opened mutant B was consistently increased (ΔG = -12.3/-16.9). It is likely that the time allowed for virus evolution was too short for some revertants to attain the most optimal configuration.

One revertant exhibits a rather unusual pathway towards viability in that two nucleotide insertions were observed (Figure 4). This is remarkable because insertions, as well as deletions, are not frequently observed in this type of experiment. Among the polyA revertants, this variant is the only variant with a base insertion, and it is even more striking that a second insertion was introduced in the same genome. It is rather amazing that both insertions coincide precisely with the position of the two bulged nucleotides that were deleted in mutant A. We can rule out contamination by the wild-type virus because the revertant contains two U-bulges, whereas the wild-type has one U- and one C-bulge. Furthermore, the G-U to G-C basepair conversion in the lower stem of mutant A was maintained in this peculiar revertant.

A similar event of precise, multiple-nucleotide reversions has been described previously for the poliovirus [56]. A temperature-sensitive poliovirus mutant reverted to the wild-type sequence with five base substitutions. Most surprisingly, this seemingly complex repair mechanism occurred at an extremely high frequency, and no intermediates were observed. These results suggest that the mutations were not introduced as sequential independent misincorporations, but rather in a genetically linked manner by e.g. an error-prone polymerase activity, an unusual RNA recombination event or a unique RNA editing process. These findings cannot be adequately explained at present, but they indicate that there may be an additional level of complexity in the field of virus genetics and evolution [56].

The combined results obtained in the analysis of polyA revertants convincingly demonstrate that a hairpin with a wild-type stability does optimally serve virus replication. Other features of the structure (position and sequence of bulges/internal loops, size of loop etc.) are not of primary importance. For instance, natural virus isolates seem to prefer a bulged nucleotide on the right hand side of the polyA stem [11], but the revertant analysis clearly indicates that this position is not critically important. We therefore think that these motifs are merely there to restrain the thermodynamic stability of the helix, which otherwise would exceed the allowable margin. A role of the thermodynamically well-balanced polyA hairpin in the process of regulated polyadenylation will be discussed in section 4.1.

3.3. THE ROLE OF THE TAR HAIRPIN IN VIRUS REPLICATION

To assess the role of the TAR hairpin in virus replication, we constructed proviral clones with the Xho+10 mutation in the 5'R or 3'R region. As a control for restoration of the function, we included one of the revertant hairpins. Upon transfection of these constructs into cells, we measured the level of intracellular HIV-1 RNA and the level of viral protein production. These values were used to calculate the translational efficiency. Virions produced by the transfected cells were analyzed for the RNA content, and T cells were infected with these virions to measure the level of reverse transcription [57]. The results obtained for the 5' TAR mutant are schematically depicted in Figure 6.

There has been some controversy concerning the transcriptional activity of truncated TAR motifs. Initial transfection studies with LTR-CAT constructs performed in COS cells indicated that a truncated TAR is fully active in transcription. However, subsequent studies in a variety of other cell types, including the T cell lines used for HIV-1 replication studies, demonstrated a significant reduction of viral gene expression [58]. This was confirmed for the Xho+10 mutant, which produced 58% RNA compared with the wild-type control (Figure 6). Since the protein production did not decrease, the mutant HIV-1 mRNA with a destabilized 5' TAR structure is apparently translated more efficiently than the wild-type transcript (Figure 6). A reduced level of packaged RNA was measured for the 5' TAR mutant (Figure 6). However, part of this reduction is directly due to the reduced level of intracellular HIV-1 RNA, suggesting that the process of RNA packaging is exquisitely sensitive to changes in the concentration of intracellular RNA. Therefore, the ratio of virion RNA to intracellular HIV-1 RNA may be a better measure of the packaging efficiency than the ratio of virion RNA to virion protein. After correction for the reduced amount of intracellular RNA, a small

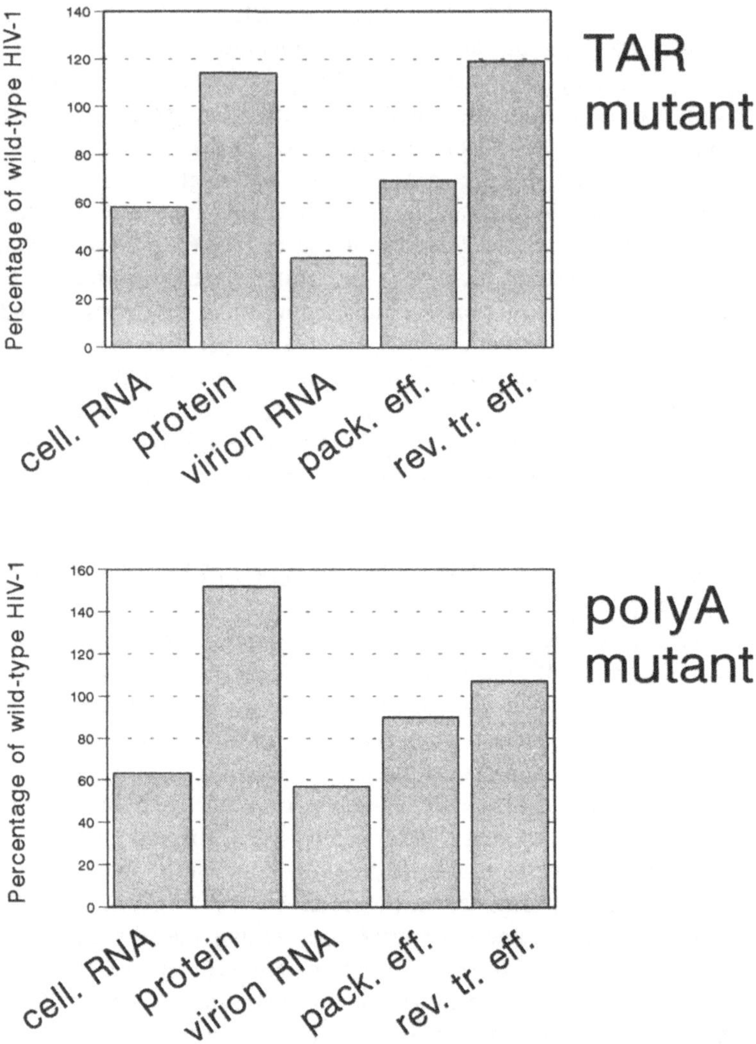

Figure 6. The R-region hairpins function at several points in the virus life cycle. An overview is presented of the effects of opening of the TAR hairpin (top panel) and polyA hairpin (bottom panel). Plotted are the amounts of intracellular HIV-1 RNA, HIV-1 protein and virion RNA produced upon transfection of cells. In addition, we calculated the packaging efficiency as the ratio of virion RNA to intracellular HIV-1 RNA, see the text for further details. Reverse transcription was measured by infecting SupT1 cells with the wild-type and mutant viruses and analysis of the reverse transcription (cDNA) products. We calculated the reverse transcription efficiency as the ratio of cDNA products to virion RNA template. All values are related to the wild-type pLAI virus, of which all parameters were set at 100%. The TAR mutant used in the assay was the 5' Xho+10 mutant, but reverse transcription was analyzed for the 5'+3' double mutant because identical R regions are required for the first strand transfer. The polyA results are the average values measured for three mutants with an opened polyA hairpin structure (mutants B, C and D in [37]).

packaging defect remained for this TAR mutant. The contribution of the 5' TAR motif to packaging is consistent with a previous study [34], and we measured a quantitatively similar contribution for the 3' TAR element [57]. Upon infection of T cells with these virions, less reverse transcription products were measured for the 5' TAR-mutated virus, but this reduction correlated precisely with the reduced level of RNA template within these virions. Thus, no net reverse transcription defect was apparent for the 5' TAR mutant (Figure 6).

3.4 THE ROLE OF THE POLYA HAIRPIN IN VIRUS REPLICATION

Proviral constructs with mutant and revertant polyA hairpins in 5'R or 3'R were tested for the expression of viral RNA and proteins in transfected cells as described for the TAR mutants in the previous section. We found that the polyA hairpin is required for efficient repression of the polyadenylation site in the 5'R [21]. Opening of the 5' hairpin triggered premature polyadenylation and resulted in the production of a short 5'R transcript [21]. Consequently, the amount of full-length viral transcripts was reduced by 40% (Figure 6). Apparently, the repressive potential of the wild-type polyA hairpin is overcome at the 3' polyA site, which is probably due to the upstream USE enhancer. However, when the stabilized mutant hairpin A was inserted in the 3'R, we found that 3' polyadenylation was completely blocked. Thus, the stability of the polyA hairpin is fine-tuned to allow efficient repression in the 5'R, yet full activation in the 3'R.

Although destabilization of the 5' polyA hairpin reduced the level of cellular HIV-1 RNA, no loss of viral protein production was measured (Figure 6). This result is very similar to that obtained for constructs with an opened 5' TAR hairpin, and may indicate that both structures negatively affect the process of mRNA translation (see section 4.1). We and others have reported previously that the 5' polyA hairpin structure contributes to packaging of the viral RNA into virion particles [35,36]. Although the packaged RNA level was indeed reduced (Figure 6), our current analysis suggests that this reduction is a direct consequence of the reduced level of intracellular HIV-1 RNA. Upon infection of T cells with these virions, reduced cDNA synthesis was measured (Figure 6), but this defect correlates with the reduced amount of template RNA genome in the mutant virions. Thus, mutation of the polyA hairpin does not significantly affect the processes of RNA encapsidation and reverse transcription.

4. Discussion

4.1. THE TAR AND POLYA HAIRPIN MOTIFS CONTROL VIRAL GENE EXPRESSION AT SEVERAL LEVELS

The combined results indicate that the TAR and polyA hairpins are pivotal in HIV-1 replication. The motifs observed in natural virus isolates will reflect the superimposed demands of multiple essential processes, which are difficult to separate experimentally. A standard mutational analysis combined with simplified assay systems will not test all these functions, and may thus not suffice to obtain a complete understanding of such complicated, multi-functional RNA motifs. The method of forced viral evolution is particularly suited because it studies these signals in the natural context of virus replication, an assay that will test all functions. As demonstrated in this study, the revertant viruses also provide useful reagents for subsequent mechanistic studies. Our current view of the most important functions of the R region hairpins is illustrated in Figure 7. We will not review the mechanism of Tat/TAR-mediated transcriptional activation here.

4.1.1. *Polyadenylation*

Retroviruses with an extended R region encode the polyadenylation (polyA) signal within the R region, such that it is present at both the 5' and 3' end of the viral transcript. This necessitates differential regulation either to repress recognition of the 5' polyA signal or to enhance usage of the 3' signal. HIV-1 has been reported to have both regulatory features (illustrated in Figure 7). Usage of the 3' polyA site is promoted by an upstream enhancer motif that is uniquely present at the 3' end of viral transcripts [6,13,18-20]. This upstream sequence element (USE) appears to stabilize binding of the cleavage polyadenylation specificity factor (CPSF) to the AAUAAA hexamer motif [7]. Repression of the 5' polyA site is thought to be mediated by several mechanisms. First, since the 5' polyA site becomes active when moved further downstream in the transcript, this polyA site may be inefficient because it is positioned too close to the transcription initiation site [14,17]. It is possible that the transcription complex engaged in synthesis of the HIV-1 leader transcript is not yet competent for polyadenylation, as it was recently shown that polyadenylation factors gain access to the nascent transcript through the RNA polymerase II complex [59]. A second repression mechanism proposes that the 5' polyA site is negatively influenced by the major splice donor signal (SD) that is uniquely present downstream of the 5' polyA site [8,16]. Mutational inactivation of the SD site in full-length HIV-1 transcripts triggered usage of the 5' polyA site [16]. This repression is mediated by binding of the U1 snRNP to the splice donor site, but it is currently unknown how this affects 5' polyA site usage [8]. We propose that the polyA hairpin structure is instrumental in this regulatory circuit.

Destabilization of the 5' polyA hairpin triggered premature polyadenylation, indicating that the wild-type structure is involved in occlusion of the 5' polyA site. In the 3' context with the upstream USE enhancer, the wild-type polyA hairpin does not interfere with efficient polyadenylation. However, 3' polyadenylation can be inhibited by further stabilization of the hairpin. Because the AAUAAA hexamer is partially occluded by basepairing, it is possible that binding of CPSF to this signal is blocked by the polyA hairpin conformation. Preliminary evidence from *in vitro* RNA-protein binding assays indicates that this initial step of the polyadenylation reaction is indeed blocked by stable RNA structure (Klasens and Berkhout, manuscript in preparation). These results suggest that the role of the polyA hairpin is to create a regulatable polyA site, which can be either repressed in the presence of

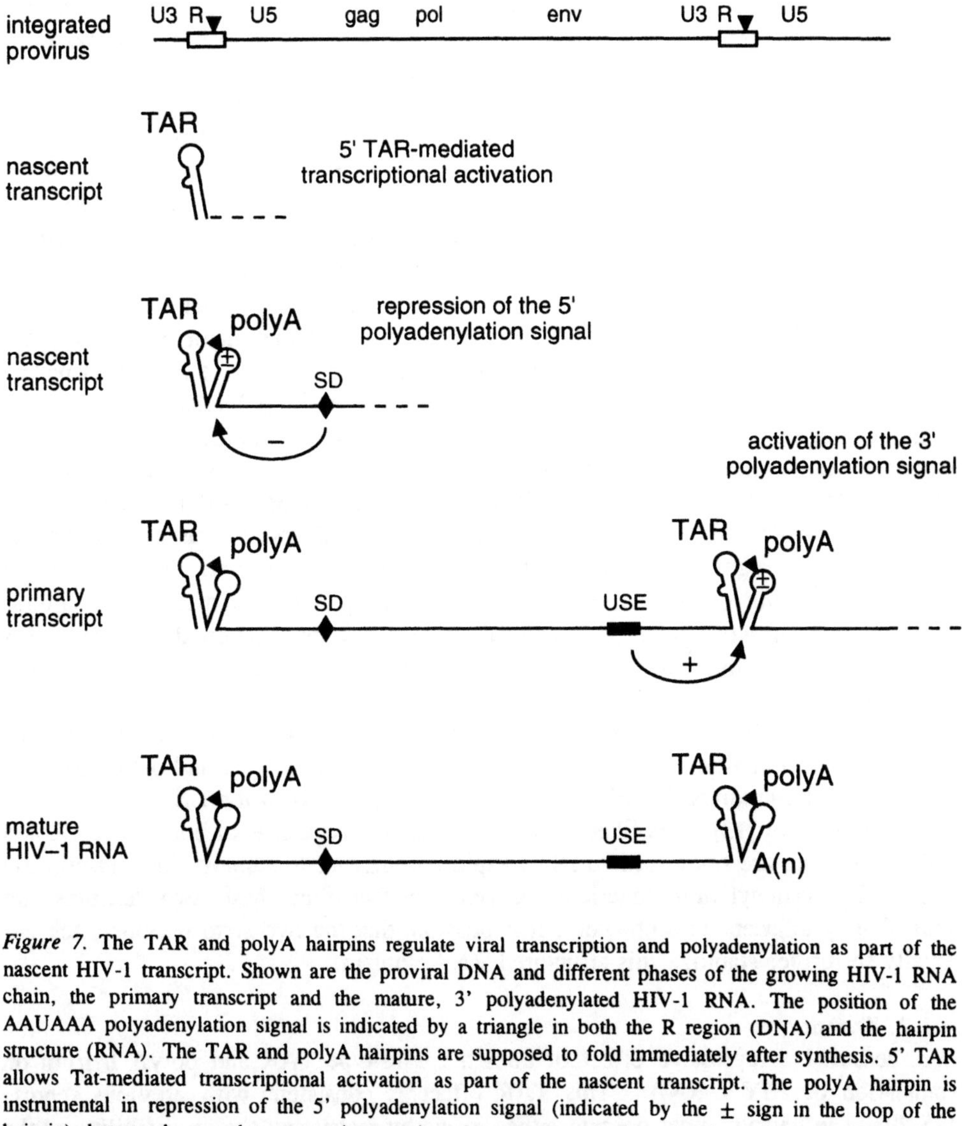

Figure 7. The TAR and polyA hairpins regulate viral transcription and polyadenylation as part of the nascent HIV-1 transcript. Shown are the proviral DNA and different phases of the growing HIV-1 RNA chain, the primary transcript and the mature, 3' polyadenylated HIV-1 RNA. The position of the AAUAAA polyadenylation signal is indicated by a triangle in both the R region (DNA) and the hairpin structure (RNA). The TAR and polyA hairpins are supposed to fold immediately after synthesis. 5' TAR allows Tat-mediated transcriptional activation as part of the nascent transcript. The polyA hairpin is instrumental in repression of the 5' polyadenylation signal (indicated by the ± sign in the loop of the hairpin), but nearly complete repression requires the presence of other signals, such as the downstream splice donor (SD). The polyA hairpin also puts the 3' polyadenylation site in an unfavourable context, but this is overcome by the presence of the USE upstream enhancer.

silencers (5' situation) or activated in the presence of an enhancer (3' situation). It seems that the thermodynamic stability of the polyA hairpin is fine-tuned to allow this on-off switching of polyadenylation. The analysis of revertants that evolved from poorly replicating virus mutants with a stabilized or destabilized hairpin demonstrated a definite drift to a thermodynamic stability comparable with that of the wild-type structure (Figure 5).

The HIV-1 polyA signal has been demonstrated to represent an inherently efficient polyadenylation sequence in reporter constructs in which the polyA site is taken out of its natural context. In the HIV-1 RNA, the partially repressive effect of the polyA hairpin structure allows nearly complete inhibition of 5' polyadenylation through the SD mechanism and promoter-proximity, and activation of 3' polyadenylation through the USE mechanism. Thus, the idea that occlusion of the HIV-1 polyA site by RNA structure plays a critical role in differential polyadenylation does not replace the existing models, but rather provides a mechanistic explanation for 5' down-regulation and 3' up-regulation. Conceivably, rapid folding of the polyA hairpin structure on the nascent viral transcript will delay the recognition by polyadenylation factors, such that sufficient time is available for recognition of the major SD by the U1 snRNP [8]. As the RNA commits itself to the splicing reaction, e.g. is transported into spliceosomes, it will become even less accessible to the polyadenylation factors. Thus, a complex interplay of polyadenylation and splicing signals, repressive RNA structure, and either enhancer or silencer motifs are involved in regulated HIV-1 polyadenylation. The polyA hairpin is well-conserved among different human and simian immunodeficiency viruses [11], suggesting that all these viruses use the same mechanism to regulate polyadenylation. This mechanism may be even more widespread among other lentiviruses and also the spumaretroviruses, as a similar RNA hairpin structure can be predicted for the bovine immunodeficiency virus (BIV), the equine infectious anemia virus (EIAV), and the human spumaretrovirus (HSRV) [21].

Interestingly, the TAR hairpin structure is also thought to perform its transcriptional function as part of the nascent viral transcript [1], and TAR has been suggested to be involved in polyadenylation as spacer between the USE enhancer and the 3' polyA site [13]. It is likely that the nucleotide sequence of both the TAR and polyA elements are optimized to allow rapid basepairing to support the transcription and anti-polyadenylation functions. Given the fact that these two hairpins are immediately adjacent to each other, it is possible that the two stem regions stack co-axially to further stabilize this structured RNA domain.

4.1.2. *Translation*

We measured a repressive effect of both the wild-type TAR and polyA hairpin on translation of HIV-1 RNAs. This TAR effect is consistent with previous reports [25-29]. The hairpins may operate either in *cis* by restricting the accessibility of the mRNA cap-structure to the translational machinery or in *trans* through induction of the PKR protein kinase system (see section 2.1). It is possible that translational inhibition is not executed by the individual hairpins, but rather by an extended structure that can be formed by coaxial stacking of these juxtaposed hairpins [31]. This may explain why translation is boosted in both the destabilized TAR and polyA hairpin mutants.

Sub-optimal translation as measured for the wild-type HIV-1 RNA may be part of a viral strategy to balance the processes of translation and packaging. It may provide the optimal amounts of genomic RNA and viral proteins, and therefore ultimately control the production of infectious virus. Previous findings with murine leukemia virus suggested that full-length viral RNA is routed to either a pool for translation or a pool for packaging, and that the RNA bound by ribosomes could not be packaged [60]. For the Rous sarcoma virus, it has been reported recently that this sorting mechanism is mediated by the viral Gag proteins [61]. Analysis of the intracellular distribution of the wild-type and mutant HIV-1 RNAs may be one way to further study this phenomenon.

4.1.3 *RNA packaging*

Besides the well-known transcriptional function of the 5' TAR hairpin, this motif was found to partially repress translation and to stimulate packaging of the viral RNA genome. The 3' TAR motif also contributed to encapsidation of the RNA in virions, and this effect was quantitatively similar to that of the 5' TAR element [57]. Combined with the results of other studies [36,62-65], it may be appropriate to consider the entire untranslated leader RNA as the packaging signal [31]. It is likely that the complete leader region is required to fold a specific tertiary RNA structure, which may form the actual packaging signal. It is also possible that the TAR motif affects RNA dimerization [33,66], which in turn may affect the process of RNA packaging [67]. Less reverse transcription products were measured with these viruses upon infection of cells, but this reduction correlates with the reduced level of RNA template within these virions. Although all reported TAR effects seem to depend on the basepaired structure of this RNA signal, it cannot be excluded that important nucleotide sequences within TAR are also critical for some of these functions. However, the finding that the TAR functions can be restored by the additional nucleotide changes observed in a revertant TAR element with a repaired stem, suggests that the actual sequence of the lower TAR stem is less critical for these functions.

4.2. THE METHOD OF FORCED EVOLUTION

4.2.1. *Other applications of forced evolution*

We and others demonstrated that the method of forced evolution is ideally suited to study RNA elements that play critical roles in the replication of HIV-1 [32,35,50,55,66,68-71]. The approach works for most RNA viruses because of their tremendous genetic variation, which results from high frequencies of nucleotide misincorporation during replication of the viral genome. For retroviruses, mutations are introduced during transcription by the cellular RNA polymerase or reverse transcription by the viral RT enzyme, as both polymerases lack an error-correcting mechanism. The method of forced evolution is particularly powerful to dissect the function of structured RNA motifs. The enormous genetic flexibility of structured RNA motifs is due to the fact that completely different sequences can form very similar basepaired structures. The forced evolution approach is also applicable for functional studies on viral DNA motifs, e.g. HIV-1 LTR promoter elements [72], and there is evidence that at least some of the HIV-1 proteins are amenable to second-site repair [73,74]. However, the genetic flexibility of proteins may be much more restricted than that of RNA motifs. For instance, we recently finished a large scale analysis of revertants of HIV-1 variants with an inactivated Tat protein due to a single pointmutation (Verhoef and Berkhout, manuscript submitted). Only one second-site Tat revertant was identified, indicating that there may be very few amino acid changes at secondary sites that can render such Tat mutants active. In addition, the presence of overlapping reading frames and underlying RNA signals (for instance signals to regulate splicing) may put additional constraints on the evolutionary flexibility of some of the HIV-1 genes. Thus, the evolutionary approach may not be an efficient method to study structure-function relationships in small proteins with a high information density. It is theoretically possible that the activity of some viral protein mutants can be restored by amino acid changes at multiple secondary sites, but the probability of finding such hypermutated variants will be remote (see also section 4.2.3).

4.2.2. *Pitfalls of forced evolution*

In the HIV-1 evolution system, genetic variation is generated either during transcription of the integrated provirus or during reverse transcription of the viral genome. Although the diversity will be random in nature, the population is shaped by selection of replication-competent viruses. Unlike current SELEX evolution experiments [75] that are started with a pool of randomized sequences, our virus evolution protocol will strongly favour the advantageous mutations that happen to occur relatively early after transfection. Novel mutations augment existing variation, so that the evolutionary search is biased by selection events that have already occurred. Thus, a virus selected in the evolution system may represent a sub-optimal solution and is likely to be one of many potential revertants. Indeed, different reversion routes were observed in independent experiments with the TAR and polyA mutants. The analysis of multiple repair routes can increase the understanding of the RNA motif under scrutiny.

To study the contribution of a structured RNA motif in virus replication, the obvious mutant to make is one in which the basepairing interaction is disrupted. At the same time, this turns out to be a rather dangerous approach, as the mutant sequence can participate in new basepairing interactions. In fact, we experienced this phenomenon in both the TAR and polyA evolution experiments (see section 3), and some other examples have been described in literature [76,77]. In case such a refolded RNA structure is inhibitory to virus replication, the repair pathway may include mutations that are meant to destroy the unwanted structure. It is obvious that such effects can considerably complicate the interpretation of RNA evolution studies [76].

In contrast to RNA mutants with a disrupted RNA structure, mutants that further stabilize existing structures are unlikely to induce structural rearrangements. However, such mutants also have a clear disadvantage, as the stabilized RNA structure may cause replication problems that are unrelated to the function of the wild-type RNA motif. For instance, an RNA stem in the untranslated leader region that is too stable can theoretically interfere with at least two steps of viral replication. Such a stable RNA structure may block the movement of scanning ribosomes over the untranslated leader [78,79], and the elongating Reverse Transcriptase (RT) enzyme during reverse transcription [52,80-82]. We analyzed the *in vitro* elongation properties of the HIV-1 RT enzyme on the wild-type RNA template and mutants thereof with a stabilized or destabilized polyA hairpin. It was found that stable RNA structure can interfere with elongation of reverse transcription (Klasens and Berkhout, submitted for publication). However, addition of the viral nucleocapsid protein (NC) to the *in vitro* assay was found to overcome such structure-induced RT stops, and no reverse transcription defect was apparent for this mutant template in virus-infected cells. NC protein has been reported to induce conformational changes in nucleic acids through altering energy barriers of duplex melting and annealing [83-85], and the observed resolution of pause sites is consistent with the idea that NC causes RNA structures to unfold more readily. These combined results suggest that retroviruses can use relatively stable RNA structures to control different steps in the viral life cycle without interfering with the process of reverse transcription. Nevertheless, revertant analysis of viruses with a stabilized RNA structure may not always provide information on the actual function of the wild-type RNA signal.

Putative second-site suppressor mutations should always be tested by subcloning in the original mutant genome, which should provide the revertant phenotype. In this way, spontaneous sequence variation that does not contribute to improved replication will be recognized. Although most of this spontaneous sequence variation is filtered out by focusing solely on the mutations that become fixated in the viral progeny, this by itself is not an absolute guarantee that the changes did cause the reversion event. For instance, fixation of new sequences is possible by non-random sampling effects during passage of the crippled virus mutant (bottle neck passage or founder effect). Alternatively, these fixated mutations may represent 'bystander mutations' that were linked to another mutation that did improve replication and therefore was the actual target for selection.

Forced evolution can provide genetic support for long-distance RNA-RNA interactions or functional interactions between the RNA and viral proteins. For instance, analysis of an HIV-1 mutant with a defect in RNA dimerization resulted in the selection of a revertant with a compensatory change in the gene encoding the viral nucleocapsid protein [86]. However, such 'second-locus' revertants may not be seen as frequently in other systems. Despite the analysis of many revertants of a Tat-mutated virus, we never observed compensatory changes in the TAR RNA hairpin motif, which is the binding site for Tat protein. *Vice versa*, TAR-mutated viruses did never yield revertants with an altered Tat protein.

It is important to realize that the replication of a defective virus mutant can be improved in a non-specific manner by mutation of an unrelated viral function. For instance, we reported recently that a translation-impaired HIV-1 mutant can dramatically improve its replication by optimizing the mechanistically unrelated Env function [87]. Another example was obtained in long-term cultures of the delta-3

HIV-1 construct with deletions in the *vpr* and *nef* genes and part of the U3 domain. This potential live-attenuated vaccine strain was able to regain replication capacity by duplication of the three Sp1 binding sites in the remaining U3 part of the LTR promoter (Berkhout *et al.*, unpublished results). Thus, to rule out the effect of genotypic changes elsewhere in the viral genome, it is essential that the effect of any mutation in a revertant virus is verified by recloning into the original virus mutant, but also into other virus mutants with unrelated defects. By doing so, it can be determined whether the reversion acts specifically to restore the replication of the original mutant, or whether replication is boosted in a more general manner.

4.2.3. *In vitro and in vivo evolution in total sequence space*

Studies on the *in vitro* evolution of virus mutants should ideally be complemented by *in vivo* studies on virus variation. This approach is particularly fertile for HIV-1 given the wealth of sequence information available on different virus isolates, the different HIV-1 subtypes, HIV-2 and the related simian immunodeficiency viruses [54]. We reported such a detailed analysis for the TAR hairpin [88], and comparison of different virus isolates provided the first evidence for the existence of the polyA hairpin structure [11]. Similar to the results obtained in *in vitro* evolution, this phylogenetic survey indicated the importance of a polyA hairpin of intermediate thermodynamic stability. The combined results of forced *in vitro* evolution and the natural phylogeny show that different RNA structures are selected by evolution to facilitate a particular function and that structural mimicry exists in the RNA world. Most importantly, such a comparative sequence analysis can provide important information on the critical sequence and structure elements within the RNA element under scrutiny.

Even the natural diversity in HIV-SIV viruses is likely to represent only a very limited section of total sequence space. Whereas one could argue that the area occupied by modern HIV-1 motifs represents the highest peak, reached after a careful evolutionary walk through all of sequence space, this seems unlikely because the genetic repertoire of natural HIV-SIV isolates is constrained by the success of a predecessor of the contemporary viruses [89]. In other words, the existing RNA motifs may well lie at local, not global, optima of sequence space. As described in this chapter, a powerful virological method to locate other sequence optima is the selection of revertant viruses in long-term infections with a replication-defective mutant, but this method is also restricted in its walk through sequence space (see section 4.2.2.). Thus, the virus evolution experiment may benefit from strategies to introduce random mutations within the function under scrutiny [90,91] or from the introduction of a genome segment with a randomized nucleotide sequence [92]. In both cases, the search for second-site revertants will occur in a much broader section of sequence space. A disadvantage of these hypermutation approaches is that a significant fraction of manipulated viral genomes with a beneficial mutation will contain additional mutations that interfere with virus replication [92].

5. Acknowledgements

We thank Bep Klaver, Bianca Klasens, Koen Verhoef and Jeroen van Wamel for technical assistance and Wim van Est for photography work. This work was supported by the Dutch Organization for the Advancement of Pure Research (NWO), the Dutch AIDS Fund (AIDS Fonds, Amsterdam) and the European Union (EC 950675).

6. References

1. Dingwall, C., I. Ernberg, M. J. Gait, S. M. Green, S. Heaphy, J. Karn, A. D. Lowe, M. Singh, M. A. Skinner, and R. Valerio. (1989) Human Immunodeficiency Virus 1 tat protein binds trans-activating-responsive region (TAR) RNA in vitro. *Proc. Natl. Acad. Sci. USA* **86**, 6925-6929.

2. Berkhout, B., R. H. Silverman, and K. T. Jeang. (1989) Tat trans-activates the human immunodeficiency virus through a nascent RNA target. *Cell* **59**, 273-282.

3. Jeang, K. T., R. Chun, N. H. Lin, A. Gatignol, C. G. Glabe, and H. Fan. (1993) In vitro and in vivo binding of human immunodeficiency virus type 1 Tat protein and Sp1 transcription factor. *J. Virol.* **67**, 6224-6233.

4. Wu-Baer, F., D. Sigman, and R. B. Gaynor. (1995) Specific binding of RNA polymerase II to the human immunodeficiency virus trans-activating region RNA is regulated by cellular cofactors and Tat. *Proc. Natl. Acad. Sci. USA* **92**, 7153-7157.

5. Wu-Baer, F., W. S. Lane, and R. B. Gaynor. (1996) Identification of a group of cellular cofactors that stimulate the binding of RNA polymerase II and TRP-185 to human immunodeficiency virus 1 TAR RNA. *J. Biol. Chem.* **271**, 4201-4208.

6. Brown, P. H., L. S. Tiley, and B. R. Cullen. (1991) Efficient polyadenylation within the human immunodeficiency virus type 1 long terminal repeat requires flanking U3-specific sequences. *J. Virol.* **65**, 3340-3343.

7. Gilmartin, G. M., E. S. Fleming, J. Oetjen, and B. R. Graveley. (1995) CPSF recognition of an HIV-1 mRNA 3'-processing enhancer: multiple sequence contacts involved in poly(A) site definition. *Genes Dev.* **9**, 72-83.

8. Ashe, M. P., L. H. Pearson, and N. J. Proudfoot. (1997) The HIV-1 5' LTR poly(A) site is inactivated by U1 snRNP interaction with the downstream major splice donor site. *EMBO J.* **16**, 5752-5763.

9. Muesing, M. A., D. H. Smith, and D. J. Capon. (1987) Regulation of mRNA accumulation by a human immunodeficiency virus trans-activator protein. *Cell* **48**, 691-701.

10. Berkhout, B. and I. Schoneveld. (1993) Secondary structure of the HIV-2 leader RNA comprising the tRNA-primer binding site. *Nucleic Acids Res.* **21**, 1171-1178.

11. Berkhout, B., B. Klaver, and A. T. Das. (1995) A conserved hairpin structure predicted for the poly(A) signal of human and simian immunodeficiency viruses. *Virol.* **207**, 276-281.

12. Klaver, B. and B. Berkhout. (1994) Comparison of 5' and 3' long terminal repeat promoter function in human immunodeficiency virus. *J. Virol.* **68**, 3830-3840.

13. Gilmartin, G. M., E. S. Fleming, and J. Oetjen. (1992) Activation of HIV-1 pre-mRNA 3' processing in vitro requires both an upstream element and TAR. *EMBO J.* **11**, 4419-4428.

14. Weichs an der Glon, C., J. Monks, and N. J. Proudfoot. (1991) Occlusion of the HIV poly(A) site. *Genes Dev.* **5**, 244-253.

15. Weichs an der Glon, C., M. Ashe, J. Eggermont, and N. J. Proudfoot. (1993) Tat-dependent occlusion of the HIV poly(A) site. *EMBO J.* **12**, 2119-2128.

16. Ashe, M. P., P. Griffin, W. James, and N. J. Proudfoot. (1995) Poly(A) site selection in the HIV-1 provirus: inhibition of promoter-proximal polyadenylation by the downstream major splice donor site. *Genes Dev.* **9**, 3008-3025.

17. Cherrington, J. and D. Ganem. (1992) Regulation of polyadenylation in human immunodeficiency virus (HIV): contributions of promoter proximity and upstream sequences. *EMBO J.* **11**, 1513-1524.

18. DeZazzo, J. D., J. E. Kilpatrick, and M. J. Imperiale. (1991) Involvement of long terminal repeat U3 sequences overlapping the transcription control region in human immunodeficiency virus type 1 mRNA 3' end formation. *Mol. Cell Biol.* **11**, 1624-1630.

19. Valsamakis, A., S. Zeichner, S. Carswell, and J. C. Alwine. (1991) The human immunodeficiency virus type 1 polyadenylylation signal: a 3' long terminal repeat element upstream of the AAUAAA necessary for efficient polyadenylylation. *Proc. Natl. Acad. Sci. USA* **88**, 2108-2112.

20. Valsamakis, A., N. Schek, and J. C. Alwine. (1992) Elements upstream of the AAUAAA within the human immunodeficiency virus polyadenylation signal are required for efficient polyadenylation in vitro. *Mol. Cell Biol.* **12**, 3699-3705.

21. Das, A. T., B. Klaver, and B. Berkhout. A hairpin structure in the R region of the HIV-1 RNA genome is instrumental in polyadenylation site selection. *J. Virol.* in press.

22. Cullen, B. R. (1986) Trans-activation of human immunodeficiency virus occurs via a bimodal mechanism. *Cell* **46**, 973-982.

23. Braddock, M., A. M. Thorburn, A. Chambers, G. D. Elliot, G. J. Anderson, A. J. Kingsman, and S. M. Kingsman. (1990) A nuclear translational block imposed by the HIV-1 U3 region is relieved by the Tat-TAR interaction. *Cell* **62**, 1123-1133.

24. Braddock, M., R. Powell, A. D. Blanchard, A. J. Kingsman, and S. M. Kingsman. (1993) HIV-1 TAR RNA-binding proteins control TAT activation of translation in Xenopus oocytes. *FASEB J.* **7**, 214-222.

25. Parkin, N. T., E. A. Cohen, A. Darveau, C. Rosen, W. Haseltine, and N. Sonenberg. (1988) Mutational analysis of the 5' noncoding region of human immunodeficiency virus type 1: effects of secondary structure. *EMBO J.* **7**, 2831-2837.

26. SenGupta, D. N., B. Berkhout, A. Gatignol, A. M. Zhou, and R. H. Silverman. (1990) Direct evidence for translational regulation by leader RNA and Tat protein of human immunodeficiency virus type 1. *Proc. Natl. Acad. Sci. USA* **87**, 7492-7496.

27. Viglianti, G. A., E. C. Rubinstein, and K. L. Graves. (1992) Role of the TAR RNA splicing in translational regulation of simian immunodeficiency virus from rhesus macaques. *J. Virol.* **66**, 4824-4833.

28. Edery, I. R., R. Petryshyn, and N. Sonenberg. (1989) Activation of double-stranded RNA dependent kinase (dsI) by the TAR region of HIV-1 mRNA: a novel translational control mechanism. *Cell* **56**, 303-312.

29. Roy, S., M. Agy, A. G. Hovanessian, N. Sonenberg, and M. G. Katze. (1991) The integrity of the stem structure of human immunodeficiency virus type 1 Tat-responsive sequence of RNA is required for interaction with the interferon-induced 68,000-Mr protein kinase. *J. Virol.* **65**, 632-640.

30. Allain, B., J.-B. Rascle, H. De Rocquigny, B. Roques, and J.-L. Darlix. (1998) cis elements and trans-acting factors required for minus-strand DNA transfer during reverse transcription of the genomic RNA of murine leukemia virus. *J. Virol.* **72**, 225-235.

31. Berkhout, B. (1996) Structure and function of the Human Immunodeficiency Virus leader RNA. *Progr. Nucl. Acid. Res. Mol. Biol.* **54**, 1-34.

32. Klaver, B. and B. Berkhout. (1994) Evolution of a disrupted TAR RNA hairpin structure in the HIV-1 virus. *EMBO J.* **13**, 2650-2659.

33. Rounseville, M. P., H. C. Lin, E. Agbottah, R. R. Shukla, A. B. Rabson, and A. Kumar. (1996) Inhibition of HIV-1 replication in viral mutants with altered TAR RNA stem structures. *Virol.* **216**, 411-417.

34. Berkhout, B., B. B. Oude Essink, and I. Schoneveld. (1993) In vitro dimerization of HIV-2 leader RNA in the absence of PuGGAPuA motifs. *FASEB J.* **7**, 181-187.

35. Hoglund, S., A. Ohagen, J. Goncalves, A. T. Panganiban, and D. Gabuzda. (1997) Ultrastructure of HIV-1 genomic RNA. *Virol.* **233**, 271-279.

36. McBride, M. S., M. D. Schwartz, and A. T. Panganiban. (1997) Efficient encapsidation of human immunodeficiency virus type 1 vectors and further characterization of cis elements required for encapsidation. *J. Virol.* **71**, 4544-4554.

37. Das, A. T., B. Klaver, B. I. F. Klasens, J. L. B. van Wamel, and B. Berkhout. (1997) A conserved hairpin motif in the R-U5 region of the human immunodeficiency virus type 1 RNA genome is essential for replication. *J. Virol.* **71**, 2346-2356.

38. McBride, M. S. and A. T. Panganiban. (1996) The human immunodeficiency virus type 1 encapsidation site is a multipartite RNA element composed of functional hairpin structures. *J. Virol.* **70**, 2963-2973.

39. Harrich, D., C. Ulich, and R. B. Gaynor. (1996) A critical role for the TAR element in promoting efficient human immunodeficiency virus type 1 reverse transcription. *J. Virol.* **70**, 4017-4027.

40. Arts, E. J., X. Li, Z. Gu, L. Kleiman, M. A. Parniak, and M. A. Wainberg. (1994) Comparison of deoxyoligonucleotide and tRNA(Lys-3) as primers in an endogenous human immunodeficiency virus-1 in vitro reverse transcription/template-switching reaction. *J. Biol. Chem.* **269**, 14672-14680.

41. Harrich, D., C. Ulich, L. F. Garcia-Martinez, and R. B. Gaynor. (1997) Tat is required for efficient HIV-1 reverse transcription. *EMBO J.* **16**, 1224-1235.

42. Cao, X. and E. Wimmer. (1996) Genetic variation of the poliovirus genome with two VPg coding units. *EMBO J.* **15**, 23-33.

43. Brown, M. D., K. L. DeYoung, and D. H. Hall. (1994) A non-directed, hydroxylamine-generated suppressor mutation in the P3 pairing region of the bacteriophage T4 td intron partially restores self-splicing capability. *Mol. Microbiol.* **13**, 89-95.

44. Olsthoorn, R. C. L., N. Licis, and J. van Duin. (1994) Leeway and constraints in the forced evolution of a regulatory RNA helix. *EMBO J.* **13**, 2660-2668.

45. Zhang, C., T. Tellinghuisen, and P. Guo. (1995) Confirmation of the helical structure of the 5'/3' termini of the essential DNA packaging pRNA of phage p29. *RNA* **1**, 1041-1050.

46. Olsthoorn, R. C. L. and J. van Duin. (1996) Evolutionary reconstruction of a hairpin deleted from the genome of an RNA virus. *Proc. Natl. Acad. Sci. USA* **93**, 12256-12261.

47. Macadam, A. J., G. Ferguson, J. Burlison, D. Stone, R. Skuce, J. W. Almond, and P. D. Minor. (1992) Correlation of RNA secondary structure and attenuation of Sabin vaccine strains of poliovirus in tissue culture. *Virol.* **189**, 415-422.

48. Gmyl, A. P., E. V. Pilipenko, S. V. Maslova, G. A. Belov, and V. I. Agol. (1993) Functional and genetic plasticities of the poliovirus genome: quasi-infectious RNAs modified in the 5'-untranslated region yield a variety of pseudorevertants. *J. Virol.* **67**, 6309-6316.

49. Jacobson, S. J., D. A. M. Konings, and P. Sarnow. (1993) Biochemical and genetic evidence for a pseudoknot structure at the 3' terminus of the poliovirus RNA genome and its role in viral RNA amplification. *J. Virol.* **67**, 2961-2971.

50. Harrich, D., G. Mavankal, A. Mette-Snider, and R. B. Gaynor. (1995) Human immunodeficiency virus type 1 TAR element revertant viruses define RNA structures required for efficient viral gene expression and replication. *J. Virol.* **69**, 4906-4913.

51. Qu, F., C. Heinrich, P. Loss, G. Steger, P. Tien, and D. Riesner. (1993) Multiple pathways of reversion in viroids for conservation of structural elements. *EMBO J.* **12**, 2129-2139.

52. Klaver, B. and B. Berkhout. (1994) Premature strand transfer by the HIV-1 reverse transcriptase during strong-stop DNA synthesis. *Nucleic Acids Res.* **22**, 137-144.

53. Zuker, M. (1989) On finding all suboptimal foldings of an RNA molecule. *Science* **244**, 48-52.

54. Myers, G., B. Korber, B. H. Hahn, K.-T. Jeang, J. H. Mellors, F. E. McCutchan, L. E. Henderson, and G. N. Pavlakis. 1995. Human retroviruses and AIDS. A compilation and analysis of nucleic acid and amino acid sequences. Theoretical Biology and Biophysics Group, Los Alamos National Laboratory, Los Alamos, New Mexico.

55. Berkhout, B., B. Klaver, and A. T. Das. (1997) Forced evolution of a regulatory RNA helix in the HIV-1 genome. *Nucleic Acids Res.* **25**, 940-947.

56. De la Torre, J. C., C. Giachetti, B. L. Semler, and J. J. Holland. (1992) High frequency of single-base transitions and extreme frequency of precise multiple-base reversion mutations in poliovirus. *Proc. Natl. Acad. Sci. USA* **89**, 2531-2535.

57. Das, A. T., B. Klaver, and B. Berkhout. (1998) The 5' and 3' TAR elements of the human immunodeficiency virus exert effects at several points in the virus life cycle. *J. Virol.* **72**, 9217-9223.

58. Verhoef, K., M. Tijms, and B. Berkhout. (1997) Optimal Tat-mediated activation of the HIV-1 LTR promoter requires a full-length TAR RNA hairpin. *Nucleic Acids Res.* **25**, 496-502.

59. McCracken, S., N. Fong, K. Yankulov, S. Ballantyne, G. Pan, J. Greenblatt, S. D. Patterson, M. Wickens, and D. L. Bentley. (1997) The C-terminal domain of RNA polymerase II couples mRNA processing to transcription. *Nature* **385**, 357-361.

60. Levin, J. G. and M. J. Rosenak. (1976) Synthesis of murine leukemia virus proteins associated with virions assembled in actinomycin-D-treated cells: evidence for persistence of viral messenger RNA. *Proc. Natl. Acad. Sci. USA* **73**, 1154-1158.

61. Sonstegard, T. S. and P. B. Hackett. (1996) Autogenous regulation of RNA translation and packaging by Rous sarcoma virus Pr76Gag. *J. Virol.* **70**, 6642-6652.

62. Aldovini, A. and R. A. Young. (1990) Mutations of RNA and protein sequences involved in human immunodeficiency virus type 1 packaging results in production of noninfectious virus. *J. Virol.* **64**, 1920-1926.

63. Clavel, F. and J. M. Orenstein. (1990) A mutant of human immunodeficiency virus with reduced RNA packaging and abnormal particle morphology. *J. Virol.* **64**, 5230-5234.

64. Harrison, G. P. and A. M. L. Lever. (1992) The human immunodeficiency virus type 1 packaging signal and major splice donor region have a conserved stable secondary structure. *J. Virol.* **66**, 4144-4153.

65. Clever, J., C. Sassetti, and T. G. Parslow. (1995) RNA secondary structure and binding sites for gag gene products in the 5' packaging signal of human immunodeficiency virus type 1. *J. Virol.* **69**, 2101-2109.

66. Berkhout, B. and J. L. B. van Wamel. (1996) Role of the DIS hairpin in replication of human immunodeficiency virus type 1. *J. Virol.* **70**, 6723-6732.

67. Fu, W., R. J. Gorelick, and A. Rein. (1994) Characterization of human immunodeficiency virus type 1 dimeric RNA from wild-type and protease-defective virions. *J. Virol.* **68**, 5013-5018.

68. Zhang, Z., S.-M. Kang, Y. Li, and C. D. Morrow. (1998) Genetic analysis of the U5-PBS of a novel HIV-1 reveals multiple interactions between the tRNA and RNA genome required for initiation of reverse transcription. *RNA* **4**, 394-406.

69. Das, A. T., B. Klaver, and B. Berkhout. (1995) Reduced replication of human immunodeficiency virus type 1 mutants that use reverse transcription primers other than the natural tRNA(3Lys). *J. Virol.* **69**, 3090-3097.

70. Berkhout, B. and B. Klaver. (1995) Revertants and pseudo-revertants of human immunodeficiency virus type 1 viruses mutated in the long terminal repeat promoter region. *J. Gen. Virol.* **76**, 845-853.

71. Liang, C., X. Li, L. Rong, P. Inouye, Y. Quan, L. Kleiman, and M. A. Wainberg. (1997) The importance of the A-rich loop in human immunodeficiency virus type 1 reverse transcription and infectivity. *J. Virol.* **71**, 5750-5757.

72. Kashanchi, F., R. Shibata, E. K. Ross, J. N. Brady, and M. A. Martin. (1994) Second-site long terminal repeat (LTR) revertants of replication-defective human immunodeficiency virus: effects of revertant TATA box motifs on virus infectivity, LTR-directed expression, in vitro RNA synthesis, and binding of basal transcription factors TFIID and TFIIA. *J. Virol.* **68**, 3298-3307.

73. Willey, R. L., E. K. Ross, A. J. Buckler-White, T. S. Theodore, and M. A. Martin. (1989) Functional interaction of constant and variable domains of human immunodeficiency virus type gp120. *J. Virol.* **63**, 3595-3600.

74. Taddeo, B., F. Carlini, P. Verani, and A. Engelman. (1996) Reversion of a human immunodeficiency virus type 1 integrase mutant at a second site restores enzyme function and virus infectivity. *J. Virol.* **70**, 8277-8284.

75. Gold, L., C. Tuerk, P. Allen, J. Binkley, D. Brown, L. Green, S. MacDougal, D. Schneider, D. Tasset, and S. R. Eddy. 1993. RNA: the shape of things to come, p. 497-509. In R. F. Gesteland and J. F. Atkins (eds.), The RNA world. Cold Spring Harbor Laboratory Press, Cold Spring Harbor, USA.

76. Berkhout, B. (1997) The primer-binding site on the RNA genome of human and simian immunodeficiency viruses is flanked by an upstream hairpin structure. *Nucleic Acids Res.* **25**, 4013-4017.

77. Banks, J. D., A. Yeo, K. Green, F. Cepeda, and M. L. Linial. (1998) A minimal avian retroviral packaging sequence has a complex structure. *J. Virol.* **72**, 6190-6194.

78. Pelletier, J. and N. Sonenberg. (1985) Insertion mutagenesis to increase secondary structure within the 5'noncoding region of a eukaryotic mRNA reduces translational efficiency. *Cell* **40**, 515-526.

79. Kozak, M. (1986) Influence of mRNA secondary structure on initiation by eukaryotic ribosomes. *Proc. Natl. Acad. Sci. USA* **83**, 2850-2854.

80. Pathak, V. K. and H. M. Temin. (1992) 5-Azacytidine and RNA secondary structure increase the retrovirus mutation rate. *J. Virol.* **66**, 3093-3100.

81. Harrison, G. P., M. S. Mayo, E. Hunter, and A. M. L. Lever. (1998) Pausing of reverse transcriptase on retroviral RNA templates is influenced by secondary structures both 5' and 3' of the catalytic site. *Nucleic Acids Res.* **26**, 3433-3442.

82. Suo, Z. and K. A. Johnson. (1997) Effect of RNA secondary structure on the kinetics of DNA synthesis catalyzed by HIV-1 Reverse Transcriptase. *Biochem.* **36**, 12459-12467.

83. Darlix, J.-L., M. Lapadat-Tapolsky, H. De Rocquigny, and B. P. Roques. (1995) First glimpses at structure-function relationships of the nucleocapsid protein of retroviruses. *J. Mol. Biol.* **254**, 537

84. Tsuchihashi, Z. and P. O. Brown. (1994) DNA strand exchange and selective DNA annealing promoted by the human immunodeficiency virus type 1 nucleocapsid protein. *J. Virol.* **68**, 5863-5870.

85. Herschlag, D. (1995) RNA chaperones and the RNA folding problem. *J. Biol. Chem.* **270**, 20871-20874.

86. Liang, C., L. Rong, M. Laughrea, L. Kleiman, and M. A. Wainberg. (1998) Compensatory point mutations in the human immunodeficiency virus type 1 Gag region that are distal from deletion mutations in the dimerization initiation site can restore viral replication. *J. Virol.* **72**, 6629-6636.

87. Das, A. T., A. P. van Dam, B. Klaver, and B. Berkhout. (1998) Improved envelop function selected by long-term cultivation of a translation-impaired HIV-1 mutant. *Virol.* **244**, 552-562.

88. Berkhout, B. (1992) Structural features in TAR RNA of human and simian immunodeficiency viruses: a phylogenetic analysis. *Nucleic Acids Res.* **20**, 27-31.

89. Domingo, E., C. Escarmis, N. Sevilla, A. Moya, S. F. Elena, J. Quer, I. S. Novella, and J. J. Holland. (1996) Basic concepts in RNA virus evolution. *FASEB J.* **10**, 859-864.

90. Siderovski, D. P., T. Matsuyama, E. Frigerio, S. Chui, X. Min, H. Erfle, M. Sumner Smith, R. W. Barnett, and T. W. Mak. (1992) Random mutagenesis of the human immunodeficiency virus type-1 trans-activator of transcription (HIV-1 Tat). *Nucleic Acids Res.* **20**, 5311-5320.

91. Martinez, M. A., J. P. Vartanian, and S. Wain-Hobson. (1994) Hypermutagenesis of RNA using human immunodeficiency virus type 1 reverse transcriptase and biased dNTP concentrations. *Proc. Natl. Acad. Sci. USA* **91**, 11787-11791.

92. Berkhout, B. and B. Klaver. (1993) In vivo selection of randomly mutated retroviral genomes. *Nucleic Acids Res.* **21**, 5020-5024.

INTERACTION OF NATIVE RNAs WITH TAT PEPTIDES

E. WYSZKO[1)], M. SZYMAŃSKI[1)], J. P. FÜRSTE[2,3)], M. GIEL-PIETRASZUK[1)], M.Z. BARCISZEWSKA[1)], P. MUCHA[4)], P. REKOWSKI[4)], G. KUPRYSZEWSKI[4)], V.A. ERDMANN[2)] and J. BARCISZEWSKI[1)]*

[1)]Institute of Bioorganic Chemistry of the Polish Academy of Sciences, Noskowskiego 12, 61794 Poznań, Poland, [2)]Institute of Biochemistry of the Free University, Thielallee 63, 14195 Berlin, Germany, [3)]Noxxon, Pharma AG, Gustav-Meyer-Allee 25, D-13355 Berlin and [4)]Faculty of Chemistry, Gdańsk University, Sobieskiego 18, 80-952 Gdańsk, Poland.

We present experimental data on the interaction of arginine-rich Tat protein of human immunodeficiency virus (HIV-Tat) and its short fragments with three different RNAs: HIV-TAR RNA, yeast $tRNA^{Phe}$ and lupin ribosomal 5S rRNA. The aim of the studies was to learn about the basis of the specificity of RNA-protein interaction. All of these RNA molecules contain the same structural motif, the single stranded nucleotide sequence UGGG and form complexes with the Tat peptide of the amino acid sequence GRKKRRQRRRA and its derivatives, where R is substituted by D-arginine, citrulline or ornithine. The structure of the RNA-Tat peptide complexes were probed with specific RNases and Pb^{2+}-induced specific cleavage of the RNA phosphodiester bond. The nucleotide sequence UGGG located in the dihydrouridine loop of $tRNA^{Phe}$ and in the loop D of plant 5S rRNA is involved in binding of Tat-peptide and in the complex it is resistant to RNases and a hydrolysis by Pb^{2+} ion. It seems that the specificity of the peptide - RNA complex formation depends on direct recognition of guanine moieties of tRNA with arginine residues of Tat peptide. The nature of the interactions are similar to those observed in many DNA - protein complexes, but are different from those previously observed for TAR RNA-Tat complexes. We also described a method for preparation of TAR RNA which uses hammerhead ribozymes, to cut off the required RNA from the *in vitro* synthesised transcript.

1. Introduction

The human immunodeficiency virus type 1 (HIV-1) Tat protein is a potent trans activator of long terminal repeat gene expression and is essential for viral replication [1]. Transactivation by the Tat protein (72 amino acids in length) is mediated through binding to the transactivation response element (TAR) RNA, found at the 5' end of all mRNA transcripts. The predicted secondary structure of 59 nucleotide TAR RNA consists of two stem regions separated by three unpaired nucleotides (bulge) and a

J. Barciszewski and B.F.C. Clark (eds.), RNA Biochemistry and Biotechnology, 277–290.

terminal loop of six nucleotides [2-4]. In various studies it has been shown that carboxy-terminal fragment of Tat spanning a basic (arginine rich) region binds specifically to a region of TAR RNA containing a trinucleotide bulge [3]. The cysteine-rich region makes Tat protein difficult to handle. Therefore several groups adopted reductionist approach, trimming the Tat proteins down to minimal RNA binding peptides in order to delineate specific residues involved in RNA recognition and facilitate structural studies [1,3,5]. A TAR RNA - argininamide complex characterised by nuclear magnetic resonance spectroscopy (NMR) reveals an RNA conformational change upon arginine addition, indicating that TAR RNA contains a specific arginine binding pocket [4,5].
A similar conformational change in TAR RNA has been observed upon binding to arginine rich peptides [2]. Many authors came to conclusion that the Tat protein, Tat peptide and arginine binding studies suggest that single amino acid in Tat is primarily responsible for specific recognition of TAR RNA and that great deal of the specificity interaction is due to the RNA structure formed by TAR [4]. On the other hand, there are reports that arginine alone cannot mimic all of the interactions in the full TAR RNA-Tat complex and a careful biochemical analysis of various TAR RNA-Tat interactions has documented a loss of binding energy and sequence specificity of truncated Tat [2]. Recent advances in understanding specific interactions in protein-RNA complexes originate mostly from studies of three-dimensional structures of the complexes of glutaminyl-, aspartyl- and seryl-tRNA synthetases with their cognate tRNAs [6-8], R17 coat protein with 19-nucleotide RNA hairpin [9] and U1A protein domain with 20-nucleotide RNA hairpin [10]. Several protein-RNA binding motifs for specific interaction with ribonucleic acid have been identified. One of them consists of a short string of basic amino acids, mainly arginine residues (arginine-rich motif, ARM) [11,12]. Similar domains have been found in many other proteins as bacterial antiterminators, ribosomal proteins, coat proteins of RNA viruses, a human Tat (HIV-Tat) and a bovine Tat (BIV-Tat) immunodeficiency virus, and Rev (HIV-Rev) proteins [13, 14].
In order to get better insight into the RNA domain which binds to synthetic arginine rich peptides with amino acid sequence corresponding to the RNA-binding domain of HIV-1 Tat, we used yeast $tRNA^{Phe}$ and plant ribosomal 5S rRNA as the model molecules. They contain the (5')-UGGG sequence in the loops identical to that occurring in the TAR RNA molecule and is involved in formation of the three dimensional fold of yeast $tRNA^{Phe}$ [15].
We found that $tRNA^{Phe}$ and 5S rRNA form complexes with these peptides. Digestion of the complexes with specific RNases, identified the dihydrouridine loop of $tRNA^{Phe}$ and the loop D of 5S rRNA to be involved in the complex formation. The RNA-protein interactions resemble very much of the guanosine - arginine recognition mode observed previously in numerous DNA-protein complexes, but are different from those proposed earlier for TAR-Tat, where the protein recognises distorted helical RNA fragment. For biochemical studies of RNA protein complexes, a large quantities of both RNA and protein are required. RNA up to ca 50-mer could be prepared either chemically [16] or enzymatically with T7 RNA polymerase [17-19]. Although the last method is very efficient, it is known that sometimes T7 RNA polymerase incorporates additional nucleotides at the ends of the transcript making a product heterogenous [17,20-22]. In this paper we describe the application of two hammerhead ribozymes acting *in cis* to

produce TAR RNA on preparative scale. This simple approach could be used *in vitro* for production of different RNA aptamers e.g. TAR decoy and other active RNA molecules of interest and also *in vivo* for generation of RNA targets to bind strongly and inactivates various protein.

2. Materials and methods

2.1. SYNTHESIS OF POLYPEPTIDES

The Tat1-4 peptides were synthesised manually on a cross linked polystyrene resin (capacity 0.68 mmol/g) by the solid-phase method, using the Boc chemistry as described previously [23].

2.2. ISOLATION AND LABELLING OF NATIVE RNA

The $tRNA^{Phe}$ was extracted from yeast and purified on 15% 7M urea polyacrylamide gel. The TAR RNA and $tRNA^{Phe}$ were labelled at the 3'-end with $[^{32}P]pCp$ and RNA ligase [24]. $[^{32}P]$-labelled tRNA was purified by electrophoresis on 10% polyacrylamide gel (PAGE) with 7M urea, eluted from the gel and renatured [25].

2.3. RNA BINDING REACTIONS AND ELECTROPHORETIC MOBILITY GEL SHIFT ASSAYS

The $tRNA^{Phe}$-Tat peptide complex formation assay was performed at 22°C for 40' in the 0.05 M Tris-HCl pH 7.5 buffer containing 0.07 M NaCl, 0.001 M EDTA, 0.1% Nonidet P-40. $1x10^{-9}$ M $tRNA^{Phe}$, 2 µg of crude tRNA and $3x10^{-9}$ M Tat-peptide in total volume of 10µl were used in the reaction. An analysis of the complexes was carried out on 0.7 % agarose gel in 0.05 M Tris/borate pH 8.3 buffer.

2.4. RNase FOOTPRINT ASSAY

In a footprint reaction of the tRNA-peptide complex, the following amounts of RNases were used T1 ($2x10^{-4}$ U), V1 ($6x10^{-2}$U) and S1 (3U) (Pharmacia). For the localisation of the Tat binding site, 4µg of tRNA and 40000 cpm of labelled $tRNA^{Phe}$ were digested with 0.04 U of T1 RNase in 20 mM sodium acetate buffer pH 4.5, 7M urea, 1mM EDTA and 0.05% xylene cyanol and analysed on 10% polyacrylamide gel with 7M urea in 0.09 M Tris/borate buffer.

2.5. METAL ION INDUCED RNA HYDROLYSIS

Labelled $tRNA^{Phe}$ was supplemented with 8 µg of non-specific RNA. Tat-1 peptide was used at final concentration 0.1; 0.5 and 0.8 mM. The complex formation reaction was performed as described (paragraph 2.3), 1 µl of 5mM $Pb(CH_3COO)_2$ was added for RNA hydrolysis. Reaction (15 min.) was stopped by mixing with 10µl of 8M urea/dyes 20 mM EDTA and loaded on 10% polyacrylamide gel.

2.6. PREPARATION OF HIV TAR RNA

The template for T7 polymerase transcription of TAR RNA was prepared by insertion into the pT7T3α-18 plasmid of an 183 bp long fragment which was obtained by chemical synthesis and ligation of six short oligodeoxynucleotides (U1-U3 and L1-L3), (manuscript in preparation).

2.7. RNA TRANSCRIPTION

RNA was synthesised by T7 RNA polymerase transcription. After linearisation of 4-40 μg of the template with Hind III, transcription was carried out in 80 mM Hepes buffer pH 7.5, containing 25 mM $MgCl_2$, 1 mM spermidine, 5 mM dithiotreitol, 0.12 mg/ml BSA, 4 mM NTPs (C, G, U), 4 mM ATP, 40 U/μl T7 RNA polymerase for 5 h at 37°C. The DNA template was digested with RNase-free DNase, and RNA was extracted with phenol, then chloroform and precipitated with ethanol [17,21]. A pellet was dried and resuspended in 10 μl of deionised formamide. RNA was separated on 12% polyacrylamide gel electrophoresis in the presence of 7M urea. The band corresponding to transcript of 35 nt was cut off, and RNA was eluted from gel with the elution buffer [17] and recovered by ethanol precipitation. RNA length markers have been prepared by chemical synthesis [16]. A ribozyme intramolecular cleavage reaction of the T7 RNA transcript was carried out in 50 mM Hepes buffer pH 7.5 containing 25 or 50 mM $MgCl_2$ and incubated at 37°C for 1h, 2h, 4h, and 12 h (over night). RNA was renatured by heating to 70°C for 3 min every 2h or 1h.

2.8. TAR RNA/TAT BINDING ASSAY

The Tat2 protein was expressed and purified from *E. coli.* It was prepared *in vitro* as the fusion protein with glutathione S-transferase (GST) and purified in a single step by binding to glutathione-Sepharose beads. The fusion proteins were cleaved with (0.5U/0.5 mg protein) thrombin (EW, unpublished). Tat2 protein binding activity to TAR RNA was assayed with 0.6×10^{-11} M (10 000 cpm) of [3'-^{32}P]-labeled TAR-RNA and $0.65\text{-}4.5 \times 10^{-12}$ M of the Tat2 protein. The reaction was carried out in 50 mM Tris-HCl buffer pH 7.5, containing 70 mM NaCl, 1mM EDTA and 0.1% Nonidet for 30 minutes at room temperature (23°C). The analysis of the complex formation was done on 0.7% agarose gel.

3. Results and discussion

We are interested in an understanding of the mechanism of protein - ribonucleic acids recognition. The important issue here is to select proper model system where detailed interactions can be identified and analysed. One of such model is a fragment RNA of HIV mRNA TAR which interacts with the Tat protein. The NMR data of the HIV TAR - Tat complex have suggested that the arginine residue of the Tat protein recognises the UCU bulge of RNA [26]. These interactions however are totally different from those observed in some DNA - protein complexes, where direct binding of guanine and arginine occurs [27]. To solve this intriguing question whether interactions within RNA

- protein and DNA - protein complexes are similar and whether the arginine residues do interact with the guanine-rich RNA loop, we turned into native tRNA^{Phe} from yeast for which various structural data are available including its three dimensional structure [15]. This RNA contains the (5') UGGG oligonucleotide in the dihydrouridine loop. The same fragment is present also in plant ribosomal 5S rRNA which is identical to that found in the TAR RNA hairpin tip (Fig. 1).

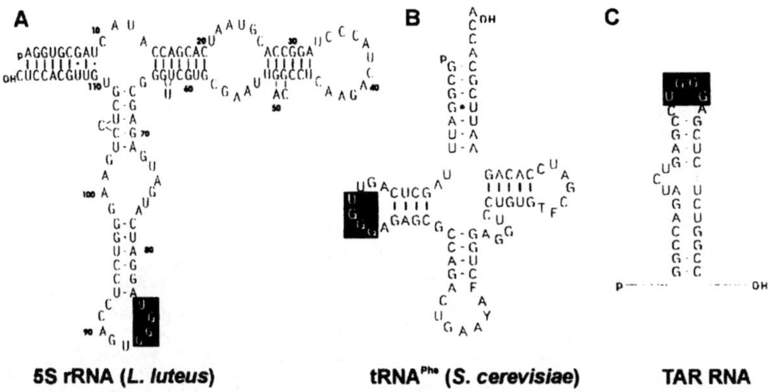

Fig. 1. The secondary structure of: A - 5S rRNA of plant [28] , B - tRNA^{Phe} of yeast [15], C - TAR RNA of HIV-1 [26]. UGGG sequence is highlighted.

For an analysis of specificity of tRNA^{Phe} and 5S rRNA Tat-binding properties, we prepared the synthetic arginine rich peptide Tat1 containing 11 amino acid residues. In addition to that, the Tat analogues of Tat2, Tat3 and Tat4 substituted at position Arg52 with D-arginine, citrulline and ornithine, respectively, have been also synthesised (Fig. 2). The amino acid sequence of these peptides is identical to the RNA binding motif of the HIV-1 Tat protein [29].

In order to check whether these peptides form complexes with tRNA^{Phe}, we carried out agarose gel shift assay. It shows that the Tat1 peptide forms very stable complex with tRNA^{Phe} at molar ratio 1:3 (Fig. 3, lanes 3-5). The Tat2 peptide (D-Arg) binds to yeast tRNA^{Phe} with lower affinity than the Tat1 (molar ratio 1:5) (lane 9-10). An excess of both the Tat1 and Tat2 peptides (lanes 1, 2, 8) causes formation of aggregates but the Tat2 at concentration of 16.4 and 21.9 nM in reaction mixture (lanes 6,7) precipitates tRNA^{Phe}. The Tat3 and Tat4 form complexes with tRNA only with large excess of peptide (23 : 1) (Fig. 3 lanes: 11, 16). To determine a peptide binding site on tRNA^{Phe}, a specific RNase hydrolysis with RNase T1 (guanosine specific), S1 (single stranded RNA specific) and V1 (double stranded RNA specific) were applied. T1 RNase digestion of tRNA^{Phe} gave two weak bands corresponding to nucleotides G18G19 (Fig. 4), not present in the digestion pattern of the tRNA-Tat complex. There are no differences in the RNase V1 hydrolysis of the complex. However, RNase S1, in contrast to T1 RNase, hydrolyses tRNA^{Phe} in the complex with the Tat1 peptide at G17G18G19, but not a free tRNA (Fig. 4). The results of specific cleavages of the complex with T1, S1, and V1 RNases clearly show that indeed the Tat1 polypeptide interacts directly with the G-rich oligonucleotide occurring in dihydrouridine loop D of tRNA^{Phe} molecule.

Tat1	Arg-Lys-Lys-	Arg	-Arg-Gln-Arg-Arg-Arg
Tat2	Arg-Lys-Lys-	D-Arg	-Arg-Gln-Arg-Arg-Arg
Tat3	Arg-Lys-Lys-	Cit	-Arg-Gln-Arg-Arg-Arg
Tat4	Arg-Lys-Lys-	Orn	-Arg-Gln-Arg-Arg-Arg

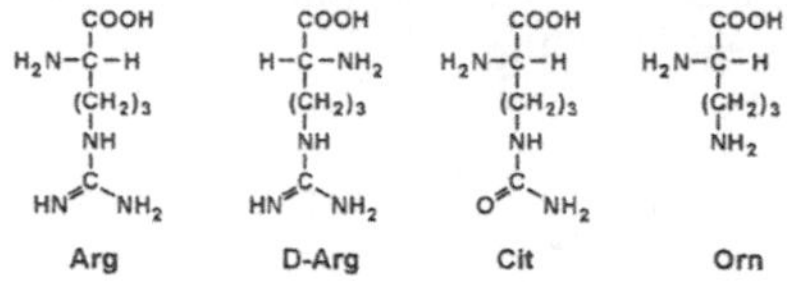

Fig. 2. The amino acid sequences of 11-amino acid long peptide Tat1 and its derivatives used in the experiments. Arginine (Arg52) in native HIV Tat peptide was substituted by D-arginine, citrulline and ornithine in Tat2, Tat3 and Tat4, respectively.

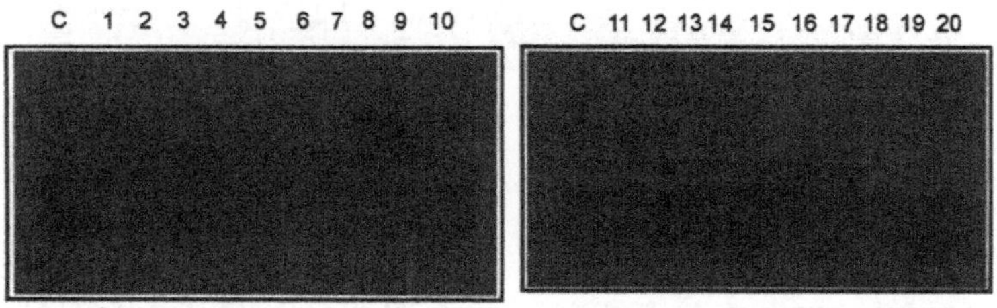

Fig. 3. An analysis of the binding of the Tat-peptides to [3'-^{32}P]tRNA^{Phe} on 0.7% agarose gel; Tat1 lanes 1–5, Tat2 lanes 6–10, Tat3 lanes 11–15, Tat4 lanes 16-20. Concentration of the peptide used in the experiments was as follows, lanes: 1, 6, 10, 16 – 21.9 nM; 2, 7, 12, 17 – 16.4 nM; 3, 8, 13, 18 – 10.9 nM; 4, 9, 14, 19 – 5,4 nM; 5, 10, 15, 20 – 2.7 nM.

For a deep elucidation of the mechanism of the interaction of Tat peptide with the D-loop of tRNA, which is involved in the tertiary structure formation, we have carried out the specific lead induced cleavage reaction. The reaction can take place only when native structure of the tRNA^{Phe} molecule is preserved (Fig. 5). The location of cleavage sites were identified with RNase T1 ladder (Fig. 5 lane T1). The Pb^{2+}-induced hydrolysis bands occur between nucleotides 17-16 and 16-15 of free tRNA^{Phe} (lane 1), but after complex formation with the Tat1 peptide (lanes 2-3) there are no cleavages at all (Fig. 5).

The lack of the lead induced hydrolysis we interpreted as a result of disturbance of the tertiary structure of tRNA effected by the Tat peptide. From the crystallographic structure of yeast tRNA^{Phe}, it is known that G18 and G19 of the loop D form hydrogen bonds with C56 and Ψ55 in the ribothymidine loop (T), respectively [15]. If so, no digestion at G18 and G19 by RNase S1 in the absence of the Tat1 and a strong hydrolysis in the presence of the polypeptide suggest conformational changes of tRNA leading to accessibility of the sugar-phosphate backbone in the complex, where the guanosine residues are protected against T1 RNase. From the previous CD-spectra analysis of Tat1-tRNA^{Phe} complex, we know that the structures of free and peptide bound tRNAs are different [30]. Therefore it is not surprising that, the nucleotides in the loop D in complexed tRNA^{Phe} are not hydrolyzed by Pb^{2+} ions, which is known as a very sensitive probe for the correct folding of tRNA. A detailed mechanism of Pb-cleavages has been proposed many years ago, in which $Pb(OH)^+$ coordinated to U59 and C60, abstracts a proton from the 2'-OH group of D17 to facilitate phosphodiester hydrolysis via a cyclic phosphodiester intermediate [31,32]. In such case, the lack of lead-induced hydrolysis of $tRNA^{Phe}$ in presence of the Tat peptide strongly suggests that the distance between the dihydrouridine loop (D) and ribothymidine loop (T) in the complex is longer than in free tRNA. The tertiary hydrogen bonds of $tRNA^{Phe}$, G19-C56 and G18-Ψ55 between D and T loops are disrupted in the complex of $tRNA^{Phe}$ with Tat-peptide. Taking this into account, as well as the results of the specific RNase hydrolysis, we propose that the peptide is located in the cleft between the loops D and T.

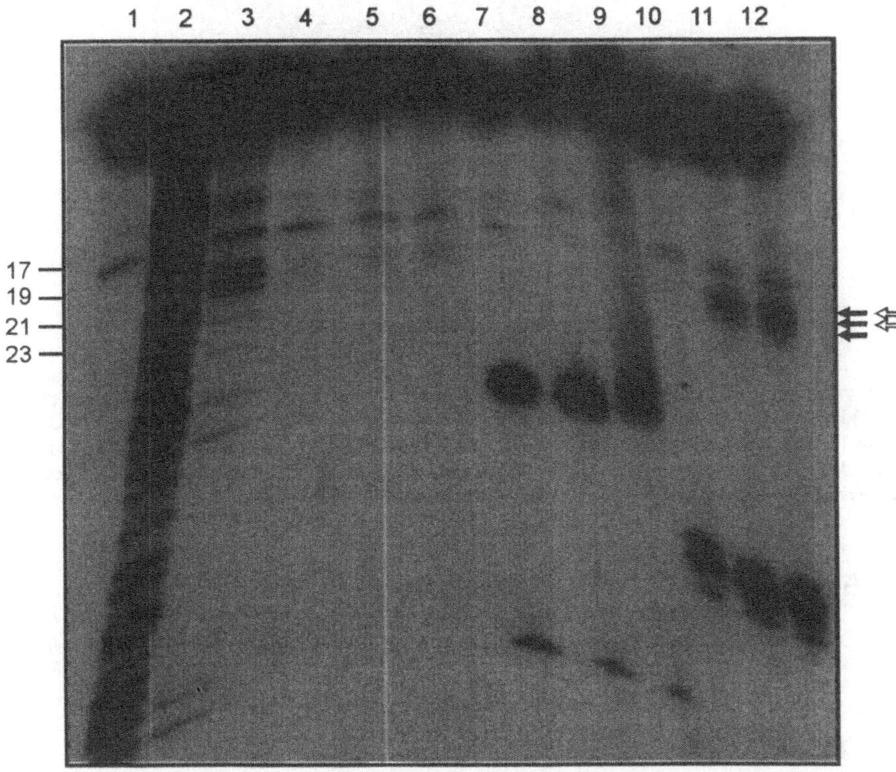

Fig. 4. Autoradiogram of 10% polyacrylamide gel with 7M urea showing hydrolysis products of [3'-^{32}P]$tRNA^{Phe}$-Tat1 complex obtained by RNase T1 (lanes 4-6), RNase V1 (lanes 7-9) and RNase S1 (lanes 10-12). Lanes: 1 - control $tRNA^{Phe}$ incubated in the reaction buffer (10'/22°C), 2 - ladder; 3 - limited hydrolysis of $tRNA^{Phe}$ with RNase T1, 4,7,10 $tRNA^{Phe}$; 5,8,11 $tRNA^{Phe}$ + 1.45 nM Tat1 peptide; 6,9,12 $tRNA^{Phe}$ + 3nM Tat1 peptide. Differences in hydrolysis pattern of free tRNA and in complex are marked by arrows on the right site (open arrows indicate the nucleotides protected from hydrolysis with RNase T1 and full arrows the sites of enhanced digestion by RNase S1), the numbers on the left side correspond to the nucleotides in primary structure of $tRNA^{Phe}$ (see Fig 1).

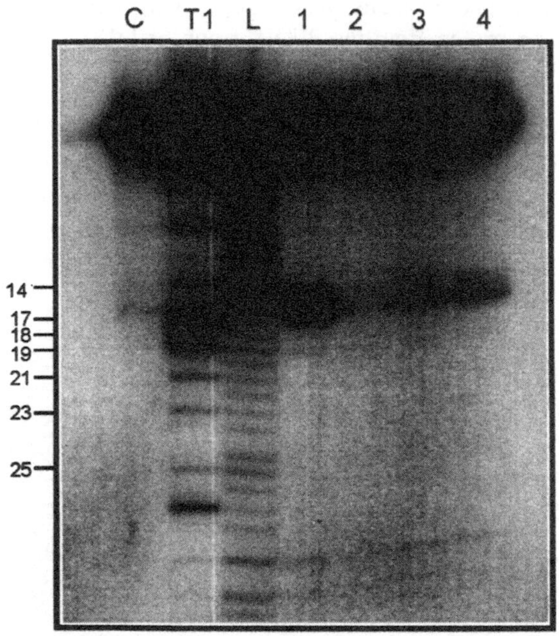

Fig. 5. Autoradiogram of 10% polyacrylamide gel with 7M urea showing Pb^{2+}-induced hydrolysis products of [3'-^{32}P]tRNA^{Phe}-Tat1 complex. Lanes: C - control tRNA^{Phe} in water, T1- limited hydrolysis of tRNA^{Phe} with RNase T1, L - alkali ladder; 1-4 - hydrolysis of tRNA^{Phe} free (lane 1) and in presence of Tat1 peptide (lanes 3-4) induced by Pb^{2+}. The conditions of the reactions are as described in Materials and Methods. The numbers on the left side correspond to nucleotides in primary structure of tRNA^{Phe} (see Fig 1).

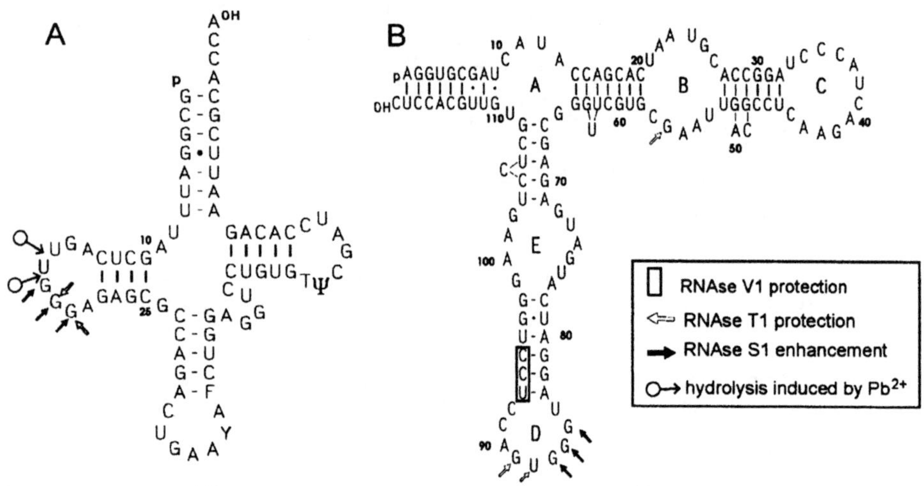

Fig. 6. RNase and Pb^{2+}-induced hydrolysis data of the yeast tRNA^{Phe} (A) and lupin 5S rRNA (B) superimposed on the secondary structures of the molecules. Open arrows indicate the nucleotides protected from hydrolysis with RNase T1 and full arrows the sites of enhanced digestion by RNase S1. Ring arrows mark the positions hydrolysed by Pb^{2+} in free molecules of tRNA^{Phe}. Box shows RNase V1 protection

We think that the Tat peptides form hydrogen bonds with $tRNA^{Phe}$ in a similar way to the Zif268 protein in complex with DNA. Most interactions take place between guanosine and arginine residues [27]. Having new data about possibility of direct and specific interaction of guanine residue with arginine moiety, we decided to carry out biochemical studies on the TAR RNA-Tat protein complex.
Up to now RNA molecules are mostly prepared by T7 RNA polymerase reaction. To avoid heterogeneity of the TAR RNA preparation we used a new approach for preparation any RNA molecules, taking advantage from RNA self-cleavage reaction *in cis* with two hammerhead ribozymes. To obtain HIV TAR RNA on preparative scale, the construct of DNA for T7 RNA polymerase transcription was synthesised. It contains the hammerhead ribozymes derived from positive strand of lucerne transient streak virusoid (+vLTSV) located upstream to the TAR RNA segment having 5' cleavage site and the second hammerhead ribozyme being a part of (+) strand of tobacco ring spot virus satellite RNA (+sTRSV) placed at the 3' end of TAR RNA sequence having 3' cleavage site [33]. In the construct there are 8 nucleotides from the 5' end of TAR RNA which form the stem I of vLTSV ribozyme, and 8 nucleotides of the 3' end of TAR RNA forming the stem III of sTRSV (Fig. 7).

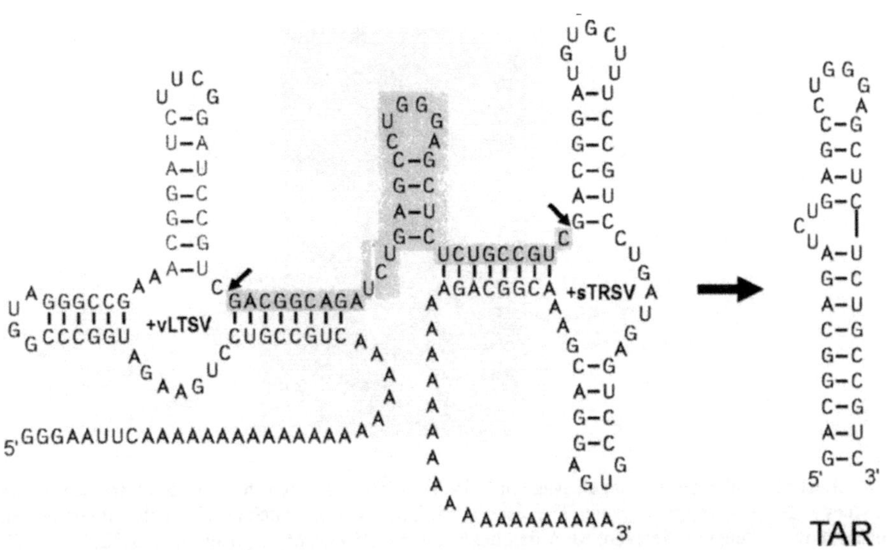

Fig. 7. The secondary structure of the T7 RNA transcript containing two hammerhead ribozymes derived from lucerne transient streak virusoid (+vLTSV) and tobacco ringspot virus satellitte RNA (+sTRSV RNA) located at the 5' and 3'end of HIV TAR RNA (35 nt) nucleotide sequence (grey box), respectively. 8 nucleotides of the 5' and 3' ends of TAR RNA element form the stem I of vLTSV and the stem III of sTRSV ribozymes, respectively. At the both termini of the hammerhead ribozymes, the oligo (A) tails were programmed. Additionally the 5' oligo (A) tail is capped with T7 RNA polymerase promotor. The cleavage sites are marked with arrows.

The DNA oligodeoxynucleotide containing TAR RNA and ribozymes was prepared by ligation of six shorter the chemically synthesised fragments (U1-U3 and L1-L3) and cloned into pT7T3α-18 plasmid. Two cleavage sites are included in single RNA chain. In fact, similar situation occurs in nature. Two of the hammerhead's helical arms of vLTSV and sTRSV are terminated by nucleotide loops and thus the ribozyme catalyses intramolecular cleavage of phosphodiester bond [33]. To prove the self-cleavage activity of the ribozymes and splitting off the TAR RNA element, the T7 RNA transcript was incubated with magnesium ions at various conditions. We showed that after transcription at 37°C for 5 h, TAR RNA has not been observed (Fig. 8 lane 6). However after incubation of the RNA primary transcript during 1 or 2 hours with 25 or 50 mM $MgCl_2$, a self-cleavage reaction occurred and TAR RNA band of 35 nt length was observed on the gel. The effects of an incubation over night and for 4 hours with heating at 70°C for 3 minutes every hour are similar (Fig. 8 lanes 1 and 4). We think that 25 mM $MgCl_2$ and 4 hr at 70°C for 3 minutes every hour are the best for slicing off TAR RNA. The reaction product of 35 nt was purified on large scale with 12 % PAGE. The selected conditions suggested that folding process of the hammerhead ribozyme active site is thermodynamically controlled. Appearance of the band corresponding to RNA chain of 35 nt identify that both ends of TAR RNA were trimmed. However on the gel there are some bands with length ca 100 nt (Fig. 8). It is reasonable to think that they are TAR RNA intermediate products having on the 5' or 3' end of the transcript of single ribozyme moiety.

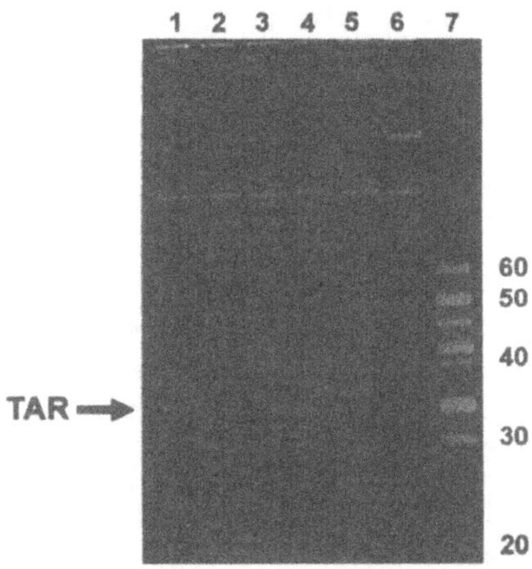

Fig. 8. Effect of incubation time on cleavage off HIV TAR RNA fragment from T7 RNA transcript with two ribozymes at the GUC target sequence. The reaction products were separated on 12% polyacrylamide gel. The numbers show the length of relevant RNA fragments. Lane 1: 25 mM $MgCl_2$ over night, 37°C; lane 2: 25 mM $MgCl_2$, 4 hr, 37°C; lane 3: 25 mM $MgCl_2$ 4 hr, 37°C with heating at 70°C for 3 min every 2 h; lane 4: 25 mM $MgCl_2$. 4 hr, 37°C with heating at 70 °C for 3 min every 1 h; lane 5: 25 mM $MgCl_2$, 2 hr, 37°C, lane 6: T7 transcript (control); lane 7: RNA length markers. The band corresponding to 35 nt of TAR RNA element is shown with an arrow.

The TAR RNA element prepared with this method forms the stem-loop structure (Fig. 7), already proposed by several groups [33-35]. If so, we checked its activity in complex formation with the HIV-2 Tat protein.

The Tat2 protein was expressed in *E. coli* as fusion with glutathione S-transferase (GST) protein (with molecular mass approximately 38 kDa). After lysis of cultures by sonication, the GST-Tat2 protein was present in the whole cell lysate. The GST-Tat2 fusion proteins were purified from cell lysates by selective binding to glutathione Sepharose beads and eluted with reduced glutathione (Fig. 9 lane 1-3). The Tat2 protein was cleaved from GST by thrombin digestion which proteolytic site was engineered between the GST and Tat moieties of the fusions and then used in binding reaction with TAR RNA.

The RNA-protein complex was clearly observed on the agarose gel in the presence of increased amount of the Tat2 protein (Fig. 10 lanes 2-7).

The results indicate that two RNA hammerhead ribozymes are sufficient for intramolecular self-cleavage off TAR RNA and that the intron molecule acquires the active conformation required for interactions with protein(s).

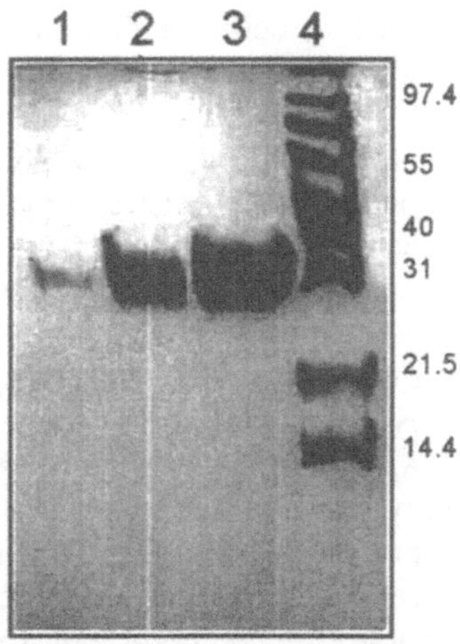

Fig. 9. 15 % polyacrylamide gel analysis of the GST-Tat2 fusion proteins expressed from pGEX2TK vector in *E. coli*. Gel was stained with Coomassie brillant blue. Lane1-3 increasing amounts of fusion protein fractions eluted from glutathione-Sepharose beads, lane 4 -molecular mass markers.

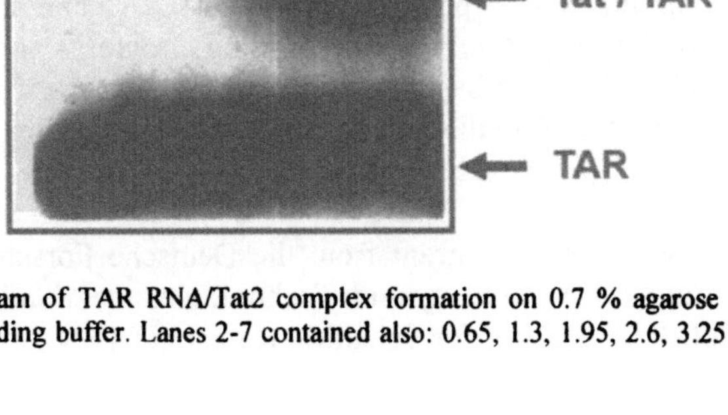

Fig. 10. An autoradiogram of TAR RNA/Tat2 complex formation on 0.7 % agarose gel. Lanes 1-7 TAR (0.6×10^{-9} M) in the binding buffer. Lanes 2-7 contained also: 0.65, 1.3, 1.95, 2.6, 3.25 and 4.5×10^{-12} M of Tat2 protein.

The method described above is based on a manipulation with the GUC hammerhead ribozymes and can be used to prepare any homologous RNA with the predetermined termini. Such approach offers a new possibility for generation *in vivo* any RNA molecule of interest. One would expect them to act as decoys to sequester some proteins e.g. Tat binding protein. It is well known that one of the big problem with a ribozymes targeting to the cell is their sensitivity towards ribonucleases. As a rather small molecules they are quickly turn down and therefore it is difficult to predict whether or not a ribozyme will arrive at the target sequence. A longer RNA molecule with strong folded structure should be generally more resistant to nuclease [36]. If so probability is much higher to get it at right site, where e.g. RNA aptamer should be cleave off.
As we found above the conditions for a ribozyme catalyzed reaction, require heating for renaturation and folding of the active conformation. It means that RNA transcript misfolds preferentially to non-ribozymic structure without a catalytic activity and a misfolding is probably caused by hairpin structure of inter-ribozymic RNA sequence. This is reasonable because it is well known that TAR RNA forms a stable stem-loop structure. If so, we ask question how T7 RNA transcript is folded. To do it, we applied the Zuker algorithm [37] to calculate putative secondary structures of the T7 RNA transcript of 183 nt long (Fig. 7). In this case, each calculated secondary structure can be easily verified, because it is known, from biochemical and crystallographic studies that efficient cleavage reaction occurs only when GU<u>C</u> target sequence of the ribozyme is in extended conformation and <u>C</u> is unpaired [33-35]. Once we obtained the required product of 35 nt long TAR RNA, obviously the ribozymes on both ends acquired the active conformation. Surprisingly none of the calculated structure with the lowest energy fulfill the structural requirement for the GUC hammerhead ribozyme activity. Generally two families of putative structures were found. These with the lowest energy contain the GUC sequence in double stranded portion of RNA, that forms an extension of the TAR-element stem. Such RNA molecules should not possess catalytic activity. In the second group, each structure contains either on the 5' or 3' ribozyme, with non-paired <u>C</u>, but not both. If so, there is a possibility to obtain precursor of the TAR RNA element with a ribozyme molecule at one terminus, only. Inspection of the gel bands (Fig. 8) suggests the longer precursors of 115 nt (3' cleavage site) or 105 nt (5' cleavage site) can be formed. This finding suggests that folding of the hammerhead ribozymes is not favorable process, probably due to the strong secondary structure of TAR. The calculations also show that there may be some variability in the structure of the stems I and III that may influence folding of the catalytic core of the ribozymes (data not shown). Finally we can conclude that computer calculations do not provide the correct secondary structure of the transcript with two active GUC hammerhead ribozymes. Therefore one has to be very careful using folding programs to predict secondary structure of long RNA and on that basis functional consequences of a molecule. On the other hand, computer structural predictions for small RNA molecules are reasonable.

Acknowledgments
This work was supported with the grant from the Deutsche Forschungsgemeinschaft, the Bundesministerium fur Forschung und Technologie, the Fonds der Chemischen Industie E.V and Polish Committee for Scientific Research (KBN).

References

1. Gait, M. and Karn, J. (1993) RNA recognition by the human immunodeficiency virus Tat an Rev proteins. *Trends Biochem. Soc.* **18**, 255-259.
2. Sundquist, W.I. (1996) Tattle tales. *Nature Struct. Biol.* **3**, 8-11.
3. Frankel, A.D. (1992) Peptide models of Tat-TAR protein-RNA interaction. *Prot. Sci.* **1**, 1539-1542.
4. Puglisi, J.D., Chen, L., Frankel, A.D. and Wiliamson, J.R. (1993) Role of RNA structure in arginine recognition of TAR RNA. *Proc. Natl. Acad. Sci. USA* **90**, 3680-3684.
5. Brodsky, A.S. and Williamson, J.R. (1997) Solution structure of the HIV-2 TAR-agininamide complex. *J. Mol. Biol.* **267**, 624-639.
6. Rould, M.A., Perona, J.J., Soll, D. and Steitz, T.A. (1989) Structure of *E. coli* glutaminyl-tRNA synthetase complexed with $tRNA^{Gln}$ and ATP at 2.8 Å resolution. *Science* **246**, 1136-1142.
7. Ruff, M., Krishnavarmy, S., Boeglin, M., Poterszman, A., Mitschler, A., Podjarny, A., Rees, B., Thierry, J.C. and Moras, D.(1991) Class II aminoacyl transfer RNA synthetases: crystal structure of yeast aspartyl-tRNA synthetase complexed with $tRNA^{Asp}$. *Science* **252**, 1682-1689.
8. Biou, V., Yaremchuk, A., Tukalo, M. and Cusack, S. (1994) The 2.9 Å crystal structure of *T. termophilus* seryl-tRNA synthetase complexed with $tRNA^{Ser}$. *Science* **263**, 1404-1410.
9. Valegard, K., Murray, J.B., Stockley, P.G., Stonehouse, N.J. and Liljas, L. (1994) Crystal structure of an RNA bacteriophage coat protein-operator complex. *Nature* **371**, 623-626.
10. Oubridge, C., Ito, N., Evans, P.R., Teo, C.H. and Nagai, K. (1994) Crystal structure at 1.92 Å resolution of the RNA-binding domain of the U1A splicesomal protein complexed with an RNA hairpin. *Nature* **372**, 432-438.
11. Mattaj, I.W. (1993) RNA-recognition: a family matter. *Cell*, **73**, 837-840.
12. Burd, C.G. and Dreyfuss, G. (1994) Conserved structures and diversity of function of RNA-binding proteins. *Science* **265**, 615-621.
13. Łazinski, D., Grządzielska, E. and Das, A. (1989) Sequence specific recognition of RNA hairpins by bacteriophage antiterminators requires a conserved arginine rich motif. *Cell* **59**, 207-218.
14. Chen, L. and Frankel, A.D. (1994) An RNA-binding peptide from bovine immunodeficiency virus Tat protein recognises an unusal RNA - structure. *Biochemistry* **33**, 2708-2715.
15. Kim, S.H., Suddath, F.L., Quingley, G.J., McPherson, A., Sussman, J.L., Wang, A.H.J., Seeman, N.C. and Rich, A. (1974) Three dimensional tertiary structure of yeast phenylalanine transfer RNA. *Science* **185**, 435-440.
16. Bald, R., Brum, K., Buchholz, B., Furste, J.P., Hartmann, R.K., Jaschke, A., Kretschmer-Kazemi Far, R., Lorentz, S., Raderschall, J., Schlegl, T., Specht, T., Zhang, M., Cech, D. and Erdmann, V.A. (1992) New possibilities in RNA research through RNA engineering, in Structural Tools for Analysis of Protein-Nucleic Acid Complexes. Advances in Life Sciences, Lilley, D.J., Heumann, H. and Suck, D. Eds., Basel-Birkhauser-Verlag, pp. 449-466.
17. Mulligan, J.F., Groebe, D.R., Witherell, G.W. and Uhlenbeck, O.C. (1987) Oligoribonucleotide synthesis using T7 RNA polymerase and synthetic DNA templates. *Nucleic Acids Res.* **15**, 8783-8798.
18. Churcher, M.J., Lamont, C., Hamy, F., Dingwall, C., Green, S.M., Lowe, A.D., Butler, J.G., Gait, M. and Karn, J. (1993) High affinity binding of TAR RNA by the human immunodeficiency virus type-1 Tat protein requires base pairs in the RNA stem and amino acid residues flanking the basic region. *J. Mol. Biol.* **230**, 90-110.
19. Hamy, F., Asseline, U., Grasby, J., Iwai, S., Pritchard, C., Slim, G., Butler, J.G., Karn, J. and Gait, M.J. (1993) Hiydrogen-bonding contacts in the major groove are required for human immunodeficiency virus type-1 Tat protein reciognition of TAR RNA. *J. Mol. Biol.* **230**, 111-123.
20. Sakamoto,T., Kawai, G., Katahira, M., Kim, M.H., Tanaka, Y., Kurihara, Y., Kohno, T., Watanabe, S., Yokoyama, S., Watanabe, K. and Uesugi, S. (1997) Haipin structure of an RNA 28-mer, which contains a sequence of the enzyme omponent of hammerhead ribozyme system:evidence for tandem G:A pairs that are not of side-by-side type. *J. Biochem.* **122**, 556-562.
21. Triana-Alonso, F.J., Dabrowski, M., Wadzack, J. and Nierhaus, K.H. (1995) Self coded 3' extention of run-off transcripts produces aberrant products during *in vitro* transcription with T7 RNA polymerase, *J. Biol. Chem.* **270**, 6298-6307.
22. Pleiss, J.A., Derrick, M.L. and Uhlenbeck, O.C., (1998) T7 RNA polymerase produces 5' end heterogeneity during in vitro transcription from certain templates. *RNA* **4**, 1313-1317.
23. Buśkiewicz I, Giel-Pietraszuk M, Mucha P., Rekowski P., Kupryszewski G., Barciszewska M.Z.(1998) Interaction of HIV Tat peptides with tRNA Phe from yeast. *Collect. Czech. Chem. Commun.* **63**, 842-850.

24. Barciszewska, M., Dirheimer, G. and Keith, G. (1983) The nucleotide sequence of methionine elongator tRNA from wheat germ. *Biochem. Biophys. Res. Commun.* **114**, 1161-1168.
25. Pieler, T. and Erdmann, V.A. (1982) Three dimensional structural model of eubacterial 5S RNA that has functional implications. *Proc. Natl. Acad. Sci. USA* **79**, 4599-4603.
26. Puglisi, D., Chen, L., Frankel, A.D. and Williamson, J.R.(1993) Role of RNA structure in arginine recognition of TAR RNA. *Proc. Natl. Acad. Sci.* **90**, 3680-3684.
27. Pavletich, N.P. and Pabo, C.O. (1991) Zinc finger - DNA recognition: crystal structure of a Zif268-DNA complex at 2.1 Å. *Science* **252**, 809-817.
28. Barciszewska, M., Huang, H.W., Marshal, A.G., Erdmann, V.A. and Barciszewski, J. (1992) Biochemical and NMR spectroscopy evidence for a new tertiary A-U base pair in lupin ribosomal 5S RNA structure. *J. Biol Chem.* **267**, 16691-16695.
29. Hormes, R., Homann, M., Oelze, I., Marschall, P., Tabler, M., Eckstein, F. and Sczakiel, G. (1997) The subcellular localisation and length of hammerhead ribozymes determine efficacy in human cells, *Nucleic Acids Res.* **25**, 769-775.
30. Giel-Pietraszuk, M., Barciszewska, M.Z., Mucha, P., Rekowski, P., Kupryszewski, G. and Barciszewski J. (1997) Interaction of HIV Tat model peptides with tRNA and 5SrRNA. *Acta Biochim. Polon.*, ***44***, 591-600.
31. Brown, R.S., Hingherty, B.E., Dewan, J.C. and Klug, A. (1983) Pb(II) catylysed cleavage of the sugar-phosphate backbone of yeat $tRNA^{Phe}$-implications for lead toxicity and self-splicing RNA. *Nature* **303**, 543-546 .
32. Brown, R.S., Dewan J.C. and Klug A.(1985) Crystallographic and biochemical investigation of the lead (II)-catalysed hydrolysis of yeast phenylalanine tRNA. *Biochemistry* **24**, 4785-1801 .
33. Symons, R.H. (1997) Plant pathogenic RNAs and RNA catalysis. *Nucleic Acids Res.* **25**, 2683-2689.
34. Birikh, K.R., Heaton, P.A. and Eckstein, F. (1997) The structure, function and application of the hammerhead ribozyme. *Eur. J. Biochem.* **245**, 1-16.
35. McKay, D.B. (1996) Structure and function of the hammerhead ribozyme: an unfinished story. *RNA* **2**, 395-403.
36. Thompson, J., Ayers, D.F., Malmstrom, T.A., McKenzie, T.L., Ganousis, L., Chowrira, B.M., Couture, L. and Stinchcomb, D.T. (1995) Improved accumulation and activity of ribozymes expressed from a tRNA-based RNA polymerase III promoter. *Nucleic Acids Res.* **23**, 2259-2286.
37. Zuker, M. and Stiegler (1981) Optimal computer folding of large RNA sequences using thermodynamics and auxiliary information, *Nucleic Acids Res.* **9**, 133-148.

BIOGENESIS, STRUCTURE AND FUNCTION OF SMALL NUCLEOLAR RNAs

Witold Filipowicz, Pawel Pelczar, Vanda Pogacic and François Dragon
Friedrich Miescher-Institut, P.O. Box 2543, 4002 Basel, Switzerland

1. Introduction

Synthesis, maturation and packaging of ribosomal RNAs (rRNAs) into ribosomal particles in eukaryotic cells takes place in the nucleolus. Ribosomal RNA genes are transcribed by RNA polymerase I into long 35/47S precursors (pre-rRNAs) which are processed into mature 18S, 5.8S and 25/28S rRNAs. The maturation process involves a large number of RNA intermediates and cleavage events which may follow alternative pathways. In addition, rRNAs are extensively modified: methylation of the 2'-hydroxyl group of sugar residues (2'-*O*-methylation) and conversion of uridines to pseudouridines (ψ) (pseudouridylation) are by far the most frequent modifications. Processing of pre-rRNA occurs concomitantly with packaging of RNA into ribonucleoprotein structures containing tens of ribosomal proteins and also nucleolar protein associating only transiently with the nascent ribosomes [for review, see 24, 25, 37, 41].

Maturation and modification of pre-rRNAs is assisted by a large number of small nucleolar RNAs (snoRNAs), which function in the form of ribonucleoprotein particles (snoRNPs). Some snoRNAs, such as vertebrate U3, U8 and U14, are required for nucleolytic cleavages of the 47S pre-rRNA or its processing intermediates [13, 24, 25, 37, 41]. However, the vast majority of snoRNAs acts as guides directing site-specific 2'-*O*-ribose methylation or ψ formation [8, 14, 15, 18, 29]. Approximately one hundred RNAs of this type have been identified to date in vertebrates and the yeast *Saccharomyces cerevisiae* [20, 38]. This large number is readily explained by the findings that one snoRNA acts as a guide usually for one or at most two modifications, and human rRNAs contain 91 pseudouridines and 106 2'-*O*-methyl residues (in yeast, these numbers are 43 and 55, respectively) [20, 24, 38]. In this article we review information about the structure and function of guide snoRNAs and discuss mechanisms of their biogenesis.

2. Structure and function of guide snoRNAs

SnoRNAs can be grouped into two major families, C/D and H/ACA, based on conserved sequence motifs [3, 20, 38] (Figure 1). Members of the box C/D family contain short sequence elements PuUGAUGA (box C) and CUGA (box D) located near their 5' and 3' ends, respectively; boxes C and D are usually flanked by short inverted repeats. Many box C/D snoRNAs contain additional C-like and D-like elements termed boxes C' and D' (Figure 2A). The box C/D snoRNAs are associated with the abundant nucleolar protein fibrillarin (known as Nop1p in yeast) [20, 25, 38].

J. Barciszewski and B.F.C. Clark (eds.), RNA Biochemistry and Biotechnology, 291–302.

Vertebrates	Yeast
Box C/D:	
U3, **U8**, **U13**, **U14**, U15, U16, U18, U20, U21, **U22**, U24, U25, U26, U27, U28, U29, U30, U31, U32, U33, U34, U35, U36, U37, U38, U39, U40, U41, U42, U43, U44, U45, U46, U47, U48, U49, U50, U51, U52, U53, U54, U55, U56, U57, U58, U59, U60, U61, U62, U63, U73, U74, U75, U76, U77, U78, U79, U80, U81	**U3**, **U14**, U18, U24, snR4, snR13, snR38, snR39, snR39b, snR40, snR41, snR45, snR47, snR48, snR190
Box H/ACA:	
U17 (E1), E2, **E3**, U19, U23, U64, U65, U66, U67, U68, U69, U70, U71, U72	snR3, snR5, snR8, snR9, **snR10**, snR11, **snR30**, snR31, snR32, snR33, snR34, snR35, snR36, snR37, snR42, snR43, snR44, snR46, snR49, snR189
Others:	
7-2/MRP	**MRP**

Figure 1. Classification of snoRNAs. The list of snoRNAs (grouped as "Box C/D", "Box H/ACA", and "Others") is given for vertebrates and yeast. SnoRNAs involved in processing reactions are indicated in bold.

Members of the H/ACA family of snoRNAs have a characteristic structure consisting of two large stem-loops. They are separated by a single-stranded sequence containing a conserved hinge or box H with a consensus ANANNA. Another conserved sequence, ACA, is present downstream of the 3'-terminal stem-loop (Figure 2B). In yeast, the H/ACA snoRNAs are associated with two essential proteins, Gar1p and Cbf5p [3, 4, 21]. Boxes C/D and H/ACA are required for formation of respective snoRNPs and for nucleolar accumulation of snoRNAs.

The 7-2/MRP RNA does not belong to either of the two families discussed above. This RNA is a component of RNase MRP, a ribonucleoprotein enzyme related in structure to RNase P. In yeast, RNase MRP is involved in the cleavage of pre-rRNA in a region upstream from 5.8S rRNA [reviewed in 28, 38]. Protein composition of the yeast RNase MRP has recently been established. The particle contains 9 different proteins, eight of them shared with RNase P [10].

As indicated in Figure 1, only very few of the C/D or H/ACA snoRNAs are involved in cleavage reactions. The box C/D U3, U13, U14 and U22 RNAs are required for 18S rRNA production, and U8 is required for processing of 5.8S and 28S rRNAs. The box H/ACA snoRNAs snR10, snR30, U17 and E3 participate in 18S rRNA production but their exact role is not understood [reviewed in 12, 37, 41].

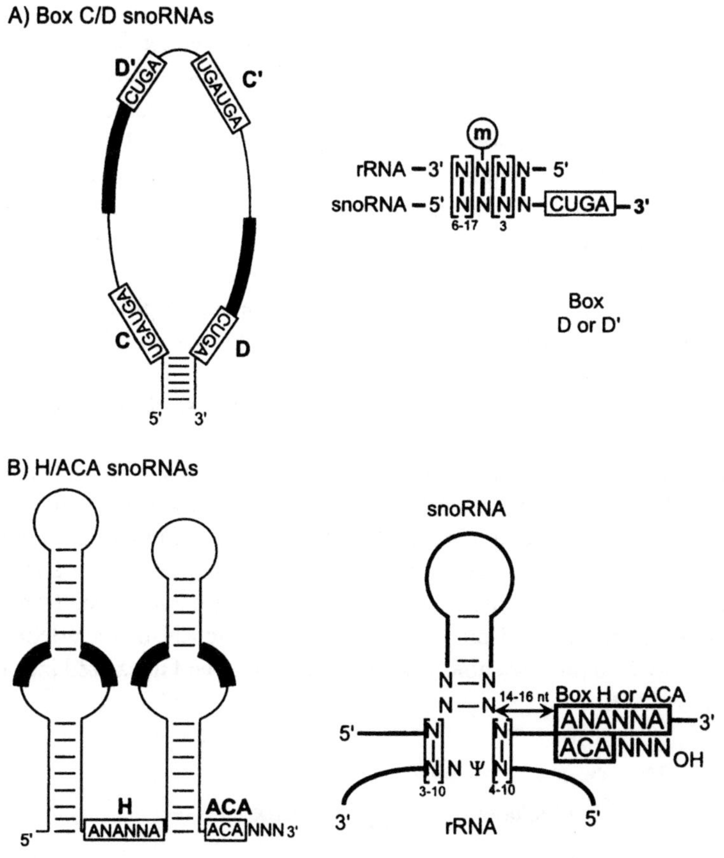

Figure 2. Structure of snoRNAs and their interaction with rRNAs. The schematic secondary structures of snoRNAs are presented on the left; the thick lines represent portions that are complementary to rRNA, and the conserved sequences are boxed and indicated in bold. Models for the selection of specific position to be modified are shown on the right. In (A), 2'-O-ribose methylation is performed on the rRNA residue that is base-paired to the fifth position upstream from box D (or D'). In (B), Ψ formation in rRNA occurs on the first unpaired U residue upstream from box H or ACA (usually at a distance of 14 to 16 nt from the box).

The vast majority of the known C/D and H/ACA snoRNAs act as guides for site-specific 2'-*O*-ribose methylation (C/D RNAs) or pseudouridylation (H/ACA RNAs). The box C/D guide snoRNAs generally contain a 10- to 21-nt-long sequence that is complementary to a region in rRNA that is a target of methylation. The sequence complementary to rRNA is always located immediately upstream of the box D or D', and the position in rRNA which is complementary to the fifth nucleotide from box D (or D') undergoes modification (Figure 2A). SnoRNAs containing sequences complementary to rRNA upstream of both the box D and D' specify two different sites of methylation. It is not yet known whether methylase activity is an integral component of snoRNPs [2, 8, 18, 34, 38, 40].

In box H/ACA snoRNAs, short (3-10 nt) sequences complementary to rRNA are present in the single stranded internal loops interrupting either one or both stem-loops characteristic of this class of snoRNAs. The U residue to undergo isomerisation is not base-paired but is flanked by two short helices formed by complementary interactions of snoRNA and rRNA. The position of pseudouridine is always 14-16 nucleotides from box H or ACA [14, 15, 29, 38] (Figure 2B). One of the protein components of the H/ACA snoRNPs in yeast, Cbf5p, is strongly homologous to the *Escherichia coli* tRNA:ψ55 pseudouridine synthase [21]. It is very likely that Cbf5p, and its mammalian counterpart NAP57, are enzymatic components of snoRNPs responsible for pseudouridylation of rRNA [21, 26]. Consistent with this is the observation that genetic depletion of Cbf5p results in impairment in pre-RNA processing and loss of rRNA pseudouridylation in yeast [21].

3. Strategies of snoRNA biosynthesis and processing of snoRNAs from introns

Eukaryotic cells use many different strategies to synthesize snoRNAs (Figure 3). Some snoRNAs, such as U3, U8 or 7-2/MRP, are individually transcribed from independent genes but many are encoded in introns of pol II transcription units. Still another strategy of expression, described for plants and yeast, involves processing of snoRNAs from polycistronic transcripts. In yeast, most of the snoRNAs are transcribed from independent genes while some are excised from introns or polycistronic RNAs. In vertebrates, the most common strategy is to process snoRNAs from the excised and linearized introns [reviewed in 5, 11, 22, 31, 38].

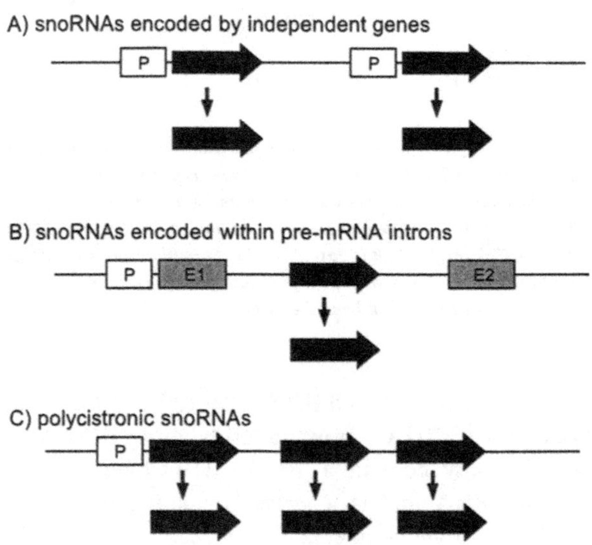

Figure 3. Expression of snoRNAs. The promoter regions (P) are shown as whites boxes, exons (E) as gray boxes, and snoRNAs as black arrows.

All known mammalian H/ACA snoRNAs and most of the box C/D snoRNAs are encoded in introns of pol II-transcribed genes [23, 25]. Using human U17 and U19 H/ACA-type snoRNAs as models, we have previously established a general mechanism for processing of snoRNAs from introns. 5' → 3' and 3' → 5' exonucleases appear to be involved in the trimming of both ends, and excised and debranched introns act as processing substrates [17] (Figure 4). The exonucleolytic trimming of introns stops at the borders of the snoRNA, most likely due to the snoRNA region being packaged into an RNP. Several lines of experimental evidence support these conclusions. In HeLa cell extracts, processing of snoRNAs from longer intron-like precursors requires free RNA termini; capping of the RNA or its circularization prevent maturation of ends. The most compelling *in vivo* evidence is the demonstration that processing is independent of sequences present upstream and downstream of the snoRNA in the intron, and that only a single snoRNA sequence can be productively processed from one intron: placement of two snoRNAs in tandem in the intron results in accumulation of dimeric snoRNAs [17].

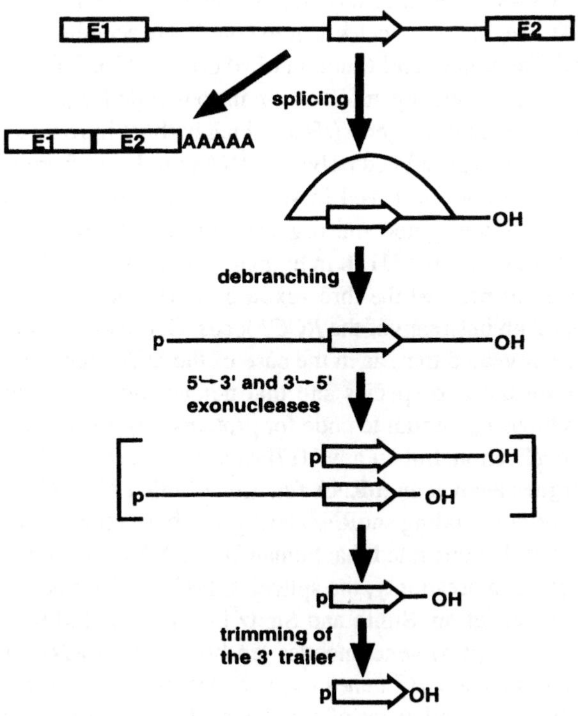

Fig. 4. A mechanism for exonucleolytic processing of snoRNAs from excised and debranched introns. The 5'-phosphate and 3'-OH ends are indicated. The snoRNA sequence (open arrow) associates with snoRNA-specific proteins prior to the final exonucleolytic processing steps.

More recent experiments performed with injected *Xenopus* oocytes [9] and with yeast [30, 32, 42] provided additional evidence supporting the exonucleotytic mechanism of snoRNA processing. In yeast, genetic inactivation of the intron debranching enzyme, Dbr1p, inhibits snoRNA processing [30, 32]. Xrn1p and Rat1p were identified as exonucleases involved in the 5' → 3' trimming of intron-encoded snoRNAs [32]. However, the mechanism of snoRNA processing outlined above has some exceptions. SnoRNAs, such as vertebrate U16 or U18, located in poorly spliced introns, are excised from the intron by endonucleases, and only the remaining extra nucleotides flanking the snoRNA are removed exonucleolytically [6, 7].

4. Host genes encoding intronic snoRNAs

Characterization of genes that act as hosts for intronic snoRNAs in vertebrates has produced many interesting findings. Most snoRNA host genes analyzed to date encode proteins essential for ribosome biogenesis or function such as ribosomal and nucleolar proteins or translational factors. This observation has led to the speculation that cotranscription of snoRNAs with mRNAs for nucleolar proteins or translational components may provide a regulatory mechanism to coordinate accumulation of different molecules required for the assembly and function of ribosomes [25, 36].

However, not all genes hosting snoRNAs in introns code for proteins. Tycowski *et al.* [39, 40] have discovered that the gene *UHG*, which harbors intronic snoRNAs U22 and U25-U31, is very unusual. The spliced poly(A)$^+$ RNAs produced from *UHG* genes in humans, mice, and frogs are not conserved in sequence and have no apparent protein coding potential. We have investigated the organization and expression of the locus encoding intronic snoRNAs U17a and U17b in human and mouse cells [31]. In humans, U17 RNAs are transcribed as parts of the three-exon transcription unit, named *U17HG*, positioned approximately 9 kb upstream of the *RCC1* locus. Comparison of the human and mouse *U17HG* genes has revealed that, as in the case of the *UHG* gene, exon sequences are not conserved between the two species and that neither human nor mouse spliced U17HG poly(A)$^+$ RNAs have a potential to code for proteins (Figure 5). The finding that, despite its cytoplasmic localization, little if any U17HG RNA is associated with polysomes in HeLa cells, further argues against an mRNA function of this RNA [31].

Two other non-protein-coding snoRNA host genes have recently been identified. Bortolin and Kiss [3a] have demonstrated that human U19 RNA is encoded in intron 2 of the multiexon U19HG gene. Interestingly, the spliced U19HG RNA has a nucleoplasmic rather than cytoplasmic localization. Smith and Steitz [35] have found that human *gas5* (growth arrest specific transcript 5) gene encodes ten box C/D snoRNAs in its introns. Comparison of human and mouse *gas5* genes has revealed no sequence conservation or significant coding potential in spliced exons of either gene. In summary, four snoRNA host genes which encode poly(A)$^+$ RNAs having no protein-coding potential have been identified to date. These genes probably act only as vehicles for expression of intron-encoded snoRNAs although it can not be entirely excluded that their spliced poly(A)$^+$ RNA products also function as regulatory or structural RNAs. RNA folding programs have not identified any obvious secondary structure elements shared by human and mouse UHG, U17HG or gas5 RNAs [31, 35, 39]. Findings that in *UHG*, *U17HG* and *gas5* genes these are snoRNA-encoding introns and not exons that are evolutionarily conserved and express functional RNAs requires a modification of the current description of exons as the main information-carrying regions of a gene.

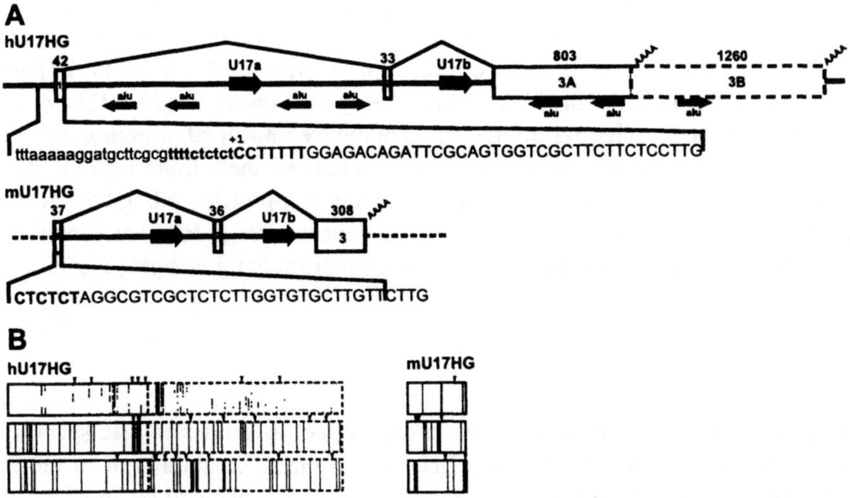

Figure. 5. Schematic structure of human and mouse U17HG genes (A) and analysis of the protein coding potential of spliced U17HG RNAs (B). Positions of exons and snoRNA regions are denoted by open boxes and black arrows, respectively. Alu sequences are shown as small black arrows. Human U17HG RNA is polyadenylated at two different sites (AAAA). The transcription-start-site-proximal sequences are shown with polypyrimidine tracts indicated in bold. (B) The coding potential of human (left panel) and mouse (right panel) U17HG RNAs. Positions of AUG codons (black triangles) and stop codons (vertical lines) are indicated for all reading frames of the spliced mouse and human U17HG RNAs. The region corresponding to the extended version of exon 3 in human RNA is drawn with dashed lines.

5. Non-protein coding snoRNA host genes are members of the 5'TOP family

Determination of the 5' terminus of the U17HG RNA revealed that transcription of the *U17HG* gene starts with a C residue followed by a polypyrimidine tract, making this gene a member of the 5'-terminal oligopyrimidine (5'TOP) family which includes genes encoding ribosomal proteins and some translation factors [31]. The polypyrimidine tracts are responsible for upregulation of translation of the 5'TOP mRNAs in response to growth factors or other conditions which require coordinated increased synthesis of proteins making up the translational apparatus [reviewed in 1, 27]. Interestingly, other known snoRNA host genes, including non-protein-coding *UHG*, *gas5* and *U19HG* genes, have features of the 5'TOP genes [3a, 31, 35, 39]. Similar characteristics of the transcription start site regions in snoRNA host and ribosomal protein genes raise the possibility that expression of components of ribosome biogenesis and translational machineries is coregulated. It is more plausible that coordination of expression of snoRNA hosts and the genes coding for translational components occurs at the transcriptional rather than the translational level. Processing of snoRNAs from introns is a nuclear event [25, 38]. Moreover, it is rather unlikely that 5'-terminal oligopyrimidine tracts have been conserved in UHG, U17HG and gas5 RNAs in order to regulate translation of these apparently non-protein-coding RNAs [31, 35, 39]. The oligopyrimidine tracts in 5'TOP genes are usually present not only downstream but also upstream of the start site C residue [1, 27] and may play a role in transcription of the 5'TOP genes. Perry and coworkers [16, 33] have reported that integrity of the +1 oligopyrimidine tracts in mouse ribosomal protein genes S16 and L30 is important for efficient and precise initiation of their transcription. Although transcriptional role for the +1 oligopyrimidine tract in non-protein-coding host genes is the most probable one, it is possible that these sequences have a different function, more directly related to the processing of snoRNAs from primary transcripts [31, 35].

6. Unanswered questions

Demonstration that most of the known snoRNAs act as guides in site-specific 2'-*O*-methylation and pseudouridylation of rRNA was a major breakthrough in the field of snoRNA/rRNA research. However, functions of modified nucleotides in rRNA remain unknown. Although modifications are clustered in evolutionarily conserved and functional regions of rRNA, multiple disruptions of most snoRNA genes in yeast have no effect on cell viability or growth [reviewed in 2, 20, 25, 38]. It is possible that modified nucleotides contribute to more global phenomena rather than act individually and deletion of tens of snoRNA genes would be required in order to see the effect. 2'-*O*-methyl groups generate more hydrophobic surfaces or may stabilize RNA stems by constraining the sugar residues into the more rigid C3'-*endo* conformation [19, 20]. On the other hand, usage of ψs increases the hydrogen bonding potential. Hence, modifications might contribute to rRNA folding or some other aspects of ribosome biogenesis and function. Alternatively, 2'-*O*-methylations might protect crucial regions of rRNA from hydrolytic degradation. In addition to elucidating functions of rRNA modifications, it will be equally important to establish the structure of guide snoRNAs and RNPs, and to understand details of the enzymatic reactions they participate in.

Questions addressing possible origins of non-protein-coding snoRNA host genes are also interesting though difficult to conclusively answer. Several scenarios can be envisaged. (I) Exons of these genes originally encoded a protein but this property was lost during evolution. (II) The host exons may have encoded a structural or regulatory RNA, whose function became obsolete with time or which we have not yet identified; (III) The genes originated from polycistronic snoRNA genes, similar to genes expressed in plants and yeast, by conversion of the inter-snoRNA spacers into spliceable exons. The last scenario would imply that units encoding poly-snoRNAs are evolutionarily old prototype genes. Such genes could have arisen by duplication of single snoRNA segments at a time when the complexity of rRNA modification was increasing. Characterization of snoRNA transcription units in additional distantly related organisms might throw more light on origins of the non-protein-coding snoRNA host genes.

References

1. Amaldi, F. and Pierandrei-Amaldi, P. (1997) TOP genes: a translationally controlled class of genes including those coding for ribosomal proteins. *Prog. Mol. Subcell. Biol.* **18**, 1-17.

2. Bachellerie, J.-P. and Cavaillé J. (1997) Guiding ribose methylation of rRNA. *Trends Biochem. Sci.* **22**:257-261.

3. Balakin, A.G., Smith, L., and Fournier, M.J. (1996) The RNA world of the nucleolus: two major families of small RNAs defined by different box elements with related functions. *Cell* **86**, 823-834.

3a. Bortolin, M.-L. and Kiss, T. (1998) Human U19 intron-encoded snoRNA is processed from a long primary transcript that possesses little potential for protein coding. *RNA* **4**, 445-454.

4. Bousquet-Antonelli, C., Henry, Y., Gélugne, J.-P., Caizergues-Ferrer, M., and Kiss, T. (1997) A small nucleolar RNP protein is required for pseudouridylation of eukaryotic ribosomal RNAs. *EMBO J.* **16**, 4770-4776.

5. Brown, J.W.S. and Shaw, P.J. (1998) Small nucleolar RNAs and pre-rRNA processing in plants. *Plant Cell* **10**, 649-657.

6. Caffarelli, E., Arese, M., Santoro, B.,Fragapane, P., and Bozzoni, I. (1994) *In vitro* study of processing of the intron-encoded U16 small nucleolar RNA in *Xenopus laevis*. *Mol. Cell. Biol.* **14**, 2966-2974.

7. Caffarelli, E., Fatica, A., Prislei, S., De Gregorio, E., Fragapane, P., and Bozzoni, I. (1996) Processing of the intron-encoded U16 and U18 snoRNAs: the conserved C and D boxes control both the processing reaction and the stability of the mature snoRNA. *EMBO J.* **15**, 1121-1131.

8. Cavaillé, J., Nicoloso, M., and Bachellerie, J.-P. (1996) Targeted ribose methylation of RNA *in vivo* directed by tailored antisense RNA guides. *Nature* **383**, 732-735.

9. Cecconi, F., Mariottini, P., and Amaldi, F. (1995) The *Xenopus* intron-encoded U17 snoRNA is produced by exonucleolytic processing of its precursor in oocytes. *Nucleic*

Acids Res. **23**, 4670-4676.

10. Chamberlain, J.R., Lee, Y., Lane, W.S., and Engelke, D.R. (1998) Purification and characterization of the nuclear RNase P holoenzyme complex reveals extensive subunit overlap with RNase MRP. *Genes Dev.* **12**, 1678-90.

11. Chanfreau, G., Rotondo, G., Legrain, P., and Jacquier, A. (1998) Processing of a dicistronic small nucleolar RNA precursor by the RNA endonuclease Rnt1. *EMBO J.* **17**, 3726-37.

12. Enright, C. A., Maxwell, E.S., Eliceiri, G.L., and Sollner-Webb, B. (1996) 5'ETS rRNA processing facilitated by four small RNAs: U14, E3, U17, and U3. *RNA* **2**, 1094-1099.

13. Filipowicz, W. and Kiss, T. (1993) Structure and function of nucleolar snRNPs. *Mol. Biol. Reports* **18**, 149-156.

14. Ganot, P., Bortolin, M.-L., and T. Kiss. (1997) Site-specific pseudouridine formation in preribosomal RNA is guided by small nucleolar RNAs. *Cell* **89**, 799-809.

15. Ganot, P., Caizergues-Ferrer, M., and Kiss, T. (1997) The family of box ACA small nucleolar RNAs is defined by an evolutionarily conserved secondary structure and ubiquitous sequence elements essential for RNA accumulation. *Genes Dev.* **11**, 941-956.

16. Hariharan, N. and Perry, R.P. (1990) Functional dissection of a mouse ribosomal protein promoter: significance of the polypyrimidine initiator and an element in the TATA-box region. *Proc. Natl. Acad. Sci. USA* **87**, 1526-1530.

17. Kiss, T. and Filipowicz. W. (1995) Exonucleolytic processing of small nucleolar RNAs from pre-mRNA introns. *Genes Dev.* **9**, 1411-1424.

18. Kiss-Lászl6, Z., Henry, Y., Bachellerie, J.-P., Caizergues-Ferrer, M., and Kiss. T. (1996) Site-specific ribose methylation of preribosomal RNA: a novel function for small nucleolar RNAs. *Cell* **85**, 1077-1088.

19. Kowalak, J.A., Dalluge, J.J., McCloskey, J.A., and Stetter, K.O. (1994) The role of posttranscriptional modification in stabilization of transfer RNA from hyperthermophiles. *Biochemistry* **33**, 7869-76.

20. Lafontaine, D.L.J. and Tollervey, D. (1998) Birth of the snoRNPs: The evolution of the modification guide snoRNAs. *Trends Biochem. Sci.* (in press).

21. Lafontaine, D.L.J., Bousquet-Antonelli, C., Henry, Y., Caizergues-Ferrer, M., and Tollervey, D. (1998) The box H+ACA snoRNAs carry Cbf5p, the putative rRNA pseudouridine synthase. *Genes Dev.* **12**, 527-537.

22. Leader, D.J., Clark, G.P., Watters, J., Beven, A.F., Shaw, P.J., and Brown, J.W.S. (1997) Clusters of multiple different small nucleolar RNA genes in plants are expressed as and processed from polycistronic pre-snoRNAs. *EMBO J.* **16**, 5742-5751.

23. Leverette, R.D., Andrews M.T., and Maxwell, E.S. (1992) Mouse U14 snRNA is a processed intron of the cognate hsc70 heat shock pre-messenger RNA. *Cell* **71**, 1215-1221.

24. Maden, B.E. (1990) The numerous modified nucleotides in eukaryotic ribosomal RNA. *Prog. Nucleic Acid Res. Mol. Biol.* **39**, 241-303.

25. Maxwell, E.S. and Fournier, M.J. (1995) The small nucleolar RNAs. *Annu. Rev. Biochem.* **64**, 897-934.

26. Meier, U.T. and Blöbel, G. (1994) NAP57, a mammalian nucleolar protein with a putative homolog in yeast and bacteria. *J. Cell Biol.* **127**, 1505-1514.

27. Meyuhas, O., Avni, D., and Shama, S. (1996) *Translational Control*, Cold Spring Harbor Laboratory Press, Cold Spring Harbor.

28. Morrissey, J.P. and Tollervey, D. (1995) Birth of the snoRNPs: the evolution of RNase MRP and the eukaryotic pre-rRNA-processing system. *Trends Biochem. Sci.* **20**, 78-82.

29. Ni, J. , Tien, A.L., and Fournier, M.J. (1997) Small nucleolar RNAs direct site-specific synthesis of pseudouridine in ribosomal RNA. *Cell* **89**, 565-573.

30. Ooi, S.L., Samarsky, D.A., Fournier, M.J., and Boeke, J.D. (1998) Intronic snoRNA biosynthesis in *Saccharomyces cerevisiae* depends on the lariat-debranching enzyme: intron length effects and activity of a precursor snoRNA. *RNA* **4**, 1096-1110.

31. Pelczar, P. and Filipowicz, W. (1998) The host gene for intronic U17 small nucleolar RNAs in mammals has no protein-coding potential and is a member of the 5'-terminal oligopyrimidine gene family. *Mol. Cell. Biol.* **18**, 4509-4518.

32. Petfalski, E., Dandekar, T., Henry Y., and Tollervey D. (1998) Processing of the precursors to small nucleolar RNAs and rRNAs requires common components. *Mol. Cell. Biol.* **18**, 1181-1189.

33. Safrany, G. and Perry, R.P. (1995) The relative contributions of various transcription factors to the overall promoter strength of the mouse ribosomal protein L30 gene. *Eur. J. Biochem.* **230**, 1066-1072.

34. Smith, C.M. and Steitz, J.A. (1997) Sno storm in the nucleolus: new roles for myriad

small RNPs. *Cell* **89**, 669-672.

35. Smith, C.M. and Steitz, J.A. (1998) Classification of gas5 as a multi-snoRNA host gene and a member of the 5' terminal oligopyrimidine gene family reveals common features of snoRNA host genes. *Mol. Cell. Biol.* (in press).

36. Sollner-Webb, B. (1993) Novel intron-encoded small nucleolar RNAs. *Cell* **75**, 403-405.

37. Sollner-Webb, B., Tycowski, K.T. and Steitz, J.A. (1996) *Ribosomal RNA: Structure, Evolution, Processing and Function in Protein Biosynthesis*, CRC Press, Boca Raton.

38. Tollervey, D. and Kiss, T. (1997) Function and synthesis of small nucleolar RNAs. *Curr. Opin. Cell Biol.* **9**, 337-342.

39. Tycowski, K.T., Shu, M.-D., and Steitz, J.A. (1996) A mammalian gene with introns instead of exons generating stable RNA products. *Nature* **379**, 464-466.

40. Tycowski, K.T., Smith, C.M., Shu, M.-D., and Steitz, J.A. (1996) A small nucleolar RNA requirement for site-specific ribose methylation of rRNA in *Xenopus*. *Proc. Natl. Acad. Sci. USA* **93**, 14480-14485.

41. Venema, J. and Tollervey, D. (1995) Processing of pre-ribosomal RNA in *Saccharomyces cerevisiae*. *Yeast* **11**, 1629-1650.

42. Villa, T., and Ceradini, F., Presutti, C., and Bozzoni, I. (1998) Processing of the intron-encoded U18 small nucleolar RNA in the yeast *Saccharomyces cerevisiae* relies on both exo- and endonucleolytic activities. *Mol. Cell. Biol.* **18**, 3376-3383.

RNA STRUCTURE MODULES WITH TRINUCLEOTIDE REPEAT MOTIFS

W.J. KRZYZOSIAK, M. NAPIERALA, M. DROZDZ
Institute of Bioorganic Chemistry Polish Academy of Sciences
Noskowskiego 12/14, 61-704 Poznań, Poland

A new mutational basis for human disease was discovered in the beginning of this decade. The genes responsible for fragile X syndrome, spinobulbar muscular atrophy and later the genes causing myotonic dystrophy and Huntington disease were shown to harbor unstable, expanding trinucleotide repeats. The dynamic nature of these mutations explained intriguing features of clinical symptoms observed in families inheriting these diseases. The increasing disease penetrance and expresivity in subsequent family generations could be correlated with the increasing number of trinucleotide repeats. Now, thirteen human hereditary neurological diseases are known to have their underlying cause in the trinucleotide repeat expansions [1] and their list is likely to grow.

A molecular cytopathogenesis of these diseases is the area of intensive research. In some of the disorders, mutant transcripts are thought to participate actively in developing the pathological phenotype (Table1). This applies to myotonic dystrophy [2-4] and to other diseases containing repeats in nontranslated sequences i.e. Friedreich ataxia [5] and fragile X syndrome [6]. The RNA repeats located in translated regions may contribute to the pathogenesis of such disorders as spinobulbar muscular atrophy, Huntington disease and spinocerebellar ataxias [7]. They may play some important regulatory or modulatory functions, involving specific repeat binding proteins. These functions may be enhanced, silenced or new functions may be gained upon the repeat expansion. However, the major pathogenic effect in these diseases seems to occur at the protein level and is most likely caused by the expanded polyglutamine tract [8].

Several years ago, we have begun to explore RNA structures formed by various trinucleotide repeats. We have been analyzing these structures within the sequence context of their natural hosts - transcripts of human genes implicated in the dynamic mutation diseases. We believe that our studies will help understanding the role played by triplet repeat regions in normal and pathological function of these RNAs.

J. Barciszewski and B.F.C. Clark (eds.), RNA Biochemistry and Biotechnology, 303–314.

TABLE 1. Human hereditary diseases caused by trinucleotide repeat expansion and selected features of the corresponding genes and transcripts, relevant to postulated RNA level effects in pathogenesis.

Disease	Gene	Repeat¶	Repeat location	Copy number		Interruptions	RNA BP‡	RNA level effect
				normal	mutant			
Fragile X syndrome	FMR1 (FRAXA)	CGG	5' UTR	6 -52 60-200 p**	200 - >2000	AGG	–	Yes
Fragile XE mental retardation	FMR2 (FRAXE)	GCC	5'UTR	7 - 35 130-150p	230-750	–	–	?
Myotonic dystrophy	DMPK	CUG	3'UTR	5 - 37	50 - >2000	–	+	Yes
Friedreich's ataxia	FRDA	GAA	intron 1	6 - 34 80 p	112 - 1700	–	–	Yes
Huntington's disease	IT15	CAG (GCC)*	ORF	6 - 39	36 - 121	–	+	?
Dentatorubral-pallidoluysian atrophy	DRPLA (B37)	CAG	ORF	6 - 35	51 - 88	–	+	?
Spinobulbar muscular atrophy	AR	CAG	ORF	11 - 33	38 - 66	–	+	?
Spinocerebellar ataxia type 1	SCA1	CAG	ORF	6 - 39	41 - 81	CAU	+	?
Spinocerebellar ataxia type 2	SCA2	CAG	ORF	14 - 31	35 - 64	CAA	+	?
Spinocerebellar ataxia type 3	SCA3 (MJD1)	CAG	ORF	12 - 41	40 - 84	–	+	?
Spinocerebellar ataxia type 6	CACNA1A	CAG	ORF	7 - 18	21 - 27	–	+	?
Spinocerebellar ataxia type 7	SCA7	CAG	ORF	7 - 17	38 - 130	–	+	?
Oculopharyngeal muscular dystrophy	PABP2	GCG	ORF	6 - 7	7 - 13	–	–	?

* accompanying polymorphic repeat located in close vicinity of CAG, ** premutation, ‡ RNA repeat binding protein known, ¶ transcript sequence shown.

1. Experimental strategy

All principal steps of the experimental procedure used in our laboratory in structural analysis of RNA repeats are shown in the diagram in Fig. 1. They include selection of the appropriate RNA fragment, synthesis of its corresponding DNA template, in vitro transcription and structure probing in solution. The standard, straightforward approach to structure probing is supplemented in this protocol with two alternative procedures. They are designed to cope with RNA conformational heterogeneity when it occurs under structure probing conditions. Some of the steps shown in the diagram are illustrated by selected experimental results in Figs. 2 and 3.

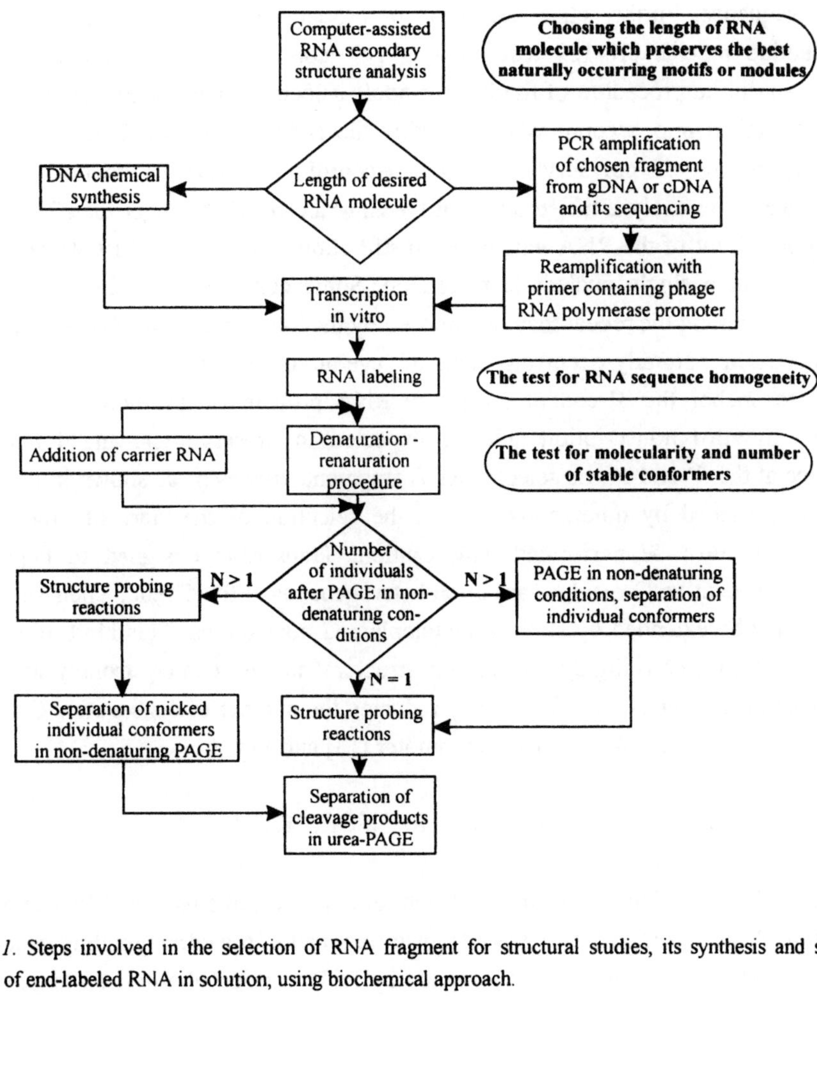

Figure 1. Steps involved in the selection of RNA fragment for structural studies, its synthesis and structure probing of end-labeled RNA in solution, using biochemical approach.

1.1. SIZE OF RNA MODULE

When approaching structure analysis of RNA motif of potential functional significance, present in a large mRNA molecule, it is important to select properly a small RNA module, which harbors this motif and is likely to preserve its putative biological function. In order to achieve that, we perform computer-assisted search of the lowest energy secondary structures [9] formed by the repeat regions of transcripts. The lengths of the repeats which are polymorphic in human population (Table 1), as well as the lengths of flanking sequences, are variables in this analysis. The known orthologous sequences from evolutionarily closest species are usually subjected to similar analysis. Their structures may be helpful in selecting the appropriate RNA module formed by human sequence. Further characterization of proteins which bind specifically to the trinucleotide repeats [10-12] will make a gel mobility shift test another valuable component in the procedure of the RNA module selection. A technical criterion, which has to be taken into account when selecting the right size of the RNA fragment, is resolving power of polyacrylamide gel. According to our experience it is not recommended to analyze the structure of the end-labeled RNA longer than 250 nt. This is the upper limit of the RNA length which still allows the details of its structure to be analyzed with a single nucleotide resolution. Some constraints on the length of the sequences flanking the repeat may be imposed by the location of short G-tracts upstream the repeat. The naturally occurring tracts of several consecutive G-residues are usually chosen to anchor the 3'-end of T7 phage RNA polymerase promoter. They ensure efficient in vitro transcription and are expected to preserve unchanged nucleotide sequence at the 5'-end of the selected RNA fragment. However, as shown in Fig. 3 and recently described by other authors [13], the selection of the start site for in vitro transcription must be performed with caution. Transcripts designed to begin with multiple consecutive G-residues are often heterogeneous at the 5'-end. Their substantial portion may be extended by several nontemplate G-nucleotides. This kind of sequence heterogeneity, which is highly undesired in structural studies, can be strongly suppressed at the cost of transcription efficiency by reducing the number of consecutive G-residues in the 3'-end of T7 RNA polymerase promoter [13] and Fig. 3.

1.2. DNA TEMPLATE FOR IN VITRO TRANSCRIPTION

A typical RNA module selected for structure analysis contains 100-200 nucleotides. Therefore PCR is used to isolate the adequate fragment of the corresponding gene. First, human genomic DNA samples are screened by PCR to select the desired length of the

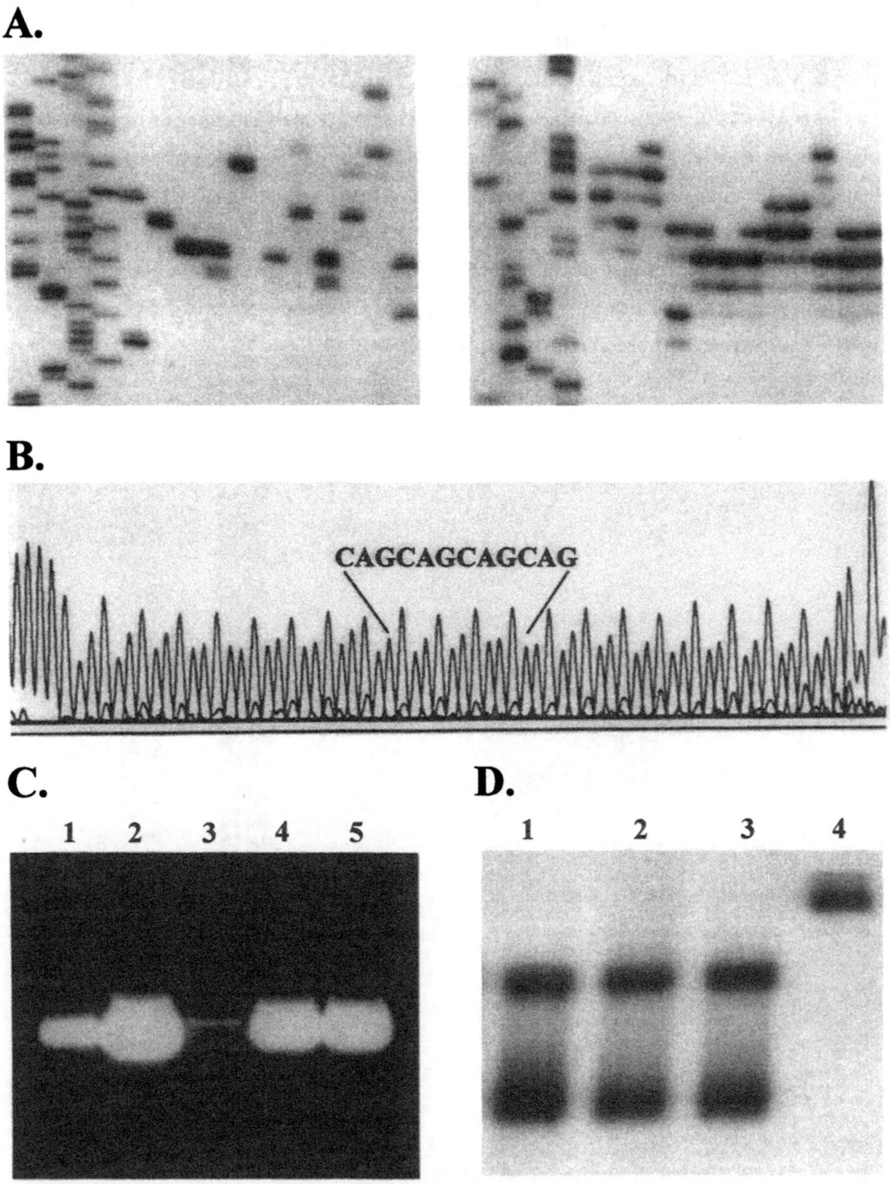

Figure 2. Selected experimental results illustrating steps preceding RNA structure probing. *(A.)* Screening of genomic DNA samples by PCR to select required polymorphic variant of the repeated sequence: AR gene (left) and SCA1 gene (right). *(B.)* Sequencing of chosen allele of DMPK gene – noncoding strand. *(C.)* Agarose gel electrophoresis of the in vitro transcription products. Lanes 1 and 2 - reference RNA of known concentration; 3 - DNA template; lanes 4 and 5 - transcripts of the repeat region of the DMPK gene. *(D.)* Polyacrylamide gel electrophoresis of the 5'-end labeled FMR-1 mRNA fragment in nondenaturing conditions. Lanes 1, 2, 3 – coexisting stable conformers of monomeric molecule, lane 4 – RNA duplex composed of two complementary strands.

polymorphic repeat. This is determined by comparing the migration of radiolabeled PCR product to M13 sequencing ladder in a denaturing polyacrylamide gel (Fig. 2A). A good practice is to sequence the selected templates (Fig. 2B) as the repeats in some genes may contain interruptions (Table 1), and localization of these interruptions needs to be known. Then, the DNA template for in vitro transcription is prepared basing on the original PCR product, by seminested or nested PCR. A new forward primer contains the sequence of T7 RNA polymerase promoter. In vitro transcription is performed under standard conditions [14, 15] and its product is usually purified on denaturing polyacrylamide gel before labeling. The labeled transcript may be analyzed for the end-sequence homogeneity at this point.

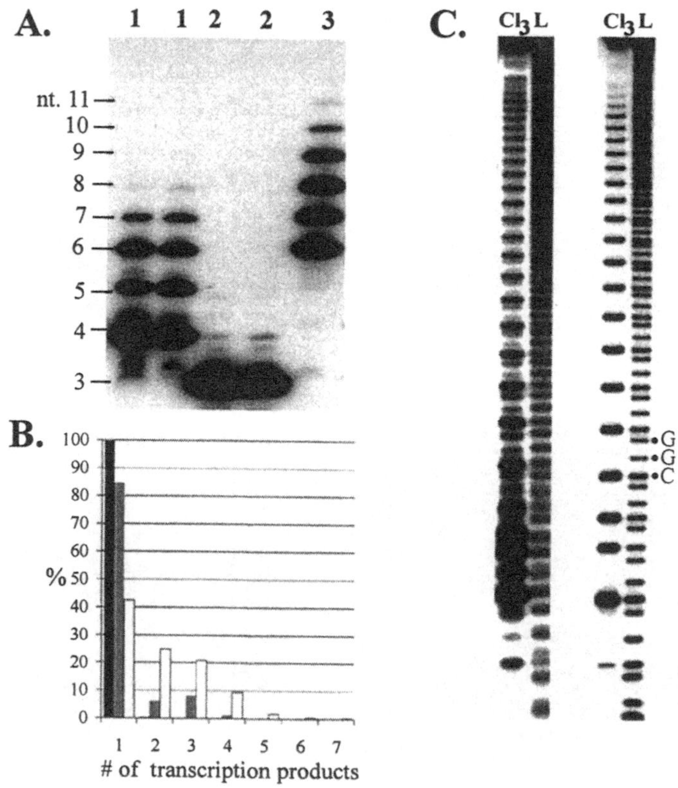

Figure 3. The 5'-end heterogeneity in in vitro transcripts designed to begin with different numbers of guanine residues. *(A.)* Analysis of the 5'-terminal oligonucleotides generated from a set of FMR-1 transcripts by complete digestion with pancreatic ribonuclease. The transcripts were designed to have the following sequences at the 5'-end: 1. **GGG**CGU...; 2. **GG**CGU...and 3. **GGGGG**CGU.... *(B.)* Quantitative representation of the results shown in panel A: transcript 1 (grey); transcript 2 (black); transcript 3 (white). *(C.)* Influence of the 5'-end heterogeneity on the quality of Cl_3 and formamide ladders generated from the 5'-end-labeled transcripts: 3 (left) and 2 (right).

1.3. TEST FOR MOLECULARITY AND NUMBER OF STABLE CONFORMERS

RNA molecules may form two or more stable conformers as well as dimers or multimers under conditions in which their structure is probed. A test for structure homogeneity is therefore strongly recommended prior to structure probing of any new RNA fragment. If a polyacrylamide gel electrophoresis in nondenaturing conditions shows two stable conformers (Fig. 2D), their structures can be established using the approaches shown in Fig. 1. In one of them, the intact conformers are first separated in native gel, and then their structure is probed after skipping the initial denaturation/renaturation step. Alternatively, structure probing is performed on the coexisting stable conformers and the partially nicked species, which migrate in native gel with the same rate as the intact conformers, are separated and analyzed in urea-containing gel. Each of these approaches has its advantages and disadvantages which are discussed elsewhere [16].

1.4. RNA STRUCTURE PROBING IN SOLUTION.

Chemical and enzymatic probing of nucleotide accessibility remains a method of choice to obtain secondary structure information rapidly, using small amounts of RNA. A number of endonucleases and chemical reagents capable of probing different features of RNA structure is available [17, 18]. Taking advantage of their distinct specificity, secondary structures of relatively large RNA molecules can be determined usually with the help of the thermodynamic method of structure prediction. There are two well established approaches to RNA structure probing in solution which have been described a decade ago [19, 20]. The first one detects chain scissions induced by nucleases or chemical reagents in end-labeled RNA. The second, relies on using reverse transcription to detect cleavage or modification sites in unlabeled RNA. In our studies the first approach is more frequently used and the preliminary structure probing experiments are usually performed with lead ions and nucleases which are highly informative and easy to handle. The advantageous feature of lead ions is the fact that the deprotonated lead-ion hydrates, which are the active species [21], are much smaller than nucleases, penetrate folded RNA easier and reveal more details of the analyzed structures [15]. The reaction conditions used in RNA structure probing by lead cleavage are basically the same as originally described [22]. Typically, RNA at about 10 μM concentration is treated with lead ions at 0.2-2 mM concentration, at pH close to neutral, in the presence of sodium and magnesium ions. In these conditions, lead ions usually differentiate between rigid double-stranded and flexible single-stranded regions of the RNA structure [23, 24]. The former are resistant and the latter susceptible to cleavage in agreement with the mechanism proposed for phosphodiester bond cleavage by lead [25].

2. First Results and Future Prospects

2.1. CUG REPEATS IN DMPK RNA

The number of CUG repeats present in normal alleles of the myotonic dystrophy protein kinase (DMPK) gene varies between 5 and 37, and the expanded repeats, found in mildly affected patients, start at 50 repeats (Table 1). For in vitro transcription and RNA structure studies we have selected the representatives of the most frequent normal alleles and one expanded allele. They contained 5, 11, 21 and 49 repeats. In all transcripts short fragments of natural sequences flanking the repeats were present, about 30 nucleotides at each end. The longest analyzed RNA module contained 212 nt. Using lead-induced cleavages and enzymatic digestion we could demonstrate [26] that $(CUG)_5$ did not form any stable secondary structure while hairpins of increasing stability were formed by $(CUG)_{11}$, $(CUG)_{21}$ and $(CUG)_{49}$. The increased stability of the stem structure in longer hairpins correlated well with the decreased efficiency of lead-induced cleavages what is shown in Fig. 4. This kind of reactivity of all internucleotide bonds in the stem indicated that the stem forms a novel, variably

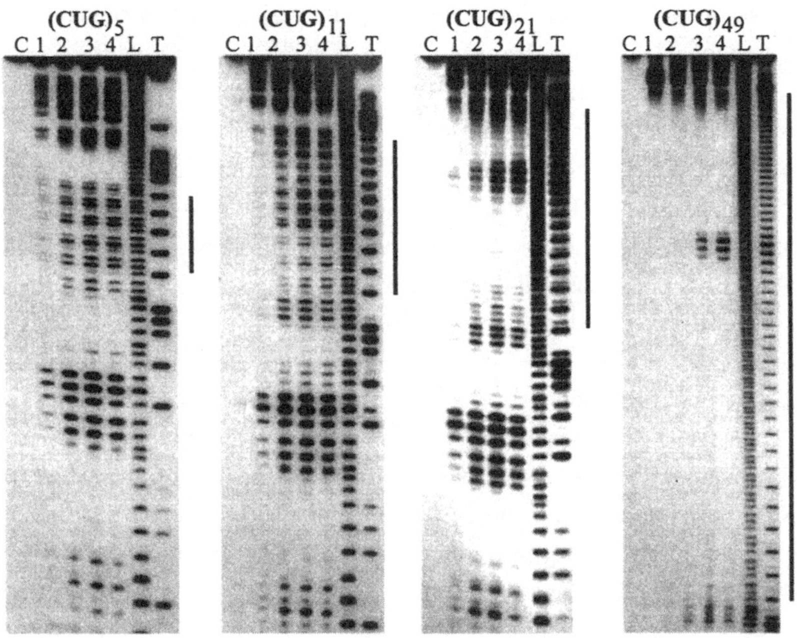

Figure 4. Susceptibility of DMPK mRNA fragments containing 5, 11, 21 and 49 CUG repeats to lead cleavage. Lanes 1-4 correspond to concentrations of lead ions increasing from 0.25 to 2.5 mM, C – incubation control, L – formamide ladder, T – limited T_1 ribonuclease digest under semidenaturing conditions. Vertical lines indicate the CUG repeat region.

relaxed type of duplex structure, rather than a structure composed of units in which two rigid base pairs C-G and G-C are followed by more flexible U-U pair. In addition to having the quasistable stem the $(CUG)_n$ hairpins were shown to be „slippery". According to the lead cleavage data, they showed several different alignments of the repeated sequence. In the coexisting variants of the $(CUG)_n$ hairpin different combinations of the central repeats are present in the hairpin loop, and different terminal repeats form protruding ends at the base of the hairpin stem (Fig. 5). The sequences flanking the repeat are single-stranded and they do not impose any constrain

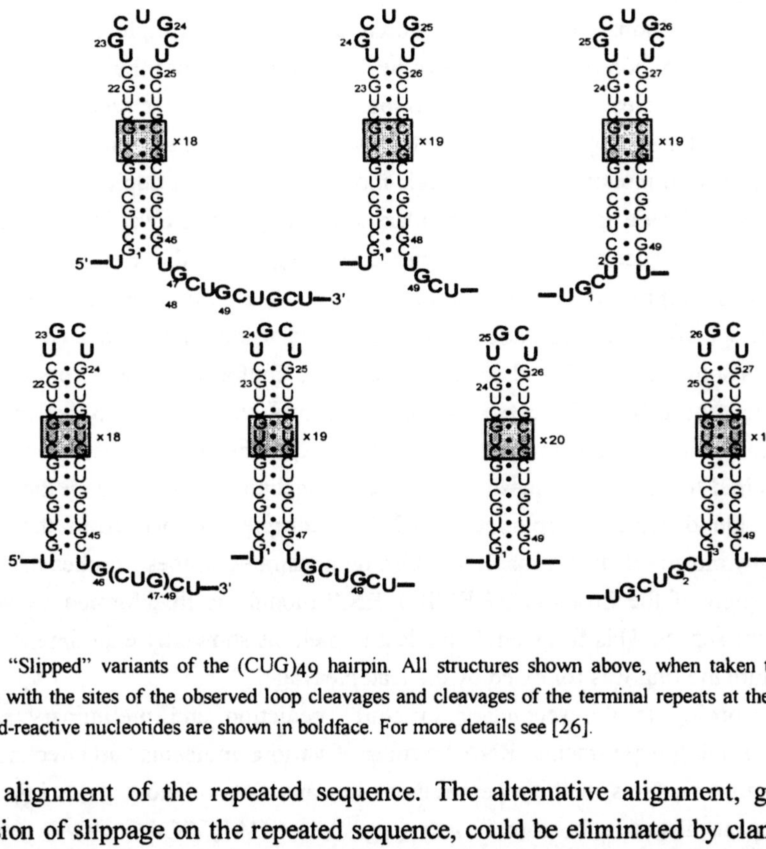

Figure 5. "Slipped" variants of the $(CUG)_{49}$ hairpin. All structures shown above, when taken together, are consistent with the sites of the observed loop cleavages and cleavages of the terminal repeats at the base of the stem. Lead-reactive nucleotides are shown in boldface. For more details see [26].

on the alignment of the repeated sequence. The alternative alignment, giving the impression of slippage on the repeated sequence, could be eliminated by clamping the repeat ends with several G-C pairs [27]. This significantly reduced the number of reactive phosphodiester bonds in the centrally located repeats forming a terminal hairpin loop.

In several recently published papers [26, 28-30] their authors hypothesized how these hairpin structures could be involved in pathogenesis of myotonic dystrophy. The experimental observations such as retention of transcripts in nucleus [31, 32], splicing defect [4] and impaired transcription of downstream genes [33, 34], can all be explained by the presence of the $(CUG)_n$ hairpins.

2.2. CGG REPEATS IN FMR-1 RNA AND CAG REPEATS

Inspection of the lowest energy secondary structures, formed by the entire FMR-1 5'UTR and its fragments of decreasing length harboring different natural variants of the repeated CGG sequence, showed that one structure of the repeat region was strongly favored. This autonomous structure containing the repeated sequence and its direct flanks was the RNA module selected for structural studies. Interestingly, some of its polymorphic variants containing the AGG interruptions migrated clearly as two distinct species in nondenaturing polyacrylamide gel (Fig. 2D). Neither different denaturation/ renaturation conditions nor different ionic conditions in the sample could significantly change the electrophoretic pattern shown in Fig. 2D. In all cases the contribution of the less prevalent species was too high to be neglected in the structure studies. The difference in rates of migration between the two species was insufficient to be explained by the presence of monomeric and dimeric molecule. The RNA duplex composed of the investigated molecule and its complementary strand migrated slower in native gel. Thus, the two discrete species formed by the repeat region of the FMR-1 mRNA were interpreted to represent two stable conformers of the monomeric molecule. The conformer-specific cleavage patterns were obtained using each of the two alternative procedures shown in Fig. 1 and structures for these conformers were proposed. One of them turned out to be a bifurcated Y-shaped structure, very similar to that predicted to be thermodynamically the most stable. Further experiments described elsewhere [35] were required to answer the question whether the second conformer is an independently folded single distorted hairpin or an RNA quadruplex formed from the Y-shaped structure by closing its two arms. It was also demonstrated in these studies that the most stable portion of the investigated FMR-1 RNA module is that formed by sequences flanking the repeat. This fragment is the last to melt, as shown by experiments in which the structure melting was followed by the lead cleavage.

According to the secondary structure prediction and preliminary results of structure probing experiments, RNA hairpins of various molecular architecture may be predominant structures of the repeat regions in transcripts of most genes implicated in triplet repeat diseases. In some cases (e.g. AR, SCA1, SCA6, SCA7 and IT15), sequences directly flanking the CAG repeats may form base-paired regions stabilizing the relaxed stem sections composed of the repeated sequence. In this respect they may resemble the FMR-1 module. In other cases (e.g. SCA3 and DRPLA), the „slippery" hairpins similar to that present in DMPK RNA may occur. Many questions regarding the details of these structures and their interactions with the repeat binding proteins need to be answered. The existence of the hairpin structures in vivo is to be demonstrated and their biological significance remains to be shown.

3. Acknowledgments

The research was partially supported by the grant 6P04B00212 from the Polish Committee for Scientific Research and by the Foundation For Polish Science in frame of program „DIAMOL". M.N. thanks the FFPS for 98' fellowship.

4. References

1. Wells, R.D. and Warren, S.T. (1998) *Genetic Instabilities and Hereditary Neurological Diseases,* Academic Press, San Diego.
2. Wang, J., Pegoraro, E., Menegazzo, E., Gennarelli, M., Hoop, R.C., Angelini, C., and Hoffman, E.P. (1995) Myotonic dystrophy: evidence for a possible dominant-negative RNA mutation, *Hum. Mol. Genet.* **4**, 599-606.
3. Korade-Mirnics, Z., Babitzke, P., and Hoffman, E.P. (1998) Myotonic dystrophy: molecular windows on a complex etiology, *Nucleic Acids Res.* **26**, 1363-1368.
4. Philips, A.V., Timchenko, L.T., and Cooper T.A. (1998) Disruption of splicing regulated by a CUG-binding protein in myotonic dystrophy. *Science* 280, 737-741.
5. Ohshima, K., Montermini, L., Wells, R.D. and Pandolfo, M. (1998) Inhibitory effects of expanded GAA.TTC triplet repeats from intron I of the Friedreich ataxia gene on transcription and replication in vivo. *J. Biol. Chem.* **273**, 14588-14595.
6. Feng, Y., Zhang, F., Lokey, L.K., Chastain, J.L., Lakkis, L., Eberhart, D., and Warren S.T. (1995) Translational suppression by trinucleotide repeat expansion at FMR1. *Science* **268**, 731-734.
7. McLaughlin, B.A., Spencer, C., and Eberwine, J. (1996) CAG trinucleotide RNA repeats interact with RNA-binding proteins. *Am. J. Hum. Genet.* **59**, 561-569.
8. Lunkes, A. and Mandel, J.L. (1997) Polyglutamines, nuclear inclusions and neurodegeneration. *Nat Med.* **3**, 1201-2.
9. Zuker, M. (1989) On finding all suboptimal foldings of an RNA molecule. *Science* **244**, 48-52.
10. Timchenko, L.T., Timchenko, N.A., Caskey, C.T., and Roberts, R. (1996) Novel proteins with binding specificity for DNA CTG repeats and RNA CUG repeats: implications for myotonic dystrophy. *Hum. Mol. Genet.* **5**, 115-121.
11. Timchenko, L.T., Miller, J.W., Timchenko, N.A., DeVore, D.R., Datar, K.V., Lin, L., Roberts, R., Caskey, C.T., and Swanson, M.S. (1996) Identification of a (CUG)n triplet repeat RNA-binding protein and its expression in myotonic dystrophy. *Nucleic Acids Res.* **15**, 4407-4414.
12. Bhagwati, S., Ghatpande, A., and Leung, B. (1996) Identification of two nuclear proteins which bind to RNA CUG repeats: significance for myotonic dystrophy. *Biochem. Biophys. Res. Commun.* **228**, 55-62.
13. Pleiss, J.A., Derrick, M.L., and Uhlenbeck, O.C. (1998) T7 RNA polymerase produces 5' end heterogeneity during in vitro transcription from certain templates. *RNA* **4**, 1313-1317.
14. Krzyżosiak, W.J., Denman, R., Nurse, K., Hellmann, W., Boublik, M., Gehrke, C.W., Agris, P.F., and Ofengand, J. (1987) In vitro synthesis of 16S ribosomal RNA containing single base changes and assembly into a functional 30S ribosome. *Biochemistry* **26**, 2353-2364.
15. Michałowski, D., Wrzesinski, J., and Krzyżosiak, W.J. (1996) Cleavages induced by different metal ions in yeast tRNA(Phe) U59C60 mutants. *Biochemistry* **35**, 10727-10734.
16. Napierała, M., Michałowski, D., and Krzyżosiak W.J. (1998) in preparation.
17. Knapp, G. (1989) Enzymatic approaches to probing of RNA secondary and tertiary structure. *Methods Enzymol.* **180**, 192-212.

18. Jaeger, J.A., SantaLucia, J. Jr, Tinoco, I. Jr. (1993) Determination of RNA structure and thermodynamics. *Annu. Rev. Biochem.* **62**, 255-287.
19. Ehresmann, C., Baudin, F., Mougel, M., Romby, P., Ebel, J.P., and Ehresmann, B. (1987) Probing the structure of RNAs in solution. *Nucleic. Acids Res.* **15**, 9109-9128.
20. Krol, A. and Carbon, P. (1989) A guide for probing native small nuclear RNA and ribonucleoprotein structures. *Methods Enzymol.* **180**, 212-227
21. Brown, R.S., Hingerty, B.E., Dewan, J.C., and Klug, A. (1983) Pb(II)-catalysed cleavage of the sugar-phosphate backbone of yeast tRNAPhe--implications for lead toxicity and self-splicing RNA. *Nature* **303**, 543-546.
22. Krzyżosiak, W.J., Marciniec, T., Wiewiorowski, M., Romby, P., Ebel, J.P., and Giege, R. (1988) Characterization of the lead(II)-induced cleavages in tRNAs in solution and effect of the Y-base removal in yeast tRNAPhe. *Biochemistry* **27**, 5771-5777.
23. Ciesiołka, J., Michałowski, D., Wrzesinski, J., Krajewski, J., and Krzyżosiak, W.J. (1998). Patterns of cleavages induced by lead ions in defined RNA secondary structure motifs. *J. Mol. Biol.* **275**, 211-220.
24. Górnicki, P., Baudin, F., Romby, P., Wiewiórowski, M., Krzyżosiak, W., Ebel, J.P., Ehresmann, C., and Ehresmann, B. (1989) Use of lead(II) to probe the structure of large RNA's. Conformation of the 3' terminal domain of E. coli 16S rRNA and its involvement in building the tRNA binding sites. *J. Biomol. Struct. Dyn.* **6**, 971-984.
25. Brown, R.S., Dewan, J.C., and Klug, A. (1985) Crystallographic and biochemical investigation of the lead(II)-catalyzed hydrolysis of yeast phenylalanine tRNA. *Biochemistry* **24**, 4785-4801.
26. Napierała, M. and Krzyżosiak, W.J. (1997) CUG repeats present in myotonin kinase RNA form metastable "slippery" hairpins. *J. Biol. Chem.* **272**, 31079-31085.
27. Napierała, M., and Krzyżosiak W.J. (1997) unpublished.
28. Mitas, M., Yu, A., Dill, J., Kamp, T.J., Chambers, E.J., and Haworth, I.S. (1995) Hairpin properties of single-stranded DNA containing a GC-rich triplet repeat: (CTG)15. *Nucleic Acids Res.* **23**, 1050-1059.
29. Mariappan, S.V., Chen, X., Catasti, P., Bradbury, E.M., and Gupta, G. (1998) Structural Studies on the Unstable Triplet Repeats in R.D. Wells and S.T. Warren (eds.), *Genetic Instabilities and Hereditary Neurological Diseases,* Academic Press, San Diego, pp. 647-676.
30. Koch, K.S., Leffert, H.L. (1998) Giant hairpins formed by CUG repeats in myotonic dystrophy messenger RNAs might sterically block RNA export through nuclear pores. *J. Theor. Biol.* **192**, 505-514.
31. Davis, B.M., McCurrach, M.E., Taneja, K.L., Singer, R.H., annd Housman, D.E. (1997) Expansion of a CUG trinucleotide repeat in the 3' untranslated region of myotonic dystrophy protein kinase transcripts results in nuclear retention of transcripts. *Proc. Natl. Acad. Sci. USA* **94**, 7388-7393.
32. Hamshere, M.G., Newman, E.E., Alwazzan, M., Athwal, B.S., and Brook JD (1997) Transcriptional abnormality in myotonic dystrophy affects DMPK but not neighboring genes. *Proc. Natl. Acad. Sci. USA* **94**, 7394-7399.
33. Klesert, T.R., Otten, A.D., Bird, T.D., and Tapscott, S.J. (1997) Trinucleotide repeat expansion at the myotonic dystrophy locus reduces expression of DMAHP. *Nat. Genet.* **16**, 402-406.
34. Thornton, C.A., Wymer, J.P., Simmons, Z., McClain, C., and Moxley, R.T. (1997) Expansion of the myotonic dystrophy CTG repeat reduces expression of the flanking DMAHP gene. *Nat. Genet.* **16**, 407-409.
35. Napierała, M., Drozdz, M., Michałowski, D., and Krzyżosiak W.J. (1998) in preparation.

PHOSPHOROTHIOATE OLIGONUCLEOTIDES AS APTAMERS OF RETROVIRAL REVERSE TRANSCRIPTASES

M. KOZIOŁKIEWICZ*, A. KRAKOWIAK, A. OWCZAREK, M. BOCZKOWSKA
Polish Academy of Sciences, Centre of Molecular and Macromolecular Studies, Department of Bioorganic Chemistry, Sienkiewicza 112, 90-363 Łódź, Poland; E-mail: mkoziol@bio.cbmm.lodz.pl

1. Introduction

Retroviral reverse transcriptases catalyze the synthesis of a double-stranded DNA copy of the RNA genome for integration into host chromosome. These enzymes possess three enzymatic activities essential for retrovirus replication: an RNA-dependent DNA polymerase, a DNA-dependent DNA polymerase which synthesizes the second strand of the proviral DNA, and an RNase H which degrades RNA template after the synthesis of the first strand of the proviral DNA. The importance of reverse transcriptase (RT) in the life cycle of retroviruses (*e.g.* HIV) makes the enzyme a preferred target for antiviral strategies [1].

Among many agents which are being considered to inhibit reverse transcriptase activity there are phosphorothioate analogues of oligonucleotides (PS-oligos) [2-6]. They can influence the RT activity by oligonucleotide sequence-independent [2, 3] and/or sequence-dependent mechanism [4-6]. A proposed mechanism for sequence-dependent antiviral activity of PS-oligos involves their hybridization to viral RNA and formation of substrates for RNase H associated with RT enzyme. This ribonuclease cleaves RNA fragments involved in the formation of RNA/PS-oligos duplexes and inhibits full-length cDNA synthesis [4-6]. For explanation of sequence-independent antiviral effect, the direct interaction of PS-oligos with reverse transcriptase has been postulated [3]. Phosphorothioate oligonucleotides bind to the enzyme with high affinity and, therefore, can competitively inhibit the synthesis of cDNA. This increased affinity is probably a result of the fact that, at least for some proteins, the dissociation rate of the PS-oligo/protein complex is much lower than that for the corresponding PO-oligo [7]. It has been suggested that the sequence-independent inhibitory effect of PS-oligos relies upon the total number of internucleotide phosphorothioate linkages rather than the oligonucleotide length or the position of modified bonds within the oligomer [8]. The sequence-independent inhibition by PS-oligos has been reported not only for reverse transcriptase but also for human DNA polymerases and human RNase H [8]. It should be noticed that in these studies PS-oligos have been used as mixtures of 2^n diastereomers, where n is a number of internucleotide phosphorothioate linkages.

Our earlier results have indicated that the direct binding of PS-oligos to the RT protein and their inhibitory effect against the enzyme can depend upon a sequence of oligonucleotide as well as upon the absolute configuration at P-atoms of internucleotide phosphorothioate bonds. We have demonstrated that the stereoregular PS-oligonucleotide d[AAG CAT ACG GGG TGT] containing phosphorothioate internucleotide functions of [R_P]-configuration effectively inhibits the AMV RT although this oligomer is not complementary to the RNA template and, therefore, cannot activate RNase H. We have suggested sequence-dependent

J. Barciszewski and B.F.C. Clark (eds.), RNA Biochemistry and Biotechnology, 315–324.

aptameric interaction of this oligonucleotide with the AMV RT enzyme [9]. The sequence-selective mode of RT inhibition by PS-oligonucleotides was earlier postulated also by Tamura *et al.* [10].

In this report we present a more detailed data on the sequence-dependent inhibition of AMV and HIV reverse transcriptases by stereoregular PS-oligos. We have examined several oligonucleotides of different nucleotide sequences. Some of them contain contiguous four G bases which can be responsible under *in vitro* conditions for the formation of tetraplex structures of oligonucleotides [11].

2. Inhibition of AMV Reverse Transcriptase

The oligonucleotides listed in Table 1 were used for the studies on the inhibition of retroviral reverse transcriptases by PS-oligos. Polyribonucleotide PO-**6** (475 nt) used as a template for RT enzymes was obtained by *in vitro* transcription using plasmid PT7-7* containing the interleukin-2 (Il-2) gene [9, 12]. Because the template PO-**6** does not contain any fragment complementary to the oligonucleotides **1-4**, the presence of these oligomers in the reaction mixture allowed us to test their ability to block the reverse transcription only by binding to the RT enzyme. At the first stage of our studies we used as potential inhibitors of the AMV RT the oligonucleotide $d[T]_{19}$ (PO-**1**) and its phosphorothioate analogues (PS-**1a-c**), while unmodified oligomer (PO-**5**) of the sequence d[AAA GGT AAT CCA TCT GTT CA] was used as a primer for the enzyme. Inhibitory effect of the oligonucleotides **1** was studied at their concentration ranging from 65 nM to 1.2 μM.

Reverse transcription of RNA template (PO-**6**) catalyzed by AMV RT (used at 5 nM concentration) gave the predicted cDNA fragment of 437 nucleotides. However, PAGE analysis of resulting products indicated that the intensity of the band corresponding to this full-length product was strongly influenced by the presence of some phosphorothioate oligonucleotides in the reaction mixture. The phosphorothioate analogues of the oligomer **1** (**1a-c**) caused 50% inhibition of the AMV RT activity at the 400 nM concentration, independently of the absolute configuration at the phosphorus atoms of their phosphorothioate internucleotide functions.

Phosphorothioate analogues of the oligonucleotides **2** and **3**, especially their [all-R_P]-isomers, appeared to be much stronger inhibitors of the AMV RT activity than the oligonucleotides **1a-c**. The oligonucleotides **2a** and **3a** have inhibited the enzyme with IC_{50} (50% of inhibition) at conc. 20-30 nM. Under the same conditions (30 nM conc. of the oligonucleotide) [all-S_P]-isomers did not show any inhibitory effect. The increase of their concentration to 130 nM gave 90% inhibition of the RT activity (Figs. 1 and 2).

Under the same conditions unmodified phosphodiester oligonucleotides PO-**2** and PO-**3** did not cause significant inhibition of the enzyme activity. This observation allows to suggest that the presence of several phosphorothioate linkages within oligonucleotide chain is a prerequisite for high affinity binding of PS-oligonucleotides to the RT protein. This is in agreement with observation of Gao *et al.* that chimeric PO/PS oligonucleotides containing 18 or 27 phosphorothioate linkages were stronger inhibitors of human DNA polymerases than the oligomers containing only 9 phosphorothioate linkages [8].

TABLE 1. PO- and PS-oligonucleotides **1-4** used for RT activity studies

Compound no.	Sequence 5'→3'	Characteristics of internucleotide bonds
PO-**1**	d[T]$_{19}$	[PO]$_{18}$[a]
PS-**1a**	d[T]$_{19}$	[R$_P$]$_{18}$[b]
PS-**1b**	d[T]$_{19}$	[S$_P$]$_{18}$[c]
PS-**1c**	d[T]$_{19}$	[mix]$_{18}$[d]
PO-**2**	d[AAC GTT GAG GGG CAT]	[PO]$_{14}$[a]
PS-**2a**	d[AAC GTT GAG GGG CAT]	[R$_P$]$_{14}$[b]
PS-**2b**	d[AAC GTT GAG GGG CAT]	[S$_P$]$_{14}$[c]
PS-**2c**	d[AAC GTT GAG GGG CAT]	[mix]$_{14}$[d]
PO-**3**	d[AAG CAT ACG GGG TGT]	[PO]$_{14}$[a]
PS-**3a**	d[AAG CAT ACG GGG TGT]	[R$_P$]$_{14}$[b]
PS-**3b**	d[AAG CAT ACG GGG TGT]	[S$_P$]$_{14}$[c]
PS-**3c**	d[AAG CAT ACG GGG TGT]	[mix]$_{14}$[d]
PO-**4**	d[TTG GGG TT]	[PO]$_7$[a]
PS-**4a**	d[TTG GGG TT]	[R$_P$]$_7$[b]
PS-**4b**	d[TTG GGG TT]	[S$_P$]$_7$[c]
PS-**4c**	d[TTG GGG TT]	[mix]$_7$[d]

[a] Oligomers PO-**1-4** contain only unmodified phosphodiester bonds (PO).
[b,c] PS-oligos **1-4a** and **1-4b** contain internucleotide phosphorothioate linkages of R$_P$ or S$_P$ configuration, respectively.
[d] Oligomers [mix] consist of a mixture of all possible diastereomers.

Unmodified oligonucleotides PO-**1-4** were prepared by the phosphoramidite method on an ABI 391 synthesizer. PS-oligos **1a-b**, **2a-b**, **3a-b** and **4a-b** were synthesized using the oxathiaphospholane method as described elsewhere [13]. Oligonucleotides **1c**, **2c**, **3c** and **4c** were synthesized by the phosphoramidite method with sulfurization of intermediary phosphites by means of S-Tetra [14]. Purification of all oligonucleotide constructs was carried out by two-step RP-HPLC (DMT-on and DMT-off) [15].

3. Inhibition of HIV-1 Reverse Transcriptase

Analogous experiments carried out with the HIV-1 reverse transcriptase (used at 32 nM conc.) have shown that the oligonucleotides **2a** and **3a** inhibit activity of this enzyme to

comparable extent as that of AMV RT: 50% inhibition of the HIV-1 reverse transcriptase by phosphorothioates **2a** and **3a** (IC_{50}) has been observed at 120 nM concentration of these oligomers (Fig. 1).
The results described above allowed us to suggest that [all-R_P]-isomers of the oligonucleotides **2** and **3** are aptameric inhibitors of the AMV and HIV-1 reverse transcriptases. High affinity of the [R_P]-phosphorothioates bearing G4 tract to AMV and HIV-1 RT can be compared to that of RNA ligands isolated by Chen and Gold by *SELEX* procedure (Table 2). These RNA ligands selected from a population of 10^{14} species contain almost 50 nucleotides and can form pseudoknots structures. They have been reported as very effective (IC_{50}=25 nM) and specific inhibitors of AMV RT [16].

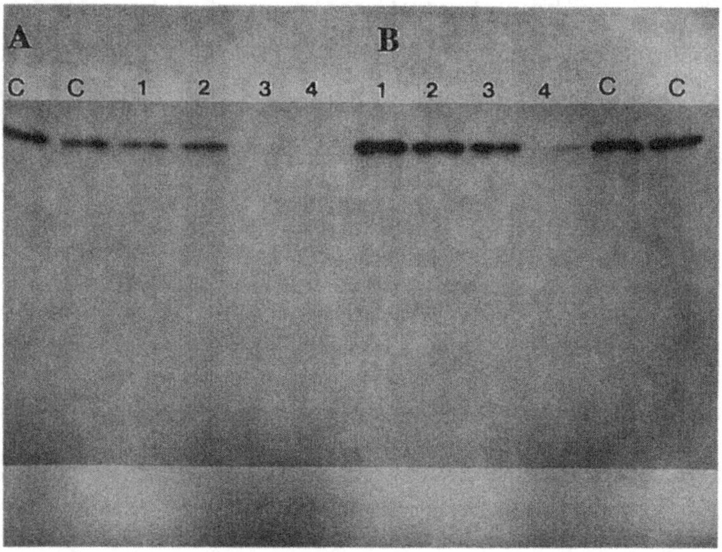

Figure 1. Inhibition of AMV and HIV-1 reverse transcriptases by oligonucleotides d[AACGTTGAGGGGCAT] (**2**) used at 130 nM concentration.
A: AMV reverse transcriptase; B: HIV-1 reverse transcriptase;
c: controls; 1: PO-**2**; 2: [mix]-PS-**2**; 3: [all-Sp]-PS-**2**; 4: [all-Rp]-PS-**2**.

RNA template (0.25 pmol) obtained by *in vitro* transcription [9], primer PO-**5** (15 pmoles) and the indicated amount of one of the oligonucleotides **1-4** were heated for 2 min at 95°C, cooled for 3 min at 0°C and then pre-incubated for 15 min at 37°C. Then to the reaction mixture were added: 5x conc. AMV RT buffer (250 mM Tris-HCl, pH 8.3, 40 mM $MgCl_2$, 250 mM NaCl, 5 mM DTT), RNasin (11 units), 7.5 nmol of each dNTP, [α-^{32}P] dCTP and AMV RT (2.5 units, 5.2 nM). This reaction mixture (total volume 20 µl) was incubated for 1 h at 37°C. Primer extension catalyzed by HIV RT was performed as above using 5x conc. HIV RT buffer (250 mM Tris-HCl, pH 8.3, 40 mM $MgCl_2$, 250 mM KCl and 5 mM DTT) and 0.5 unit (32 nM) of HIV reverse transcriptase. The products were analyzed on a 7% polyacrylamide gel. The autoradiograms were scanned using an LKB Ultrascan XL densitometer.

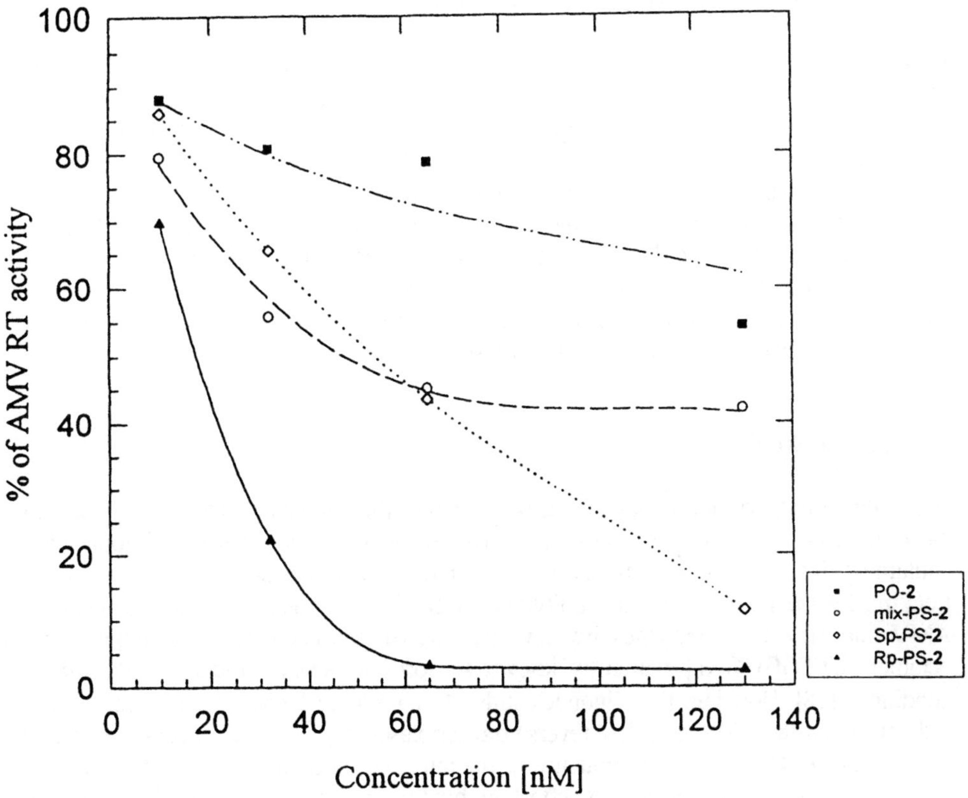

Figure 2. Inhibition of AMV reverse transcriptase by phosphorothioate analogues of the oligonucleotide d[AACGTTGAGGGGCAT] (**2**).

TABLE 2. Comparison of the inhibitory effects possessed by RNA ligand [16] and [all-R_P]-isomer of PS-oligo **2**

	Reaction conditions		Inhibitory effect: IC_{50} [nM]
	AMV RT [nM]	Template/primer [nM]	
RNA ligand	3.2	10	25
[all-R_P]-isomer of PS-oligo **2**	5.2	12.5	20

4. Klenow Fragment-Catalyzed Polymerization

In the light of above findings it was necessary to estimate a specificity of these phosphorothioate aptamers *i.e.* to find if they have any inhibitory effect on the activity of other polymerases. It has been reported that some polymerases (HIV-1 RT, Klenow fragment of DNA pol I, eukaryotic polymerase ß and T7 RNA polymerase) share a common catalytic mechanism centered around the highly conserved carboxylic acid residues [17]. We have chosen for our studies the large fragment of bacterial DNA polymerase I (Klenow enzyme). It has been found that the stereoregular oligonucleotides **3** used at 30, 130, 700 and 1200 nM concentration do not visibly inhibit the Klenow enzyme activity (data not shown). This result indicates that the phosphorothioate oligonucleotides **2** and **3** inhibit the AMV and HIV-1 RT with high efficiency and high specificity.

5. Structural Studies

It should be noticed that the nucleotide sequences of the oligonucleotides **2** and **3** share the same motif *i.e.* contiguous four G bases. There are many reports that the oligonucleotides containing this motif can form under *in vitro* conditions tetraplex structures [10, 11, 18, 19]. Looking for the reason of strong inhibitory effect of the oligomers **2a** and **3a** against the AMV and HIV-1 RT enzymes we have also tested phosphorothioate analogues of the octamer d[TTGGGGTT] which is known to form tetraplex structure under *in vitro* conditions [18]. However, the oligonucleotides **4a-c** used at 65 nM - 1.2 μM concentration did not inhibit the AMV and HIV reverse transcriptases (Fig. 3). These results indicate that the presence of G4 tract in the sequence of the oligonucleotide which interacts with the RT molecule is not sufficient for its effective binding to the protein. Probably for tight interaction between PS-oligo and RT protein suitably long oligonucleotide chain (at least 15-16 bases) is necessary.

On the other hand, Tamura *et al.* observed that phosphorothioate analogues of the octanucleotides d[TTGGGGTT] and d[GCGGGGTA] used at 0.13-0.25 μM concentration have influenced the AMV RT activity [10]. Because of the lack of detailed description of the experiment (*e.g.* lack of the enzyme concentration) we could not discuss our results with those reported by Tamura *et al.*

Since stereodependent inhibition of the RT activity by the phosphorothioate oligonucleotides **2** and **3** may be correlated with their ability to form tetraplex structures, we have studied formation of higher order structures by the PS-oligos using CD spectroscopy. The CD spectra of the phosphodiester and phosphorothioate oligonucleotides **2** were characteristic for parallel tetraplex reported by others [19] with a positive band at 264 nm and a negative band at 243 nm. The spectra recorded at 25°C for the samples of oligonucleotides **2** dissolved in water have shown that for [all-R_P]-isomer value $\Delta\varepsilon=5.7$ ($\lambda=264$ nm) is higher than that for PO-**3** ($\Delta\varepsilon=4.6$). For [all-S_P]-isomer and [mix]-form of oligonucleotide **2** $\Delta\varepsilon$ values are 3.8 and 4.2, respectively (Fig. 4). These $\Delta\varepsilon$ values may indicate differences in amount of tetraplex present in the samples, or reflect differences in spectral characteristics and conformation of tetraplexes formed by different diastereomers (M. Boczkowska - manuscript in preparation). After thermal denaturation of the samples at 95°C over 15 minutes decomposition of the tetraplex was only partial with $\Delta\varepsilon$=2.8 for oligomer **2a**, 1.7 for **2b** and 1.9 for **2c**. It should be mentioned that CD spectra were recorded at micromolar concentration of the oligonucleotides **2**, while their inhibitory effect was observed at nanomolar concentration. Although the tetraplex structure is thermodynamically less stable at the concentration below 1 μM, the extremely slow dissociation rate of this structure, as reported by Wyatt *et al.* [19], permits to assume that the stereodifferentiated tetraplex structures observed in stock solutions of the oligonucleotides **2** (at micromolar concentrations) are also preserved during the 1 hour-period of the primer extension experiment, when the oligonucleotides are diluted to nanomolar concentration. If [all-R_P]-isomers of the PS-oligonucleotides containing G4 tracts (**2a** and **3a**) actually form more abundant tetraplexes compared to their [all-S_P]-counterparts and [mix]-forms, it may explain their stronger inhibitory effect towards retroviral reverse transcriptases. On the other hand, both [all-R_P]- and [all-S_P]-isomers of PS-oligo **2** are better inhibitors of the RT activity than the [mix]-form (**2c**), although this latter seems to form the tetraplex structure as abundant as that formed by [all-S_P]-isomer.

6. Discussion

The use of several PS-oligonucleotides of different sequences and stereochemistry allowed us to conclude that their ability to bind to the RT enzymes (AMV and HIV-1) depends upon nucleotide sequence, length of oligonucleotide and the absolute configuration at P-atoms of each internucleotide phosphorothioate bond.

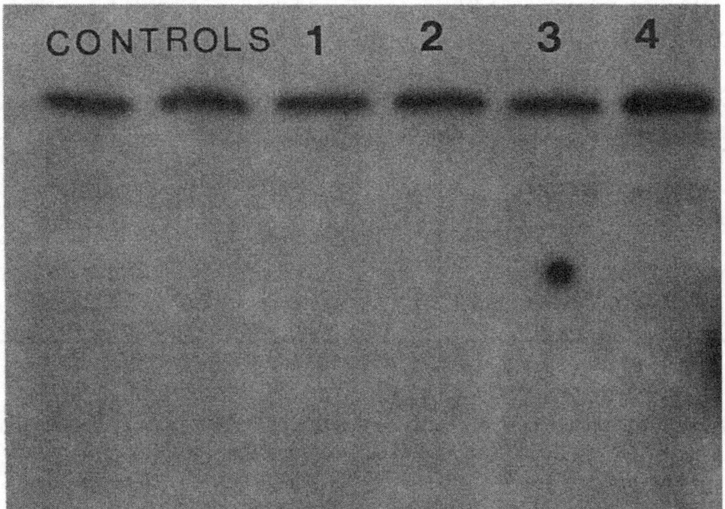

Figure 3. Activity of HIV-1 reverse transcriptase in the presence of phosphorothioate analogues of the oligonucleotide d[TTGGGGTT] (**4**) used at 750 nM concentration.
1: PO-**4**; 2: [mix]-PS-**4**; 3: [all-Sp]-PS-**4**; 4: [all-Rp]-PS-**4**.

Evident inhibitory effect (IC_{50}=20 nM) has been observed for two phosphorothioate oligonucleotides of R_P-configuration bearing contiguous four G bases (G4 tract).
The IC_{50} values found for these oligomers are very close to the IC_{50} values reported by Cheng and Gold for the RNA high-affinity ligands selected by *SELEX* procedure [16].
However, it should be underlined that PS-oligonucleotides are much more stable against nucleolytic action than oligoribonucleotides. Considering the potential application of aptamers to inhibit retroviral reverse transcriptases under *in vivo* conditions, the biological stability of PS-oligos may be of special importance.
Although the existance of tetraplex structure has not been demonstrated *in vivo*, many proteins have been found to bind G4-containing nucleic acids or oligonucleotides. It has been demonstrated that thrombin and HIV-1 integrase bind with high affinity phosphodiester oligonucleotides which are composed of deoxyguanosine and thymidine [20, 21]. They can form intramolecular tetraplex structure what has been confirmed by NMR studies [22, 23].

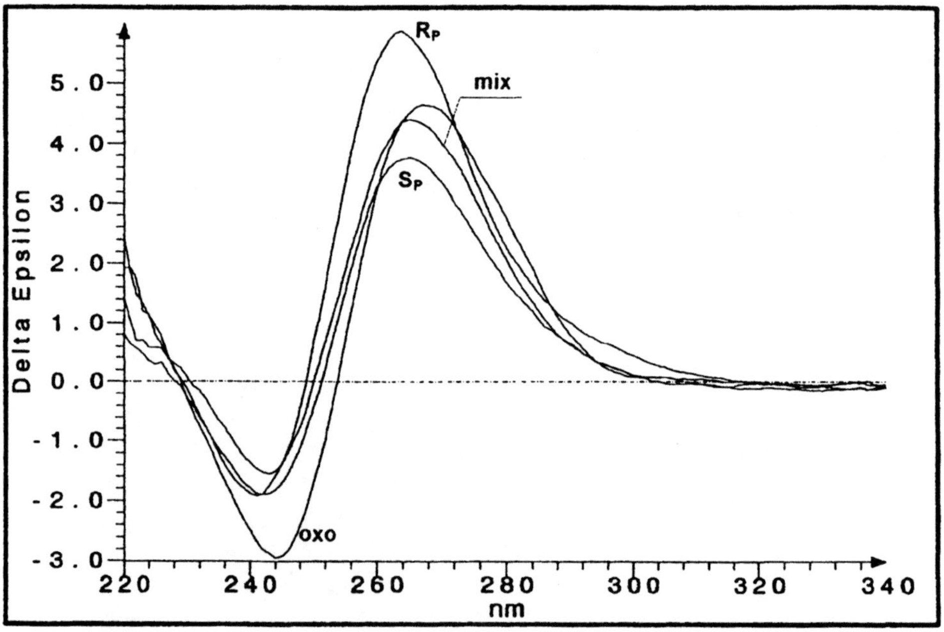

Figure 4. CD spectra for oligonucleotide **2** and its phosphorothioate analogues at concentration 6.5 μM at 25°C in H_2O, before denaturation.

The CD spectra were recorded at 25°C using a Jobin Yvon CD6 dichrograph with a cuvette of 0.5 cm pathlength. The first set of spectra was taken after the samples of oligonucleotides **2** were dissolved in 900 μl of water to give the final concentration of 99 μM per base (6.5 μM of the oligonucleotide). Then the samples were denatured at 95°C over 15 minutes and rapidly cooled in ice. After their short incubation at the temp. 25°C 5x conc. AMV RT buffer was added to each sample to final concentration of 50 M Tris-HCl (pH 8.3), 4 mM $MgCl_2$, 50 mM NaCl and 6 μM of one of the oligomers **2** (90 μM per base) and the second set of spectra was recorded.

Moreover, both the number of intramolecular G quartets and the sequence of the loops between the quartets are important for optimal activity of integrase aptamer [21]. It has been also reported that the phosphorothioate octamer d[TTGGGGTT] binds to the V3 loop domain of gp120 of human immunodeficiency virus. The concentration of the oligonucleotide required for 50% inhibition of virus-induced cytopathic effect (IC_{50}) is 0.3 μM [18]. It was postulated that the presence of sulfur atoms and the tetraplex structure of this octamer are required for its binding to the protein [18]. Presented here phosphorothioate oligonucleotides contain internucleotide bonds exclusively of R_P-configuration and can form intermolecular parallel tetraplexes. Their properties described above allow to include them to the growing list of tetraplex-forming oligonucleotides which can bind with high affinity to specified proteins (thrombin, gp120 protein, HIV-1 integrase, reverse transcriptases).

Acknowledgments

Authors wish to thank Professor Wojciech J. Stec for helpful discussion and careful reading of the manuscript. Stereoregular phosphorothioate oligonucleotides **1-4** were synthesized by Dr. Andrzej Okruszek and Bolesław Karwowski. This project has been financially assisted by the State Committee for Scientific Research, grant no. 4 P05F 023 10 (to Prof. W.J. Stec), grant no. 6 P04B 014 15 (to Dr. M. Koziołkiewicz) and, in part, by grant K-1007 (Principal Investigator - Prof. H. Takaku).

References

1. DeVico, A.L. and Sarngadharan, M.G. (1992) Reverse transcriptase - a general discussion, *J. Enzyme Inhibition* **6**, 9-34.
2. Matsukura, M., Shinozuka, K., Zon, G., Mitsuya, M., Reitz, M., Cohen, J.S. and Broder, S. (1987) Phosphorothioate analogs of oligodeoxynucleotides: inhibitors of replication and cytopathic effects of human immunodeficiency virus, *Proc.Natl.Acad.Sci.USA.* **84**, 7706-7710.
3. Majumdar, Ch., Stein, C.A., Cohen, J.S., Broder, S. and Wilson, S.H. (1989) Stepwise mechanism of HIV reverse transcriptase: Primer function of phosphorothioate oligonucleotide, *Biochemistry* **28**, 1340-1346.
4. Boiziau, C., Moreau, S. and Toulme, J-J. (1994) A phosphorothioate oligonucleotide blocks reverse transcriptase via an antisense mechanism, *FEBS LETT.* **340**, 236-240.
5. Boiziau, C., Larrouy, B., Sproat, B.S. and Toulme, J.-J. (1995) Antisense 2'-O-alkyl oligoribonucleotides are efficient inhibitors of reverse transcription, *Nucleic Acids Res.* **23**, 64-71.
6. Boiziau, C., Tarrago-Litvak, L., Sinha, N.D., Moreau, S., Litvak, S. and Toulme, J.-J. (1996) Antisense oligonucleotides inhibit in vitro cDNA synthesis by HIV-1 reverse transcriptase, *Antisense and Nucleic Acid Drug Dev.* **6**, 103-109.
7. Stein, C.A. (1996) Phosphorothioate antisense oligodeoxynucleotides: questions of specificity, *Trends Biotechnol.* **14**, 147-149.
8. Gao, W.-Y., Han, F.-S., Storm, Ch., Egan, W. and Cheng, Y.-Ch. (1991) Phosphorothiate oligonucleotides are inhibitors of human DNA polymerases and RNase H: Implications for antisense Technology, *Mol. Pharmacology* **41**, 223-229.
9. Krakowiak, A. and Koziołkiewicz, M. (1998) Influence of P-chirality of phosphorothioate oligonucleotides on the activity of AMV reverse transcriptase, *Nucleosides & Nucleotides* **17**, 1823-1834.
10. Tamura, N., Iwatani, W., Shoji, Y., Shimada, J. and Mizushima, Y. (1995) Aptameric inhibition of in vitro DNA polymerization by phosphorothioate oligonucleotides, *Nucleic Acids Symp. Series* **34**, 93-94.
11. Basu, S. and Wickstrom, E. (1997) Temperature and salt dependence of higher order structure formation by antisense c-myc and c-myb phosphorothioate oligodeoxyribonucleotides containing tetraguanylate tracts, *Nucleic Acids Res.* **25**, 1327-1332.
12. Koziołkiewicz, M., Krakowiak, A., Kwinkowski, M., Boczkowska, M. and Stec, W.J. (1995) Stereodifferentiation -

The effect of P-chirality of oligonucleoside phosphorothioate)s on the activity of bacterial RNase H, *Nucleic Acids Res.* **23**, 5000-5005.

13. Stec, W.J., Grajkowski, A., Kobylańska, A., Karwowski, B., Koziołkiewicz, M., Misiura, K., Okruszek, A., Wilk, A., Guga, P. and Boczkowska, M. (1995) Diastereomers of 3'-O-(2-thio-1.3.2-oxathia(selena)phospholanes): building blocks for stereocontrolled synthesis of oligo(nucleoside phosphorothioate)s, *J.Am.Chem.Soc.* **117**, 12020-12029.
14. Stec, W.J., Uznański, B., Wilk, A., Hirschbein, B.L., Fearon, K.L. and Bergot, B.J. (1993) Bis-(O,O-diisopropoxy phosphinothioyl)disulfide - A highly efficient sulfurizing reagent for cost-effective synthesis of oligo(nucleoside phosphorothioate)s, *Tetrahedron Lett.* **34**, 5317-5320.
15. Zon, G. and Stec, W.J. (1991) Phosphorothioate oligonucleotides. In: *Oligonucleotides and Analogs: A Practical Approach*, Eckstein, F.(ed), IRL Press, Oxford, 87-108.
16. Chen, H. and Gold, L. (1994) Selection of high-affinity RNA ligands to reverse transcriptase: inhibition of cDNA synthesis and RNase H activity, *Biochemistry* **33**, 8746-8756.
17. Pelletier, H., Sawaya, M.R., Kumar, A., Wilson, S.H. and Kraut, J. (1994) Structures of ternary complexes of rat DNA polymerase β, a DNA template-primer and ddCTP, *Science*, **264**, 1891-1903).
18. Wyatt, J.R., Vickers, T.A., Robertson, J.L., Buckheit, R.W., Klimkait, T., DeBaets, E., Davis, P.W., Rayner, B., Imbach, J.-L. and Ecker, D.J. (1994) Combinatorially selected guanosine -quartet structure is a potent inhibitor of human immunodeficiency virus envelope-mediated cell fusion, *Proc.Natl.Acad.Sci.USA.* **91**, 1356-1360.
19. Wyatt, J.R., Davis, P.W. and Freier, S.M. (1996) Kinetics of G-quartet-mediated tetramer formation, *Biochemistry* **35**, 8002-8006.
20. Bock, L.C., Griffin, L.C., Latham, J.A., Vermaas, E.H. and Toole, J. (1992) Selection of single-stranded DNA molecules that bind and inhibit human thrombin, *Nature* **355**, 564-566.
21. Mazumder, A., Neamati, N., Ojwang, J.O., Sunder, S., Rando, R.F. and Pommier, Y. (1996) Inhibition of the human immunodeficiency virus type 1 integrase by guanosine quartet structures, *Biochemistry* **35**, 13762-13771.
22. Macaya, R.F., Schultze, P., Smith, F.W., Roe, J.A. and Feigen, J. (1993) Thrombin binding DNA aptamer forms a unimolecular quadruplex structure in solution, *Proc.Natl.Acad.Sci.USA.*, **90**, 3745-3749.
23. Wang, K.Y., McCurdy, S., Shea, R.G., Swaminathan, S. and Bolton, P.H. (1993) A DNA aptamer which binds to and inhibits thrombin exhibits a new structural motif for DNA, *Biochemistry* **32**, 1899-1904.

OXATHIAPHOSPHOLANE METHOD OF THE STEREOCONTROLLED SYNTHESIS OF PHOSPHOROTHIOATE ANALOGUES OF OLIGONUCLEOTIDES

A. OKRUSZEK
Polish Academy of Sciences, Centre of Molecular and Macromolecular Studies, Department of Bioorganic Chemistry, Sienkiewicza 112 90-363 Łódź, Poland, e-mail: okruszek@bio.cbmm.lodz.pl

1. Introduction

Oligonucleotide analogues have recently found wide application in biochemistry and molecular biology as valuable tools for studying interactions of DNA/RNA with other biomolecules [1-3] and as potential candidates for therapeutics in the antisense/antigene or ribozyme strategy [4-6]. Among oligonucleotide congeners most widely used for such a studies are those modified within the internucleotide phosphate group including phosphorothioates, phosphorodithioates, methanephosphonates, phosphoramidates, phosphotriesters etc. The phosphorothioate modification, in which one of the nonbridging oxygen atoms of internucleotide phosphate is substituted by sulphur [7], is most frequently employed in physicochemical and enzymatic studies [8,9]. Phosphorothioate oligodeoxyribonucleotides are also most promising candidates for antisense drugs against several viral and cancer diseases as indicated by their use in numerous clinical trials [10-11], and by recent FDA approval of phosphorothioate 22-mer (*Fomivirsen*) for treatment of CMV retinitis in AIDS patients (Aug.26, 1998).

2. The Synthesis of Phosphorothioate Analogues of Oligonucleotides

Along with vast application of phosphorothioate analogues of oligodeoxyribonucleotides, their chemical synthesis is well documented in the literature. The first automated synthesis of oligo(deoxyribonucleoside phosphorothioate)s was accomplished by Stec *et al.* [12,13] by modified phosphoramidite approach with a stepwise sulphurization of intermediate phosphite linkage. The solution of elemental sulphur, originally employed for sulphurization [12,13], was later substituted by more efficient reagents [14]. An alternative approach was proposed by Froehler [15], who modified the H-phosphonate method by introducing final one-step sulphurization with a solution of elemental sulphur. The modifications of phosphotriester method found only limited application for the synthesis of oligo(deoxyribonucleoside phosphorothioate)s [16].

In contrast to DNA analogues, the synthesis of oligo(ribonucleoside phosphorothioate)s was described only in a few cases, usually by H-phosphonate method [17,18].

An intrinsic property of oligo(nucleoside phosphorothioate)s, as depicted in *Figure 1*, is the formation of a new centre of chirality at each internucleotide phosphorus, leading to diastereomers, designated as R_P and S_P. With an increasing length of

J. Barciszewski and B.F.C. Clark (eds.), RNA Biochemistry and Biotechnology, 325–334.

oligonucleotide, the number of diastereomers (*m*) grows exponentially ($m = 2^n$ for *n* phosphorothioate bonds).

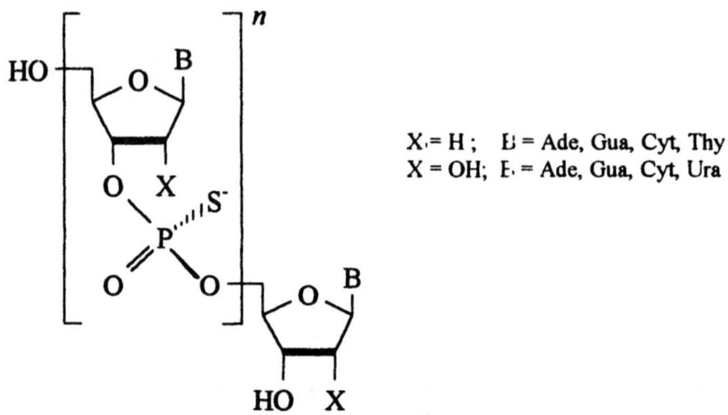

Figure 1. Schematic representation of oligo(deoxyribonucleoside phosphorothioate)s (X=H) and oligo(ribonucleoside phosphorothioate)s (X=OH).Example shows [All-S_P] configuration.

As an important consequence of aforementioned polydiastereomerism of oligo (nucleoside phosphorothioate)s, stereodifferentiated interactions could occur between particular diastereomers and other chiral biomolecules such as DNA, RNA, proteins, carbohydrates or lipids. This could in turn lead to stereodifferentiated: uptake, cellular trafficking, stability in biological media, pharmacokinetics, and toxicity. As a result, particular diastereomers may have different therapeutic properties when applied in antisense, antigene or ribozyme strategy.

The synthesis of oligo(deoxyribonucleoside phosphorothioate)s by most commonly used solid support phosphoramidite/sulphurization approach leads to a more or less random mixture of all possible diastereomers. The detailed studies have shown, however, that every individual coupling step occurs with some stereoselectivity, and the formation of phosphorothioate centre with R_P configuration is preferred (usually 52-62% of R_P) [19]. Higher stereoselectivity (up to 85% of R_P) was observed for the synthesis of short oligo(ribonucleoside phosphorothioate)s by the solid-support H-phosphonate metho-dology [18]. By application of phosphotriester approach, with diastereomerically pure nucleoside 3'-*O*-(*S*-alkyl-*O*-*p*-nitrophenyl)phosphorothioates as monomers, it was possible to synthesize diastereomerically pure phosphorothioate analogues of trithymidine [20] and triuridine [21]. However, the method could not be used under conditions of solid-support synthesis.

The enzymatic template-directed processes, utilizing DNA or RNA polymerases of different origin and [S_P]-nucleoside 5'-*O*-α-thiotriphosphates as substrates [7] have been employed for the stereocontrolled synthesis of several phosphorothioate analogues of DNA [22,23] or RNA [24]. An important limitation of this methodology is, that it can only be applied for the synthesis of [All-R_P] diastereomers of phosphorothioate DNA or RNA fragments.

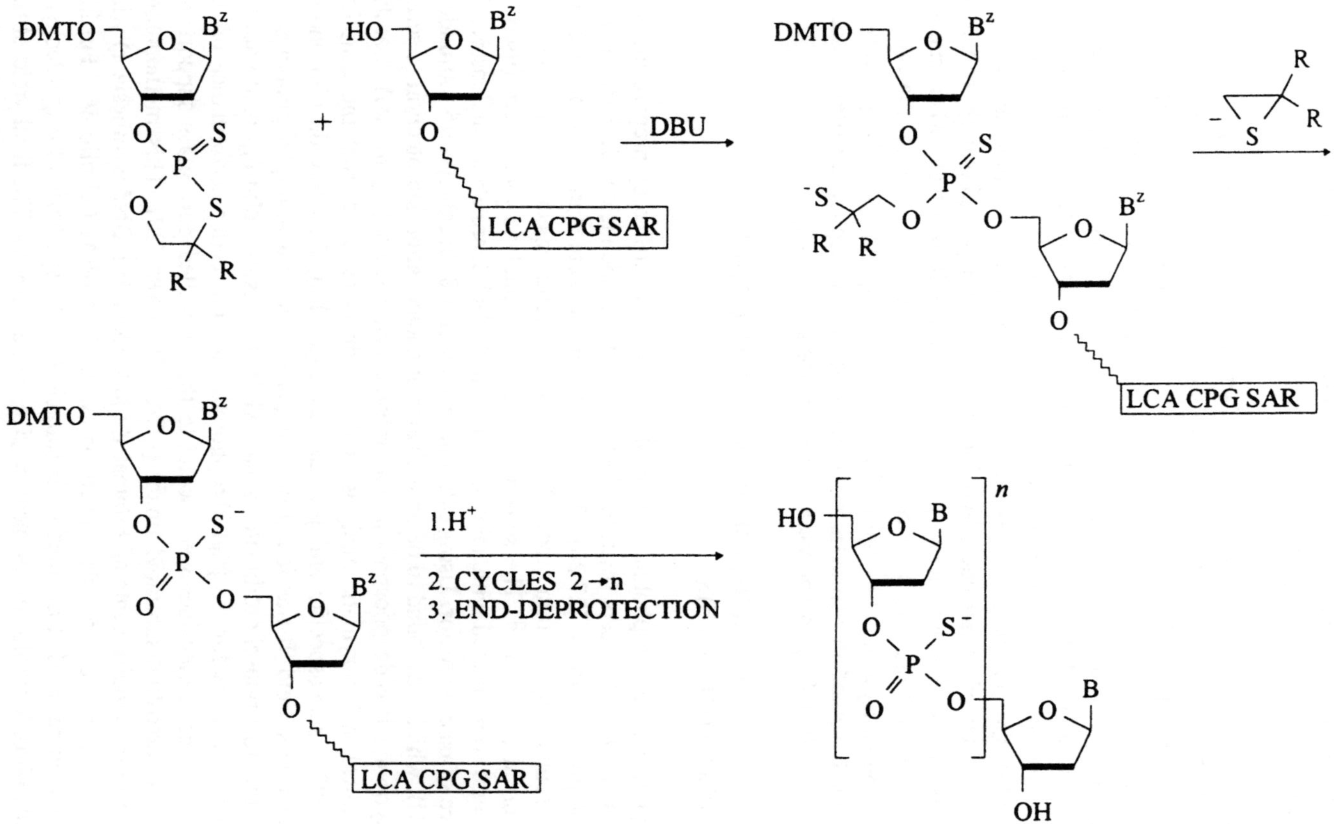

Figure 2. Oxathiaphospholane synthesis of oligo(deoxyribonucleoside phosphorothioate)s.
a:R = H, *b*:R = CH_3, *c*:R = - $(CH_2)_5$- ; B^z = Ade^{Bz}, Gua^{iBu}, Cyt^{Bz}, Thy.

3. Stereocontrolled Synthesis of Oligo(nucleoside phosphorothioate)s by Oxathiaphospholane Approach

5'-*O*-Dimethoxytrityl (DMT)-thymidine 3'-*O*-(2-oxo-1,3,2-oxathiaphospholane) was found to be participating as a reactive intermediate in a stereoselective conversion of diastereomerically pure 5'-*O*-DMT-thymidine 3'-*O*-(*O*-*p*-nitrophenyl phosphorothioate) into thymidine 3'-*O*-[^{16}O,^{17}O,^{18}O]phosphate by reaction with styrene [^{18}O]oxide in the presence of [^{17}O]water [25]. Uridine 3'-*O*-(2-oxo-1,3,2-oxathiaphospholane) was also found as reactive intermediate in a stereospecific conversion of [Sp] uridine cyclic 3',5'-phosphorothioate into uridine cyclic 2',3'-[^{18}O]phosphate by means of styrene [^{18}O]oxide [26]. On this basis a hypothesis was formulated, and then proved, that appropriately protected nucleoside 3'-*O*-(2-thio-1,3,2-oxathiaphospholane)s can react with alkohols in a similar manner to produce in a stereoselective way (when starting from separated diastereomers) corresponding nucleoside 3'-*O*-phosphorothioate diesters. The pursuing of this idee led to the development by Stec *et al.* [27] of a novel stereocontrolled method of synthesis of oligo(nucleoside phosphorothioate)s, usually referred to as *oxathiaphospholane method* (see *Figure 2*).

3.1. STEREOREGULAR OLIGO(DEOXYRIBONUCLEOSIDE PHOSPHOROTHIOATE)S

The concept of oxathiaphospholane methodology has been verified by preparation of a series of base-protected (except thymine) 5'-*O*-DMT-deoxyribonucleoside 3'-*O*-(2-thio-1,3,2-oxathiaphospholane)s [27,28]. These compounds were further separated by column chromatography into individual diastereomeric species, with opposite configuration at phosphorus. The separated oxathiaphospholane monomers (used in 10-fold excess) were reacted with solid support bound 5'-*O*-deprotected nucleosides in the presence of base activator. Using 300-fold excess of 1,8-diazabicyclo[5.4.0]undec-7-ene (DBU), which was found to be an activator of choice, after deprotection, a series of di(deoxyribonucleoside phosphorothioate)s were synthesized (*Figure 2a*). Careful physicochemical and enzymatic analysis of the products revealed, that the coupling process is >98% stereospecific and provides corresponding dinucleoside phosphorothioates in *ca* 95% yield. It was found that the „fast"-eluting oxathiaphospholanes are precursors for [S_P]-phosphorothioate bond, while the „slow"-eluting oxathiaphospholanes are precursors for the [R_P]-phosphorothioate. For better performance of the method, the 3'-end nucleosides were bound to the controlled pore glass support *via* DBU-resistant sarcosinyl-succinoyl linker (LCA CPG SAR) [29]. This methodology was adapted to the requirements of automated mode (ABI 391 DNA synthesizer) with „capping" of unreacted 5'-OH functions with DMAP/Ac_2O/Lutidine. Further experiments confirmed, that the method is suitable for the synthesis of oligo(deoxyribonucleoside phosphorothioate)s of medium size, e.g. 8-16-mers. The final yield of the purified product was lower than that of phosphorothioates obtained by the phosphoramidite-sulphurization process [12,13], however, the purity of oligomers produced by both methods was comparable. The most important difference was that now, by using diastereomerically pure oxathiaphospholanes, the chirality at each stereogenic phosphorothioate centre can be fully controlled.

The diastereomeric purity of stereoregular oligo(deoxyribonucleoside phosphorothioate)s was estimated on the basis of their diastereoselective degradation with Nuclease P1 (S_P-specific enzyme), and a mixture of snake venom phosphodiesterase and *Serratia marcescens* endonuclease (R_P-specific nucleases) [30].

Further developments of the oxathiaphospholane method were aimed mainly to improve the chromatographic separability of the monomers, which was achieved by introducing substituents at position 4 of the oxathiaphospholane ring. The introduction of two methyl groups (*Figure 2b*), although did not improve too much chromatographic separation of diastereomers [30], allowed to obtain a crystalline deoxycitidine derivative and to assign its absolute configuration by X-ray crystallography [31]. This assignment permitted to perform the stereochemical analysis of oxathiaphospholane ring opening condensation and allowed to conclude that it proceeds *via* an „adjacent" mechanism, with the participation of a pentacoordinate phosphorus intermediate [31].

Considerable improvement of the chromatographic separability of oxathiaphospholane monomers was achieved by introducing „spiro"-pentamethylene substituent at position 4 of the oxathiaphospholane ring (*Figure 2c*). For these monomers one passage through a silica gel column gave *ca* 70% recovery of the applied material in a diastereomerically pure form. The 5'-*O*-DMT-deoxyribonucleoside 3'-*O*-(2-thio-4,4-„spiro"-pentamethylene-1,3,2-oxathiaphospholane)s were chemically more stable , and using these monomers it was possible to optimize the automated oligonucleotide synthesis protocol [32]. With this protocol, which allows to obtain a consistent 92-95% step-yield of coupling, several oligo(deoxyribonucleoside phosphorothioate)s of medium size (10-25 nucleotides) were synthesized with pre-determined absolute configuration at stereogenic phosphorus centres. The preliminary results of studies of the effect of phosphorus chirality on the physicochemical and biological properties of oligo(deoxyribonucleoside phosphoro-thioate)s will be discussed in section 4 of this paper.

It is worthwhile to mention that the introduction of „spiro"-pentamethylene system allowed to synthesize 2-oxo monomers, which upon application to ring-opening condensation procedure are precursors of non-modified phosphodiester linkage, on the way compatible with oxathiaphospholane synthesis of oligo(deoxyribonucleoside phosphorothioate)s. Thus, it is now possible to synthesize chimeric oligonucleotides, possessing, at preselected positions, phosphodiester groups and phosphorothioate linkages of predetermined chirality [32,33].

3.2. STEREOREGULAR OLIGO(RIBONUCLEOSIDE 3',5'-PHOSPHOROTHIOATE)S

In order to check the applicability the oxathiaphospholane method to the stereocontrolled synthesis of oligo(ribonucleoside 3',5'-phosphorothioate)s, appropriate oxathiaphospholane monomers were synthesized with 2'-hydroxyl function protected by *tert*-butyldimethylsilyl (TBDMS) group [34]. Protected oxathiaphospholane monomers were separated chromatographically into individual diastereomers and were

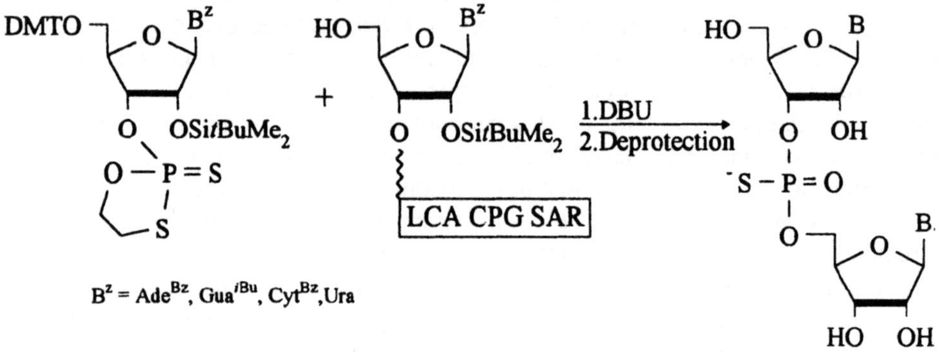

Figure3. Oxathiaphospholane synthesis of diribonucleoside 3',5'-phosphorothioates.

reacted with support-bound ribonucleosides under DBU activation (*Figure 3*). After cleavage from the support, deprotection, and HPLC purification, the resulting homo-diribonucleoside 3',5'-phosphorothioates were analyzed by spectroscopic and enzymatic methods. It was found that the syntheses proceeded with a stereospecificity >98%, however, only for the case of diuridine analogues the yields were in the range acceptable for solid-support oligonucleotide synthesis (96-98%). For other homo-diribonucleotide analogues lower yields (66-83%) were observed. Attempts to increase the yields by extension of the coupling time or increasing of DBU concentration were unsuccessful. So far, oxathiaphospholane method is applicable only to the stereo-controlled synthesis of diribonucleoside 3',5'-phosphorothioates [34].

3.3. STEREOREGULAR OLIGO(ADENOSINE 2',5'-PHOSPHOROTHIOATE)S

Phosphorothioate analogues of short 2',5'-oligoadenylates have promising antiviral and anticancer properties and were recently prepared as diastereomerically separated compounds [35,36]. In order to check the feasibility of stereocontrolled synthesis of these derivatives by oxathiaphospholane approach N^6-benzoyl-5'-*O*-DMT-2'-*O*-TBDMS-adenosine 3'-*O*-(2-thio-1,3,2-oxathiaphospholane) was synthesized and chromatographically separated into diastereomers with opposite configuration at phosphorus [37]. The separated monomers were reacted under the conditions of oxathiaphospholane synthesis, with DBU activation, into LCA CPG SAR-bound N^6-benzoyl-2'-*O*-TBDMS-adenosine. The repetition of the procedure, after removal of 5'-*O*-DMT group, led to corresponding 2',5'-tri-, and then to 2',5'-tetra-adenosine analogues, which were cleaved from the support, deprotected, and identified by physicochemical and enzymatic methods. The step-yields 94-95% were achieved for attachment of consecutive adenosine 2'-phosphorothioate moieties.

The fact, that oxathiaphospholane method can be employed for 5'-elongation of oligonucleotide prepared by a solid-support phosphoramidite approach, allows to use this methodology in a novel, RNase L-mediated approach to antisense therapy [38]. This would require the synthesis of chimeric oligomers, possessing at the 3'-end the „normal" oligodeoxyribonucleotide (phosphodiester or phosphorothioate), and at the 5'-end (connected directly or through a linker) the tetraadenosine 2',5'-phosphorothioate. The 3'-end part of the construct would act as an antisense probe while the 5'-end fragment, prepared by stereocontrolled way as more active [All-R_P] isomer [35,36], would play a role of site-directed activator of RNase L [38].

4. Biological Consequences of Phosphorus Chirality of Stereoregular Phosphorothioate Oligonucleotides

Theoretical considerations suggested that duplexes (with complementary DNA template) formed by [All-S_P] oligo(dexyribonucleoside phosphorothioate)s should be more stable than those formed by [All-R_P] isomers [39,40]. This rule, however, was not confirmed by experimental data, because the relative stability of duplexes was found to depend primarily on the nucleotide sequence of oligonucleotides, and not on the chirality of stereoregular phosphorothioate component [30,32]. On the other hand, the thermal stability of heteroduplexes formed by stereoregular oligo(deoxyribonucleoside phosphorothioate)s with complementary RNA matrix strongly depends on phosphorus chirality and it was found, that [All-R_P] diastereomers of oligo(dexyribonucleoside phosphorothioate)s form more stable complexes than their [All-S_P] counterparts [30,41,42]. As a consequence, the heteroduplexes formed by [All-R_P] diastereomers were found to be better substrates for bacterial RNase H than those prepared from [All-S_P] isomers [41].

The stereoregular oligo(deoxyribonucleoside phosphorothioate)s were also studied as substrates for human plasma 3'-exonuclease. It was found, that this exonuclease is an [R_P]-specific enzyme, and readily degrades [All-R_P] diastereomers, while not cleaving [All-S_P] isomers [43]. Moreover, it has been observed that certain types of normal and tumor human cells (HeLa, HL60, HUVEC) contain another 3'-exonuclease that is also [R_P]-specific [44].

Hexadeca(deoxyribonucleotide phosphorothioate)s complementary to a fragment of human PAI-1 mRNA were studied in cultured HUVEC cells as sequence-dependent inhibitors of PAI-1 expression [42]. The activity of random mixture of diastereomers has been compared with that of isosequential, stereoregular [All-R_P] and [All-S_P] isomers. The highest inhibitory effect on PAI-1 synthesis was observed with the [All-S_P] diastereomer. Lower inhibitory activity of [All-R_P] isomer is most probably due to its lower stability against cellular nucleases.

The P-chirality of oligo(deoxyribonucleoside phosphorothioate)s was also found to affect the activity of AMV-reverse transcriptase (AMV-RT) and terminal deoxyribonucleotidyl transferase (*TdT*). AMV-RT is inhibited to a higher extent by [All-R_P] diastereomer of pentadeca(deoxyribonucleoside phosphorothioate) of the sequence 5'- AAG CAT ACG GGG TGC (IC_{50}=30 μM), than by its [All-S_P] counterpart (IC_{50}>100 μM) [45].

The elongation of oligo(deoxyribonucleoside phosphorothioate)s, used as primers for *TdT* depends upon the absolute configuration at the phosphorus atom of the

internucleotide bond located between the second and the third nucleoside from the 3'-end. The presence of [S_P]-linkage in this position strongly reduces enzyme activity, while the presence of [R_P]-phosphorothioate allows for effective and fast elongation [46].
The aforementioned examples show, that interactions of stereoregular oligo(deoxyribonucleoside phosphorothioate)s with nucleic acids and/or proteins (enzymes) may be highly stereodependent. However, it was also found that phosphorothioate analogues of T_{19} and d(TC)$_9$T, prepared as [All-R_P] and [All-S_P] diastereomers, bind to several proteins such as: basic fibroblastic growth factor, recombinant soluble CD4, laminin, and fibronectin in a P-chirality-independent manner [47].

Acknowledgment

Part of the work presented in this paper was financially assisted by the State Committee for Scientific Research (KBN) grant no. 4.P05F.023.10 (to W.J.Stec). The author wishes to thank Professor Wojciech J.Stec for encouragement and helpful discussion.

References

1. Koziołkiewicz, M., Uznański, B. Stec, W.J. and Zon, G. (1986) P-Chiral analogues of oligodeoxyribonucleotides: synthesis, stereochemistry and enzyme studies, *Chemica Scripta* **26**, 251-260.
2. Koziołkiewicz, M. and Stec, W.J. (1992) The application of backbone-modified oligonucleotides in the studies on Eco RI endonuclease mechanism of action, *Biochemistry* **31**, 9460-9466.
3. Beaucage, S.L. and Iyer, R.P. (1993) The synthesis of modified oligonucleotides by the phosphoramidite approach and their applications, *Tetrahedron* **49**, 6123-6194.
4. Zon,G. (1988) Oligonucleotide analogs as potential chemotherapeutic agents, *Pharm.Res.* **5**, 539-549.
5. Stull, R.A. and Szoka, F.C. (1995) Antigene, ribozyme and aptamer nucleic acid drugs: progress and prospects, *Pharm.Res.* **12**, 465-483.
6. Agrawal, S. (ed.) (1996) *Antisense Theraputics*, Humana Press Inc., Totowa, NJ.
7. Eckstein, F. (1985) Nucleoside phosphorothioates, *Annu.Rev.Biochem.* **54**, 367-402.
8. Cohen, J.S. (1993) Phosphorothioate oligonucleotides, in: S.T.Crooke and B.Lebleu (eds.), *Antisense Research and Applications*, CRC Press, Boca Raton, FL, pp. 205-221.
9. Brautigam, C.A. and Steitz, T.A. (1998) Structural principles of the inhibition of the 3'-5' exonuclease activity of *Escherichia coli* DNA polymerase I by phosphorothioates, *J.Mol.Biol.* **227**, 363-377.
10. Glaser, V. (1996) Oligonucleotide therapies move toward efficiacy trials to treat HIV, CMV, cancer, *Genetic Eng.News*, Feb.1, pp. 1, 16, 17, 21.
11. Wickstrom, E. (ed.) (1998) *Clinical Trials of Genetic Therapy with Antisense DNA and DNA Vectors*, Marcel Dekker, Inc., New York, NY.
12. Stec, W.J., Zon, G., Egan, W. and Stec, B. (1984) Automated solid-phase synthesis, separation, and stereochemistry of phosphorothioate analogues of oligodeoxyribonucleotides, *J.Am.Chem.Soc.* **106**, 6077-6079.
13. Zon, G. and Stec, W.J. (1991) Phosphorothioate oligonucleotides, in: F.Eckstein (ed.), *Oligonucleotides and Analogues. A Practical Approach*, IRL Press, Oxford, pp. 87-108.
14. Guga, P., Koziołkiewicz, M., Okruszek, A. and Stec, W.J. (1998) Oligo(nucleoside phosphorothioate)s, in: C.A.Stein and A.M.Krieg (eds.), *Applied Antisense Oligonucleotide Technology*, Wiley-Liss, Inc., New York, NY, pp. 23-50.
15. Froehler, B.C. (1986) Deoxynucleoside H-phosphonate diester intermediates in the synthesis of internucleotide phosphate analogs, *Tetrahedron Lett.* **27**, 5575-5578.
16. Barber, I., Imbach, J.-L. and Rayner, B. (1995) Solution phase synthesis of phosphorothioate oligonucleotides by the phosphotriester method, *Antisense Res.& Dev* **5**, 39-47.

17. Agrawal, S. and Tang, J.-Y. (1990) Efficient synthesis of oligoribonucleotide and its phosphorothioate analogue using H-phosphonate approach, *Tetrahedron Lett.* **31**, 7541-7544.
18. Almer, A., Stawiński, J. and Strömberg, R. (1996) Solid support synthesis of All-R_P-oligo-(ribonucleoside phosphorothioate)s, *Nucleic Acids Res.* **24**, 3811-3820.
19. Wilk, A. and Stec, W.J. (1995) Analysis of oligo(deoxyribonucleoside phosphorothioate)s and their diastereomeric composition, *Nucleic Acids Res.* **23**, 530-534.
20. Leśnikowski, Z.J. and Jaworska, M.M. (1989) Studies on stereospecific formation of P-chiral internucleotide linkage. Synthesis of (R_PR_P)- and (S_PS_P)-thymidylyl (3',5')thymidylyl(3',5')thymidine di(*O,O*-phosphorothioate) using *o*-nitrobenzyl group as a new S-protection, *Tetrahedron Lett.* **30**, 3821-3824.
21. Leśnikowski, Z.J. (1992) The first stereocontrolled synthesis of thiooligoribonucleotide: (R_PR_P)- and (S_PS_P)-$U_{PS}U_{PS}U$, *Nucleosides & Nucleotides* **11**, 1621-1638.
22. Hacia, J.G., Wold, B.J. and Dervan, P. (1994) Phosphorothioate oligonucleotide-directed triple helix formation, *Biochemistry* **33**, 5367-5369.
23. Tang, J., Roskey, A., Li, Y. and Agrawal, S. (1995) Enzymatic synthesis of stereoregular All-[R_P] oligonucleotide phosphorothioate and its properties, *Nucleosides & Nucleotides* **14**, 985-990.
24. Ueda, T., Tohda, H., Chikazumi, N., Eckstein, F. and Watanabe, K. (1991) Phosphorothioate-containing RNAs show mRNA activity in the prokaryotic translation system *in vivo*, *Nucleic Acids Res.* **19**, 547-552.
25. Okruszek, A., Guga, P. and Stec, W.J. (1987) Novel approach to the synthesis of isotopomeric monoalkyl [^{16}O,^{17}O,^{18}O]phosphates. The stereospecific one-pot conversion of [R_P]-thymidine 3'-(4-nitrophenyl)phosphorothioate into [R_P]-thymidine 3'-[^{16}O,^{17}O,^{18}O]phosphate, *J.Chem.Soc., Chem.Commun.*, 594-595.
26. Okruszek, A.,Guga, P. and Stec, W.J. (1991) Stereochemistry of the reaction of ribonucleoside cyclic phosphorothioates with oxiranes, *Heteroatom Chem.* **2**, 561-568.
27. Stec, W.J., Grajkowski, A., Koziołkiewicz, M. and Uznański, B. (1991) Novel route to oligo(deoxy ribonucleoside phosphorothioate)s. Stereocontrolled synthesis of P-chiral oligo(deoxyribonucleoside phosphorothioate)s, *Nucleic Acids Res.* **21**, 5883-5888.
28. Stec, W.J. and Wilk, A. (1994) Stereocontrolled synthesis of oligo(nucleoside phosphorothioate)s, *Angew.Chem.Int.Ed.Engl.* **33**, 709-722.
29. Brown, T., Pritchard, C.E., Turner, G. and Sailsbury, S.A. (1989) A new base-stable linker for solid-phase oligonucleotide synthesis, *J.Chem.Soc., Chem.Commun.*, 891-893.
30. Stec, W.J., Grajkowski, A., Kobylańska, A., Karwowski, B., Koziołkiewicz, M., Misiura, K., Okruszek, A., Wilk, A., Guga, P. and Boczkowska, M. (1995) Diastereomers of nucleoside 3'-*O*-(2-thio-1,3,2-oxathia(selena)phospholane)s: building blocks for stereocontrolled synthesis of oligo (nucleoside phosphorothioate)s, *J.Am.Chem.Soc.* **117**, 12019-12029.
31. Stec, W.J., Karwowski, B., Guga, P., Misiura, K., Wieczorek, M.W. and Błaszczyk, J. (1996) Stereochemistry of DBU-assisted reaction of nucleoside 3'-*O*-(2-thiono-1,3,2-oxathiaphospholanes) with 5'-hydroxynucleosides, *Phosphorus, Sulfur & Silicon.* **109/110**, 257-260.
32. Stec, W.J., Karwowski, B., Boczkowska, M., Guga, P., Koziołkiewicz, M., Sochacki, M., Wieczorek, and Błaszczyk, J. (1998) Deoxyribonucleoside 3'-*O*-(2-thio- and 2-oxo-"spiro"-4,4-pentamethylene-1,3,2-oxathiaphospholane)s: monomers for stereocontrolled synthesis of oligo(deoxyribonucleoside phosphorothioate)s and chimeric PS/PO oligonucleotides, *J.Am.Chem.Soc.* **120**, 7156-7167.
33. Karwowski, B., Guga, P., Kobylańska, A. and Stec, W.J. (1998) Nucleoside 3'-*O*-(2-oxo-"spiro"-4,4-pentamethylene-1,3,2-oxathiaphospholane)s: monomers for stereocontrolled synthesis of oligo (nucleoside phosphorothioate/phosphate)s, *Nucleosides & Nucleotides* **17**, 1747-1759.
34. Sierzchała, A., Okruszek, A. and Stec, W.J. (1996) Oxathiaphospholane method of stereocontrolled synthesis of diribonucleoside 3',5'-phosphorothioates, *J.Org.Chem.* **61**, 6713-6716.
35. Charubala, R. and Pfleiderer, W. (1992) Syntheses and characterization of phosphorothioate analogues of (2'-5')adenylate dimer and trimer and their 5'-*O*-monophosphates, *Helv.Chim.Acta* **75**, 471-479.
36. Sobol, R.W., Henderson, E.E., Kong, N., Shao, J., Hitzges, P., Mordechai, E., Reichenbach, N.L., Charubala, R., Schirmeister, H., Pfleiderer, W. and Suhadolnik, R.J. (1995) Inhibition of HIV-1 replication and activation of RNase L by phosphorothioate/phosphodiester 2',5'-oligoadenylate derivatives, *J.Biol.Chem.* **270**, 5963-5978.
37. Yang, X.-B., Sierzchała, A., Misiura, K., Niewiarowski, W., Sochacki, M., Stec, W.J. and Wieczorek, M.W. (1998) The first stereocontrolled solid-phase synthesis of di-, tri-, and tetra[adenosine (2',5') phosphorothioate]s, *J.Org.Chem.* **63**, 7097-7100.

38. Torrence, P.F., Xiao, W., Li, G., Cramer, H., Player, M.R. and Silverman, R.H. (1997) Recruiting the 2-5A system for antisense therapeutics, *Antisense & Nucl.Acid.Drug Dev.* **7**, 203-206.
39. Zon, G., Sumers, M.F., Gallo, K.A., Shao, K.-L., Koziołkiewicz, M., Uznański, B. and Stec, W.J. (1987) Stereochemistry of phosphorothioate phosphotriesters, in: *Biophosphates and Their Analogues - Synthesis, Structure, Metabolism and Activity*, K.S.Bruzik and W.J.Stec (eds.), Elsevier, Amsterdam, pp. 165-178.
40. Jaroszewski, J.W., Sui, J., Majzel, J. and Cohen, J.S. (1992) Towards rational design of antisense DNA: molecular modelling of phosphorothioate DNA analogues, *Anti-Cancer Drug Design.* **7**, 253-262.
41. Koziołkiewicz, M., Krakowiak, A., Kwinkowki, M., Boczkowska, M and Stec, W.J. (1995) Stereodifferentiation - the effect of P-chirality of oligo(nucleoside phosphorothioate)s on the activity of bacterial RNase H, *Nucleic Acids Res.* **23**, 5000-5005.
42. Stec, W.J., Cierniewski, C.S., Okruszek, A., Kobylańska, A., Pawłowska, Z., Koziołkiewicz, M., Pluskota, E., Maciaszek, A., Rębowska, B. and Stasiak, M. (1997) Stereodependent inhibition of Plasminogen Activator Inhibitor Type 1 by phosphorothioate oligonucleotides: proof of sequence specificity in cell culture and *in vivo* rat experiments, *Antisense & Nucl.Acids Drug Dev.* **7**, 567-573.
43. Koziołkiewicz, M., Wójcik, M., Kobylańska, A., Karwowski, B., Rębowska, B., Guga, P. and Stec, W.J. (1997) Stability of stereoregular oligo(nucleoside phosphorothioate)s in human plasma: diastereoselectivity of plasma 3'-exonuclease, Antisense *& Nucl.Acid. Drug Dev.* **7**, 43-48.
44. Koziołkiewicz, M., Gendaszewska, E. and Maszewska, M. (1997) Stability of stereoregular oligo(nucleoside phosphorothioate)s in human cells. Diastereoselectivity of cellular 3'-exonuclease,*Nucleosides & Nucleotides* **16**, 1677-1682.
45. Krakowiak, A. and Koziołkiewicz, M. (19980 Influence of P-chirality of phosphorothioate oligonucleotides on the activity of AMV-reverse transcriptase, *Nucleosides & Nucleotides* **17**, 1823-1834.
46. Koziołkiewicz, M., Maciaszek, A., Stec, W.J., Semizarov, D., Victorova, L. and Krayevsky, A. (1998) Effect of P-chirality of oligo(nucleoside phosphorothioate)s on the activity of terminal deoxyribonucleotidyl transferase, *FEBS Letters* **434**, 77-82.
47. Benimetskaya, L., Tonkinson, J.L., Koziołkiewicz, M., Karwowski, B., Guga, P., Zeltser, R., Stec, W.J. and Stein, C.A. (1995) Binding of phosphorothioate oligonucleotides to basic fibroblast growth factor, recombinant soluble CD4, laminin and fibronectin is P-chirality independent, *Nucleic Acids Res.* **23**, 4239-4245.

TOWARDS IMPROVED APPLICATIONS OF CELL-FREE PROTEIN BIOSYNTHESIS – THE INFLUENCE OF mRNA STRUCTURE AND SUPPRESSOR tRNAS ON THE EFFICIENCY OF THE SYSTEM

MICHAEL GERRITS, HELMUT MERK, WOLFGANG STIEGE AND VOLKER A. ERDMANN

Institut für Biochemie, Freie Universität Berlin, Thielallee 63, D-14195 Berlin

1. Abstract

The cell-free protein biosynthesis has the potential to become a powerful technology for the biochemical research in particular in the determination of the structure and function of proteins. The number of possible applications is rising with the obtainable yields and with the expanded feasibility of introducing modified amino acids into proteins. Here we describe the influence of two RNA translation components, the mRNA and the suppressor tRNA, on the efficiency of protein biosynthesis.

It is shown that the rate limiting factor of the cell-free translation of the two proteins dihydrofolate reductase (DHFR) and fatty acid binding protein (FABP) is not the initiation or termination step. The efficiency of peptide bond formation in the nascent protein varies between the two genes but is independent on the size of the coding sequences. The poor translation of DHFR can be improved when its coding sequence is fused with a part of the more efficiently translated FABP gene.

We compared different amber suppressor tRNAs on the level of translational efficiency and aminoacylation capacity. Our results show that in most cases the aminoacylation rate of the tRNAs is not the limiting factor of suppression. An *E. coli* $tRNALeu_{CUA}$ exhibits the highest translational efficiency of the examined tRNAs. So this $tRNALeu_{CUA}$ may be a starting point to construct more efficient tRNAs for the introduction of unnatural amino acids into proteins in the *in vitro* translation system by eliminating the synthetase mediated aminoacylation of the tRNA.

J. Barciszewski and B.F.C. Clark (eds.), RNA Biochemistry and Biotechnology, 335–345.

2. Introduction

One important aspect of the RNA-technologies is the development of an efficient *in vitro* synthezising system for the synthesis of proteins. The methodology for the *in vitro* synthesis of biological active proteins has been known for more than 30 years and we could demonstrate that 1. the production of cytotoxic, regulatory or unstable proteins which can not be (over)expressed in living cells is possible [20]. Further advantages are 2. the proteins synthezised can be selectively labeled with isotopes in order to facilitate their detection or to study their structure and function by such methods as NMR spectroscopy. 3. The protein products are more easily isolated from *in vitro* systems [7] and 4. they show in general a superior biological activity. 5. The system can also be used for the synthesis of biological active single chain antibodies or 6. even for the *in vitro* evolution of proteins with selected biological properties. 7. Moreover, purified mRNA can be used as a template and 8. it is possible to create proteins with improved or even new biolocical activities by introducing unnatural amino acids into specific positions of a protein. It is only recently that interests have been turning to the *in vitro* translational system as a potential method for the synthesis of large amounts of biological active proteins.

In protein biosynthesis the most important interactions between the translational components take place on the level of RNA molecules. The three species of RNA namely the ribosomal RNA (rRNA), the messenger RNA (mRNA) and the transfer RNA (tRNA) are the key components during the course of translation. Here we present some interesting results concerning the mRNA and the tRNA. In the first part we focus on the optimization of templates for the efficient protein biosynthesis in a cell-free *E. coli* system [20, 12] and in the second part we report the development of an improved system, in which amber suppressor tRNAs can be used for the introduction of a desired (modified) amino acid into a certain protein.

3. Materials and Methods

Construction of plasmids

For the construction of plasmids, which served as templates for the *in vitro* transcription, we followed standard protocols [18]

In vitro transcription

The mRNA and tRNA transcripts were obtained by *in vitro* runoff transcription from linearized plasmids with T7 RNA polymerase (Stratagene) following the protocol of Triana-Alonso et al. [21] with some modifications.

Purification of tRNA transcripts

Suppressor tRNAs were purified by standard methods using denaturing polyacrylamide gel electrophoresis according to Sampson and Uhlenbeck [19] with some modifications.

In vitro translation

We used an optimized procaryotic lysate which was prepared by the method of Cronenberger and Erdmann [1] with some modifications and composed of components described by Merk et al. [12] The determination of suppression efficiencies were carried out with the addition of *in vitro* transcribed suppressor tRNAs as indicated in the text. The synthesized proteins where labeled with L-[U-^{14}C]-leucine with a specific activity of 304 mCi/mmol (Amersham). After translation protein quantification was done by measuring the incorporated radioactivity present in trichloro acetic acid-precipitated aliquots of the reaction mixtures. The proteins were analyzed by autoradiography after separation on 15% polyacrylamide gels according to Laemmli [11].

Aminoacylation of tRNA

Aminoacylation was performed under translation conditions omitting the ribosomal fraction. The four ribonucleotide triphosphates were used in final concentrations of 1 mM (ATP and GTP) and 0,5 mM (CTP and UTP).

4. Results and Discussion

4.1 INFLUENCE OF mRNA STRUCTURE ON TRANSLATIONAL EFFICIENCY

Initiation and termination of the translation are known to be rate limiting steps. The initiation in procaryotes depends on the presence of a ribosome binding site and on an appropriate spacing to the initiation codon [4, 16]. The extend of folding of the mRNA in some cases regulates the accessibility to the translational components [2, 6] as well as its own stability against ribonucleases [3]. The frequency of rare codons also can be important [5, 8], especially if eucaryotic templates are used in procaryotic systems. Pronounced folding of the mRNA might result in low translational efficiency or in premature termination of translation at least *in vivo*. This is in contrast to an increased protein synthesis due to an increased half-life of the mRNA. The translational efficiency is also dependent on the size of the coding region or the whole mRNA, respectively.

In order to analyze the influence of the translational initiation and termination as well as the protein size on the efficiency of the translation, we have constructed four plasmids coding for monomers and dimer fusion proteins of FABP (14,7 kDa) and DHFR (21,6 kDa) (Fig. 1). If the initiation and the termination are predominantly limiting the protein synthesis, the molar yields of the dimers should hardly be influenced in comparison to the monomers. Our results, however, show that the molar yields of the dimers are reduced to about half the amount of the monomers (Fig. 2a). On the other hand the overall mass synthesis was hardly influenced (Fig. 2b). Moreover, the homogenity of the products is not influenced as the ^{14}C-labeled proteins appear as clear bands in the autoradiogram of the 15% Laemmli gel (Fig. 3). As the mass yields of monomers and dimer fusion proteins are quite similar, we conclude that the elongation of the nascent peptide chain is the overall limiting factor rather than initiation and termination.

Figure 1: Schematic illustration of the fusions of FABP and DHFR

To figure out whether the poorly expression of the DHFR gene might be improved by fusion with different regions of the more efficiently translated FABP sequence, we have constructed three fusion genes: A) DHFR fused in front of FABP; B) DHFR fused in the middle of FABP and C) construct like B with shortened sequence of FABP (Fig. 1).

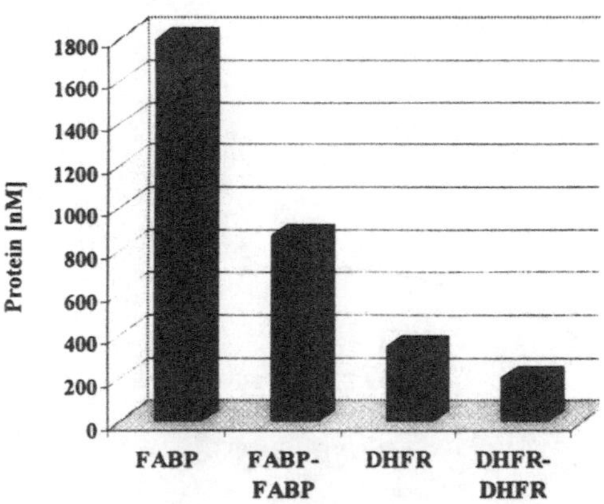

Figure 2a: Maximum molar yields from translation of monomer and dimer forms of FABP and DHFR

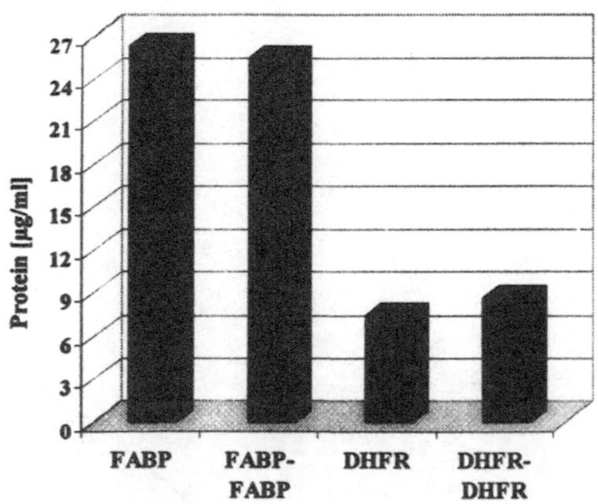

Figure 2b: Maximum mass yields from translation of monomer and dimer forms of FABP and DHFR

The results of the translation of the corresponding mRNAs derived from these constructs show that with construct B, which starts with 111 amino acids of FABP the mass yield was raised to about 240%. With construct C, starting with 11 amino acids of FABP the yield was hardly influenced. The construct A showed a negative effect (Fig. 4). So the position of DHFR in the fusion protein is critical for the translational efficiency. It seems, in this case, that the kind and size of the (coding) initiation region might be rate limiting for the protein biosynthesis and that the longer leader region of FABP in product B indeed has a positive effect on the translation of DHFR.

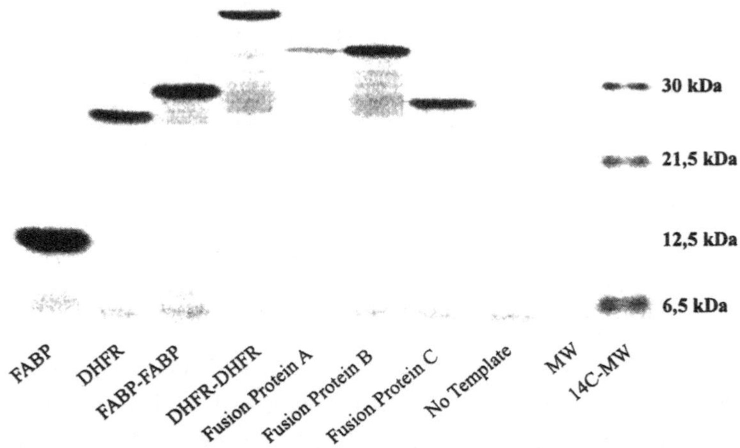

Figure 3: Autoradiogram of SDS-Gel after cell-free translation

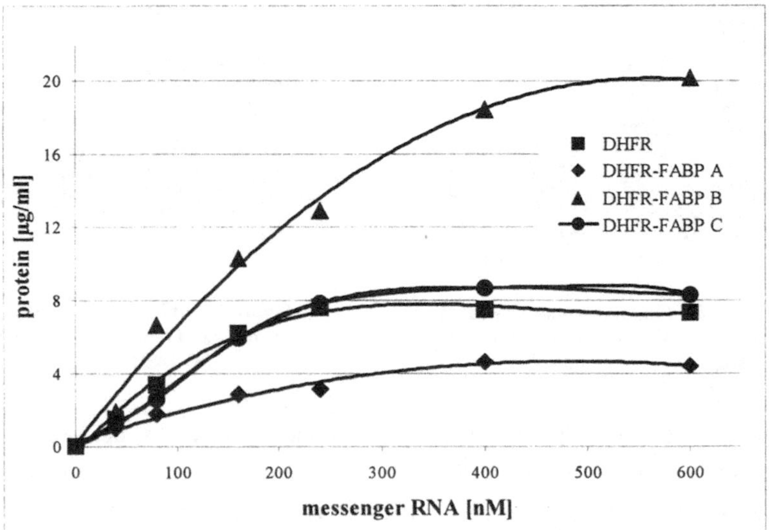

Figure 4: Mass yields from translation of DHFR and its fusion proteins with FABP

4.2. THE TRANSLATIONAL EFFICIENCY OF AMBER SUPPRESSOR tRNAs

The introduction of unnatural amino acids into proteins can be achieved by the use of an amber suppressor tRNA, that has been chemically acylated with the desired (modified) amino acid and that is not a substrate for the natural aminoacyl tRNA synthetases (for example: Robertson et al. [17]). Addition of the charged amber suppressor tRNA to the protein biosynthesis reaction results in specific incorporation of the amino acid in the protein. The amount of mutant protein generated by this method is limited (1) by the low stability of the hydrolytically labile amino acyl linkage under reaction conditions and (2) by the suppression efficiency of the charged aminoacyl tRNA. So only a small portion of the desired amino acid may be incorporated in the protein while the most part of the activated amino acid gets lost due to spontaneous deacylation of the aminoacyl tRNA. A way to increase the yield of mutant protein is to find tRNAs with higher suppression efficiencies. Our approach involves both the determination of the aminoacylation rate in the translation system and the comparison of the suppressor tRNAs on the level of translational efficiency.

In order to test the efficiency of suppressor tRNAs we used a system in which the tRNAs are aminoacylated by their cognate synthetases. We have constructed plasmids containing the genes of the following different *E. coli* amber suppressor tRNAs under the control of a T7 promotor: $tRNAAla_{CUA}$, $tRNAPhe_{CUA}$, $tRNAHis_{CUA}$, $tRNALys_{CUA}$ [10] and $tRNALeu_{CUA}$ [13]. After runoff transcription with T7 RNA polymerase and purification, the tRNAs were compared on the level of aminoacylation capacity and translational efficiency. The ability of the tRNAs to be aminoacylated was performed with the S100 fraction of the cell-free translation system, that is nucleic acid-free.

The amber suppressor tRNAs for Ala, His and Leu show aminoacylation rates on a comparable high level while in the case of $tRNALys_{CUA}$ and $tRNAPhe_{CUA}$ we could not detect any aminoacylation activity under the conditions used (Fig. 5). The missing aminoacylation of $tRNAPhe_{CUA}$ was not unexpected because the Phenylalanyl tRNA synthetase is known to use the anticodon of the tRNA as a recognition element. The tRNAPhe transcript with the CUA anticodon is reduced in catalytic activity by at least

1000-fold in comparison to the "wildtype" transcript (Peterson and Uhlenbeck, 1992).

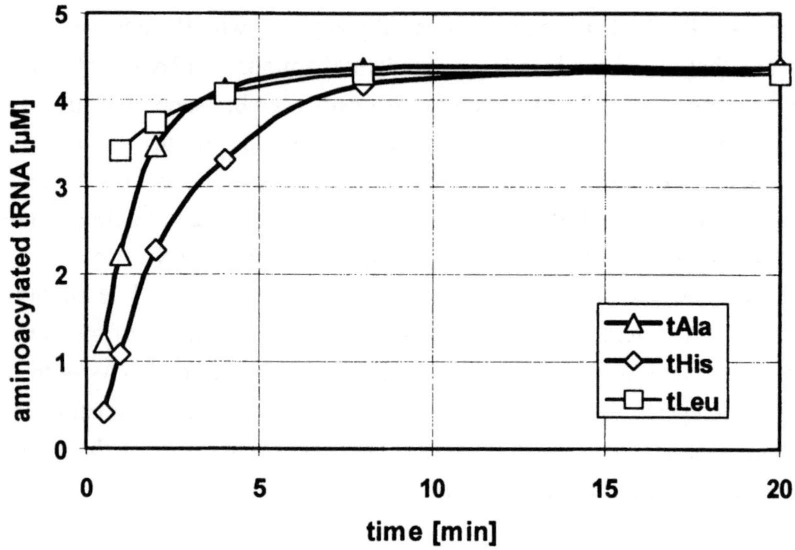

Figure 5: Aminoacylation of amber suppressor tRNA transcripts (5µM) in the S100 fraction of the translation system

In order to analyze the ability of the amber suppressor tRNA transcripts to suppress an amber codon in an *in vitro* transcription/translation reaction we have changed the codon GAC (Asp) of the gene encoding the FABP (14,7 kDa) to UAG (amber) (Fig. 6). The translation products of the corresponding mRNAs are labeled with ^{14}C-leucine. The truncated amber fragment translated in the absence of an amber suppressor is too small to appear on the autoradiogram or to be detected by TCA-precipitation while the full-length product generated by suppression of the amber codon should appear on the autoradiogram of a SDS-Gel as a clear single band.

Figure 6: Part of the sequence of the mutant FABP gene

The addition of the amber suppressor tRNA transcripts for Leu, His, Ala and Lys resulted in detectable yields of full-length FABP (Fig. 7) while no full-length product appeared in the autoradiogram in the absence of a suppressor (Fig. 7, line K). The expression of the "wildtype" FABP was not affected in the presence of the amber suppressor tRNA transcripts (data not shown). The $tRNAPhe_{CUA}$ transcript was not able to generate a detectable amount of suppression product (Fig. 7). This finding correlates with the missing aminoacylation activity of the transcript in the S100 fraction (see above), but is in contrast to the *in vivo* conditions, where the phenylalanyl tRNA amber suppressor has been shown to be a good suppressor [14]. So suppression *in vivo* and *in vitro* seems to be quite different.

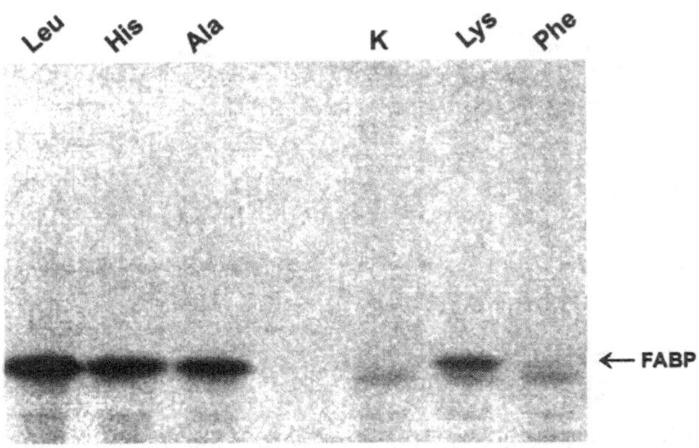

Figure 7: Autoradiogram of an SDS-Gel. Translation reactions were performed – in the absence of an amber suppressor tRNA (K); - in the presence of $tRNALeu_{CUA}$ transcript (Leu); $tRNAHis_{CUA}$ transcript (His); $tRNAAla_{CUA}$ transcript (Ala); $tRNALys_{CUA}$ transcript (Lys); $tRNAPhe_{CUA}$ transcript (Phe) and separated on an 15% Laemmli gel.

The ability of the $tRNALys_{CUA}$ transcript to generate detectable amounts of suppression products was surprising due to the missing aminoacylation activity of the tRNA in the S100 fraction. It seems likely that this finding is based on a very low level of aminoacylated $tRNALys_{CUA}$ transcripts which can not be detected in the S100 fraction. In the translation system the labile aminoacyl linkage of the aminoacyl tRNA can be stored in the very stable peptide linkage of the generated peptide chain while in the S100 fraction aminoacylation competes with spontanous deacylation of the aminoacyl tRNA.

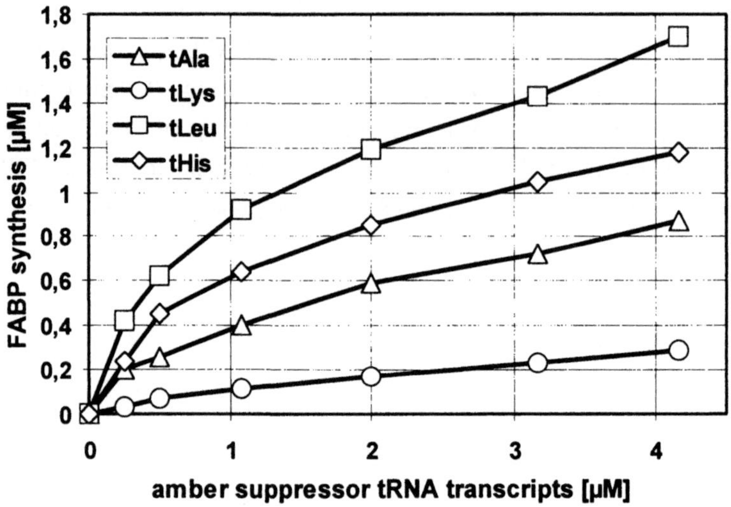

Figure 8: Translation of mutant FABP in the presence of different amber suppressor tRNA transcripts

The yield of suppression product generated by the amber suppressor tRNA transcripts for Leu, His and Ala lies between 0.8 and 1.7 µM protein in a 90-minute translation reaction (Fig. 8) while the aminoacylation capacity of the corresponding aminoacyl tRNA synthetases was shown to be higher than 4 µM aminoacyl tRNA in 10 minutes in the S100 fraction of the translation system (Fig. 5). For this reason we assume that at every time of the translation reaction the added suppressor tRNA transcripts are fully aminoacylated and that the amount of suppression product is not influenced by the charging state of the amber suppressor tRNAs. So the ability of the $tRNALeu_{CUA}$ transcript to generate two times more suppression product than $tRNAAla_{CUA}$ (Fig. 8) seems not to be a result of a higher aminoacylation capacity of its corresponding aminoacyl tRNA synthetase. The improved yield of suppression product is more likely the result of the higher suppression efficiency of the $tRNALeu_{CUA}$ transcript.

One of the best tRNAs reported to date for the introduction of unnatural amino acids into proteins is an *E.coli* $tRNAAla_{CUA}$ with a U70C mutation in the acceptorarm that is not a substrate for reactivation by the *E. coli* alanyl-tRNA synthetase [9]. Our results show that an *E. coli* $tRNALeu_{CUA}$ exhibits a substantial higher suppression efficiency than "wildtype" *E. coli* $tRNAAla_{CUA}$. So this *E. coli* $tRNALeu_{CUA}$ may be a starting point to construct more efficient tRNAs for the introduction of unnatural amino acids into proteins by eliminating the synthetase mediated aminoacylation of the tRNA.

Acknowledgements: This project has been supported in part by the Bundesministerium für Bildung, Wissenschaft, Forschung und Technologie (BMBF, No. 0311 302, a project coordinated by Boehringer Mannheim GmbH), the Deutsche Forschungsgemeinschaft (SFB 344) and the Fonds der Chemischen Industrie e.V.

References

1. Cronenberger, J.H. and Erdmann, V.A. (1975). Stimulation of polypeptide polymerization by blocking of free sulphydryl groups in Escherichia coli ribosomal proteins. *J. Mol. Biol.* **95**, 125-137
2. De Smit M.H. and Van Duin J. (1990). Control of prokaryotic translational initiation by mRNA secondary structure. *Progr. Nucl. Acid Res. Mol. Biol.* **38**, 1-35
3. Fuchs, U., Stiege, W., Erdmann, V.A. (1997). Ribonucleolytic activities in the Escherichia coli in vitro translation system and in its separate components. *FEBS Lett.* **414**, 362-364
4. Gold L., Pribnow D. Schneider T., Shinedling S., Singer B.S., Stormo G. (1981). Translational initiation in prokaryotes. *Annu. Rev. Microbiol.* **35**, 365-405
5. Grosjean H., Sankoff D., Jou W.M., Fiers W., Cedergren R.J. (1978). Bacteriophage MS2 RNA: a correlation between the stability of the codon: anticodon interaction and the choice of code words. *J. Mol. Evolution* **12**, 113-119
6. Hall M.N., Gabay J., Débarbouillé M., Schwartz M. (1982). A role for mRNA secondary structure in the control of translation initiation. *Nature* **295**, 616-618
7. Haukanes B.I., Kvam C. (1993). Application of magnetic beads in bioassays. *Biotechnology* **11**, 60-63
8. Ikemura T. (1981). Correlation between the abundance of Escherichia coli transfer RNAs and the occurrence of the respective codons in its protein genes: a proposal for a synonymous codon choice that is optimal for the E. coli translational system. *J. Mol. Biol.* **151**, 389-409
9. Karginov V.A., Mamaev S.V., Hecht S.M. (1997). In vitro suppression as a tool for the investigation of translation initiation. *Nucleic Acid Research* **25**, 3912-3916
10. Kleina L.G., Masson J.M., Normanly J., Abelson J., Miller J.H. (1990). Construction of Escherichia coli amber suppressor tRNA genes. II. Synthesis of additional tRNA genes and improvement of suppressor efficiency. *J. Mol. Biol.* **213**, 705-717
11. Laemmli U.K. (1970). Cleavage of structural proteins during the assembly of the head of bacteriophage T4. *Nature* **227**, 680-685
12. Merk H., Stiege W., Tsumoto K., Kumagai I., Erdmann V.A., Cell-free expression of two single–chain monoclonal antibodies against lysozyme – Effect of domain arrangement on the expression. *J. Biochem.*, in press
13. Normanly J., Ogden R.C., Horvath S.J., Abelson J. (1986a). Changing the identity of a transfer RNA. *Nature* **321**, 213-219
14. Normanly J., Masson J.M., Kleina L.G., Abelson J., Miller J.H. (1986b). Construction of two Escherichia coli amber suppressor genes: tRNAPheCUA and tRNACysCUA. *Proc. Nat. Acad. Sci.* **83**, 6548-6552
15. Peterson ET, Uhlenbeck OC (1992). Determination of recognition nucleotides for Escherichia coli phenylalanyl-tRNA synthetase. *Biochemistry* **31**, 10380-10389
16. Ringquist S., Shinedling S., Barrick D., Green L., Binkley J., Stormo G.D., Gold L. (1992). Translation initiation in Escherichia coli: sequences within the ribosome-binding site. *Mol. Microbiol.* **6**, 1219-1229
17. Robertson S.A., Ellmann J.A., Schultz P.G. (1991). A general and efficient route for chemical aminoacylation of transfer RNAs. *J. Am. Chem. Soc.* **113**, 2722-2729
18. Sambrook J., Fritsch E.F., Maniatis T. (1989). Molecular cloning. *Cold Spring Harbour Laboratory Press*
19. Sampson J.R., Uhlenbeck O.C. (1988). Biochemical and physical characterization of an unmodified yeast phenylalanine transfer RNA transcribed in vitro. *Proc. Nat. Acad. Sci.* **85**, 1033-1037
20. Stiege W., Erdmann V.A. (1995) The potentials of the in vitro protein biosynthesis system. *J. Biotechnol.* **41**, 81-90
21. Triana-Alonso F.J., Dabrowski M., Wadzack J., Nierhaus K.H. (1995) Self-coded 3'-extension of run-off transcripts produces aberrant products during in vitro transcription with T7 RNA polymerase. *J. Biol. Chem.* **270**, 6298-6307

RNA ON THE WEB

M. SZYMAŃSKI[1)], B.F.C. CLARK[2)] and J. BARCISZEWSKI[1)]
[1)]Institute of Bioorganic Chemistry of the Polish Academy of Sciences, Noskowskiego 12, 61704 Poznań, Poland and [2)]Institute of Molecular and Structural Biology, Aarhus University, Gustav Wieds Wej 10C, DK-8000 Aarhus C, Denmark

1. Introduction

Life involves transmission and transformation of the information that is either stored within the genome or acquired from the environment. Genetic information is a basic factor necessary to construct and manage a living organism. The molecular processes in living cells are responsible for transformation of the genetic information into physical reality. The aim of biochemistry and molecular biology is to obtain insight into the finest details of these processes and to understand the underlying logic of life. However, the answers obtained in laborious, slow and costly research process are often far from being complete.

On the other hand, such studies can be complemented by evolutionary studies that are based on an assumption that the organisation and basic mechanisms within the living cells are redundant. Thus in the course of evolution, some molecular processes were reused in different functional contexts and in different organisms. The same function performed by biological macromolecules was found to be often (but not always) associated with similar primary, secondary and tertiary structures. From the comparison of available data one can draw general conclusions about sequence and structure requirements for specific function. The importance of biological databases is well recognised by a majority of molecular biologists. The Internet-based sequence analysis tools together with on-line sequence databases allow quick comparison and identification of new primary structures resulting from research. There is now a very strong link between the workbench and the computer, and the results of computer analyses are no longer considered as accessory data. The invention and development of the Internet gave us a new tool for quick access to the data deposited within the databases for molecular sequences. The user-friendly interface offered by the World Wide Web makes it available to anyone possessing a minimal knowledge of computers. Searching, comparison, alignment and retrieval as well as the submission of new sequences to the databases is now very fast and the speed of information exchange allows the results to be brought to the computer screen within seconds. Currently, most of the basic sequence analysis procedures are available to anyone having the computer connected to the Internet and the WWW browser. They include similarity searches, multiple sequence alignments, predictions of the secondary structure of proteins and RNA as well as the generation of the 3D structure models of proteins. They do not

J. Barciszewski and B.F.C. Clark (eds.), RNA Biochemistry and Biotechnology, 347–352.

require the installation of the programmes and calculations and analyses are performed by the software on remote servers so the results are sent either to the browser in the interactive mode or to the user's e-mail address.

The prerequisite for such analyses is a collection of sequence and structural data. Shortly after publication of the first DNA, RNA and protein sequences, molecular biologists started to organise them into various databases. These efforts resulted in creation of large international and collaborative investments, whose aim was to collect and distribute all available protein and nucleic acids sequence data for the scientific community. Currently there are four main databases that contain all known nucleotide and amino acid sequences. These are EMBL Nucleotide Sequence Database, GenBank, DNA Data Bank of Japan and SwissProt.

In the last decade, the DNA sequencing technique improved dramatically. The introduction of the polymerase chain reaction and automated DNA sequencing allowed the simplification of some of the traditional methods in DNA cloning and sequence analysis. This in turn resulted in the increase of the rate and the decrease of the costs at which the new sequences can be obtained. Because of this progress, scientists undertook projects of sequencing whole genomes of several organisms, including human. The main goal of these analyses is to gain an insight into organisation of the genetic material as well as better understanding of the mechanisms, which govern its expression. The outcome of those efforts is a growing list of sequenced microbial and organelle genomes and the complete yeast genome. Furthermore, the human, *Drosophila melanogaster*, *Caenorhabditis elegans*, *Arabidopsis thaliana* and several crop genome sequencing projects are well on the way to completion.

The sequences derived from the analysis of genome sequences as well as the accessory databases of Expressed Sequence Tags (EST) and Sequence-Tagged Sites (STS) made substantial contributions to the growth of nucleotide sequence databases in the past few years. Computer analysis of the genome sequences is a crucial step in an attempt to understand the structure and function of biological macromolecules. Comparative analyses of the sequenced microbial chromosomes will allow us to determine the minimal set of the genes that is necessary for cell function. For most of the novel genes identified in the genomic sequences, no specific function is known.

The rise of bioinformatics is the result of the development of experimental techniques, which allowed generation of large amounts of sequence data. Its primary role was storage and management of data that could be used to assist experimental projects. More recently, bioinformatics became an independent scientific discipline located on the border between molecular biology and computer sciences. In fact the research projects in bioinformatics have their roots either in wet biochemistry or in computer science problems. They are merged into an interdisciplinary field, whose objective is finding answers to biologically relevant questions, using the intellectual power of computer sciences.

The accumulation of DNA and protein sequence data often makes searches of large databases difficult. There is a tendency to create new, specialised databases that contain well defined subsets of the general data banks (EMBL, SwissProt). In contrast to the general databases, which cover the whole variety of biological macromolecules, the specialised collections are devoted to specific, usually homologous, molecules or specific biological activity. This allows them to be better annotated and they provide more detailed information with an emphasis being put on the most important features of the described sequences. It also makes the access and retrieval of data much easier and straightforward.

Another advantage of the specialised molecular sequences databases is that they usually do not contain redundant data, that are often found in the general databases, which has resulted from multiple submissions of the same sequences, that occur under different accession numbers. While collecting the data for the database, these multiple entries can be sorted out and eliminated, reducing the "information noise".

2. RNA resources on the web

The discovery of various roles of RNA as a carrier of genetic information and an accessory factor in its decoding as well as a biocatalyst, was a main reason for increased interest in this type of biological molecule for the last 15 years. The recognition of the potential of RNA molecules led to a rapid growth of data. This was accompanied by a constantly growing number of the web sites related to RNA research.

The RNA resources on the World Wide Web (Table 1) fall into three general categories:

1. sequence and structure databases,
2. on-line sequence analysis tools,
3. miscellaneous information sources.

The first category comprises databases of RNA and RNA-related molecules. These databases are available on line, and in addition to the raw data they usually provide some accessory information related to their subject. They involve collections of sequences of ribosomal RNAs, guide RNAs, transfer RNAs, small nuclear RNAs, ribonuclease P RNAs etc., as well as of proteins that are known to interact with RNA and that are key factors in decoding the genetic information, e.g. aminoacyl-tRNA synthetases. The content of these databases is an invaluable source of information for molecular phylogeny and structural studies. Another category of the databases related to the research of RNA function is made up by the databases of RNA mutations and editing sites. These databases significantly contribute to the understanding of functions performed by certain RNA molecules.

Another group of databases comprises the collections of the secondary structures, inferred from the evolutionary analyses assisted by the results from the experimental, structural studies. Currently, such databases are available for ribosomal RNAs, ribonuclease P and group I self-splicing introns.

The ultimate goal of structural studies in protein as well as in nucleic acids research is the determination of the three-dimensional structure. In the case of proteins, it proved to be much easier, and resulted in a rapid growth of the protein structure database. Currently it contains 8771 entries for 8137 protein, 618 nucleic acid and 12 carbohydrate structures. Due to the accumulation of the structural data obtained for nucleic acids, in 1992 a new project, the Nucleic Acid Database, started. Its objective is to collect and distribute structural information about nucleic acids and their complexes with proteins. The data are available via WWW (see table for the net addresses).

TABLE 1. Selected RNA-related sites on the World Wide Web

SEQUENCE DATABASES
Ribosomal RNAs database *http://rrna.uia.ac.be/*
Ribosomal Database Project *http://www.cme.msu.edu/RDP*
5S rRNA *http://www.man.poznan.pl/5SData/5SRNA.html*
tRNA and tRNA genes database *http://www.uni-bayreuth.de/departments/biochemie/trna/tRNA*
Guide RNA Database *http://www.biochem.mpg.de/~goeringe/*
Ribosomal RNA Mutation Databases *http://www.fandm.edu/Departments/Biology/Databases/RNA.html*
RNA Aptamer Sequence Database *http://splatter.chem.indiana.edu/sequences/database.html*
RNA editing site *http://www.lifesci.ucla.edu/RNA/index.html*
RNA editing site, Trypanosome U insertion/deletion *http://www.lifesci.ucla.edu/RNA/trypanosome/index.html*
Small RNA Database *http://mbcr.bcm.tmc.edu/smallRNA/smallrna.html*
Signal Recognition Particle Database *http://psyche.uthct.edu/dbs/SRPDB/SRPDB.html*
RNA Modification Database *http://www-medlib.med.utah.edu/RNAmods/RNAmods.html*
snoRNA database *http://rna.wustl.edu/snoRNAdb/*
RNase P Database *http://jwbrown.mbio.ncsu.edu/RNaseP/home.html*
tmRNA Databases *http://psyche.uthct.edu/dbs/tmRDB/tmRDB.html* *http://sunflower.bio.indiana.edu/~kwilliam/tmRNA/home.html*
uRNA Database *http://psyche.uthct.edu/dbs/uRNADB/uRNADB.html*
mRNA-like non-coding RNAs *http://www.man.poznan.pl/5SData/ncRNA/index.html*
Viroid and viroid-like RNA sequences database *http://www.callisto.si.usherb.ca/~jpperra*
Aminoacyl-tRNA synthetase database *http://www.man.poznan.pl/aars/*

TABLE 1. (continued)

SECONDARY AND TERTIARY STRUCTURE DATABASES
RNA secondary structures of group I introns, 16S rRNA, 23S rRNA *http://pundit.icmb.utexas.edu*
Nucleic Acids Database Project *http://ndbserver.rutgers.edu*
Brookhaven Protein Data Bank *http://www.pdb.bnl.gov*
3D models of RNase P *http://jwbrown.mbio.ncsu.edu/RNaseP/RNA/threeD/threeD.html*
RiboWeb Project - 3D models of 30S ribosomal subunit and 16S rRNA *http://www-smi.stanford.edu/projects/helix/ribo3dmodels/index.html*
RNA ANALYSIS TOOLS
MFOLD - RNA seconadry structure analysis WWW server *http://www.ibc.wustl.edu/~zuker/rna/fold1.cgi* *http://BiBiServ.TechFak.Uni-Bielefeld.DE/mfold/*
RNA secondary structure prediction server *http://www.genebee.msu.su/services/rna2_reduced.html*
RNAfold program (Viena package) WWW server *http://www.tbi.univie.ac.at/cgi-bin/RNAfold.cgi*
RNA secondary structure analysis (ESSA) *http://www-bia.inra.fr/T/essa/Doc/essa_home.html*
tRNAscan - search for tRNA sequences *http://genome.wustl.edu/eddy/tRNAscan-SE/*
SOFTWARE SOURCES
MFOLD *http://www.ibc.wustl.edu/~zuker/*
RNAdraw *http://rnadraw.base8.se*
RNAstructure *http://128.151.176.70/RNAstructure.html*
RnaViz *http://rrna.uia.ac.be/rnaviz/*
Vienna RNA Package *http://www.tbi.univie.ac.at/~ivo/RNA/*
XRNA *ftp://fangio.ucsc.edu/pub/XRNA*
A Guide to RNA Folding Software *http://hornet.mmg.uci.edu/~hjm/projects/biocomp/rna.folding.html*

In the second category of the web sites dedicated to RNA research there are a variety of tools, which allow analysis of the RNA sequences. The majority of these sites is devoted to RNA folding. For many years, the problem of the formation of the secondary structure of RNA molecules was a key question in RNA research. Various attempts to solve this problem resulted in several algorithms, which can be used to predict the secondary structure of RNA. There are several web sites, which offer software that can perform calculations of RNA secondary structure prediction. This can also be accomplished by using the web interfaces for MFOLD and Viena RNAfold (see table). Another useful tool, available on-line is a tRNA-Scan server, which performs analysis of nucleotide sequences in search for potential tRNA genes.

The third group of web sites, that may be interesting for RNA research comprises the sources of RNA analysis and secondary structure visualisation software and miscellaneous information. A great repository of links to other RNA-related web sites can be found on the RNA-World home page at the IMB Jena site (*http://www.imb-jena.de/RNA.html*). This site also contains an up to date list of events, conferences and courses on RNA.

From exploring the Internet in search for information on RNA, one can see that with the amount of data, that can be obtained from that source makes research much easier, than it was a few years ago. The advantages of computer sciences and the development of the global network, as well as the need for rapid exchange of information make the Internet one of the most important sources of scientific knowledge and ideas.

HOW RISKY IS DIRECT DEMOCRACY FOR BASIC SCIENCE?

PETER MANI
Gene Technology & Society
ETHZ, Rämistrasse 101, 8092 Zurich, SWITZERLAND
tecrisk@smile.ch

1. Abstract

A threatful national referendum which aimed at prohibiting gene technology in Switzerland to a large extend by prohibiting transgenic animals, deliberate release of genetic modified organisms, the patenting of animals and plants as well as their parts and products and by demanding a general reversal of burden of proof in research, has initiated the discussion about the necessity to rethink public communication about new technologies and risk communication both on the side of science as on the side of industry. The paper proposes the installation of three new institutions: first of all the implementation of a Competence Center for Biosafety which fulfills a new form of risk communication as well as the cantonal execution of federal ordinances, second a new institution for establishing strong links between Universities and High Schools to improve information transfer of basic science to the public and third the installation of an information system on the internet.

2. Introduction

Switzerland is governed by seven ministers and a two chamber system with a small chamber - the Council of States - containing two representatives from each of its 23 Cantons and a large chamber - the National Council - consisting of 200 members elected proportionally to the inhabitants of each Canton. All members are elected for a period of four years. Officials and clergymen are excluded from the large chamber. Entitled to vote are about 4.6 mio Swiss over 18 years, living in about 3000 political communities. There are four large political parties: the social democrats SPS (54/5[1]), the radical democrats FDP (45 /17), the christian democrats CVP (34/5) and the traditionalistic party SVP (29/ 5). These four parties occupy 81% in the large chamber and 94% in the small chamber, the rest is taken by a handful of minor parties. The parliament meets four times a year for three weeks. The seven ministers were elected by the 246 members of parliament by an informal agreed 2 SPS/2 FDP/2 CVP/1 SVP formula and re-elected every fourth year; they head the federal administration.

Citizens trying to enforce a change in the constitution must be able to collect 100'000 signatures of inhabitants which are entitled to vote within a deadline of 18 months. In

---

1. First figure corresponds to number of seats in large (total of 200) and second figure to the number of seats in small chamber (total of 46) respectively.

J. Barciszewski and B.F.C. Clark (eds.), RNA Biochemistry and Biotechnology, 353–361.

case of success the so-called initiative - an order for a national referendum - is given to the federal department which is supposed to be involved strongest into the according matter. This allocation is made by all seven ministers in agreement. The administration is then preparing a message which is signed by the corresponding minister. This message is regarded as the official opinion and has to be advocated by all officials working on that business. There are three possibilities for the message: it could contain a direct counter proposal favoring a different change of the national constitution in order to solve the given problem in an alternative way, it could otherwise contain an indirect counter proposal to solve the problem on the level of laws and ordinances in order to make the change in the constitution obsolete, or as a last possibility it could just recommend acceptance or rejection of the initiative without any counter measures.

3. History of the regulation of gene technology in Switzerland

Opposition against gene technology in Switzerland is driven to a large extent by the animal right organizations. In 1981 a first initiative to prohibit vivisection on vertebrates was launched; this initiative was proposed to be rejected without any counter proposal by the government. It was clearly rejected by the population with 71%. 1986 the second initiative intended to prohibit animal experimentation in general, allowing for some exceptions only. It was also proposed to reject without counter proposal and in the following vote it was rejected with 56%. Despite these two unsuccessful referendums against animal experimentation, a third initiative for the prohibition of animal experimentation was initiated in 1991 and without counter proposal given for voting where it was again rejected in 1992.

In the same year a newspaper called „Der Beobachter" started an initiative demanding an amendment to the Swiss constitution regarding the protection of human beings against misuse of biotechnology and reproductive medicine. This time the government and the parliament could build a consensus and formulate an alternative proposition which was even stronger than the initiative. Namely, not only the initiative was esteemed but it was found that the new article in the federal constitution should also contain a short paragraph covering the extra human nature.

At the same time, animal right organizations and part of the anti-nuclear-power organization who had just succeeded in preventing the construction of a planned nuclear power station near Basel and therefore had some free capacity, advocated a solution where the extra human nature should be protected with the same intensity than human beings. However, this opinion was not supported by the majority of the parliament and as a consequence the so-called „Gene Protection Initiative" was launched. In the same year the „Beobachter-Initiative" was secluded by its initiants and the counter proposal formulated by the parliament was accepted in the following votation by an impressing majority of 72%.

4. The „Gene Protection Initiative"

The text of the paragraph 24 ^{novies} which has been accepted in May 1992 as a result of the Beobachter-initiative goes as follows:

Article 24 ^{novies} paragraph 1 and 2 regulates gene technology and reproduction medicine on human in great detail, only paragraph 3 which is relevant for the further discussion is cited here:

> *Article 24 ^{novies} Paragraph 3:*
>
> *The Confederation regulates work with germ and genetic material of animals, plants and other organisms. It accounts for the dignity of creature as well as for safety of men, animal and environment and protects genetic diversity of animal and plant species.*

The new article 24 ^{decies} as proposed by the Gene Protection Initiative repeats this paragraph accordingly in its first paragraph and regulates gene technology of animals, plants and micro-organisms in the next three paragraph in very detail:

> *The Federal Constitution will be amended as follows:*
>
> *Article 24 ^{decies} (new)*
>
> *1. The Confederation issues regulations against the abuse and dangers arising from genetic modification of the genome of animals, plants and other organisms. It thereby takes into account the dignity and integrity of living beings, the conservation and utilization of genetic diversity as well as the safety of human beings, animals and the environment.*
>
> *2. It is forbidden to*
>
> *a. Produce, purchase and transfer genetically modified animals;*
> *b. Release genetically modified organisms into the environment;*
> *c. Patent genetically modified animals and plants, as well as their constituents, the procedures employed thereby and the products obtained.*
>
> *3. The legislation specifically regulates*
>
> *a. Production, purchase and transfer of genetically modified plants:*
> *b. The industrial production of compounds using genetically modified organisms;*
> *c. The research on genetically modified organisms which may constitute a risk to human health or the environment.*
>
> *4. Legislation specifically requires from an applicant the proof of usefulness and safety, the lack of alternatives as well as an explanation of ethical responsibility.*

5. Interpretation of the text

In view of basic research the prohibition of transgenic animals would have caused a problem: as there are no exceptions allowed all basic research which uses transgenic animals would be forbidden. This is of concern especially for medical research and developmental biology.

The prohibition of deliberate release would have made research with transgenic plants obsolete as this kind of projects usually end up in a deliberate release. It would be ethically inconsistent to produce transgenic plants and live vaccines in order to release them abroad only. Also practical research on environmental risk assessment - e.g. field experiments - would not be possible any more.

The impact of the prohibition of patenting was difficult to assess for life scientists. However, it has to be kept in mind that more and more public research institutes will have to procure their own money in future. In this context patenting will also become important for universities. It would also have been a problem for spin-off companies from universities.

Proofing usefulness and the absence of alternatives would have been a very restrictive barrier. Usefulness of a technical solution has always to be compared with the acceptability and cost of the renunciation. This can only be done experimentally. The demand for the proof of absence of an alternative could never be fulfilled as there is always an alternative: the renunciation.

In general it could be stated that the three prohibitions do not need any interpretations by the legislator nor do they allow for any kind of margin. Interpretation of paragraph 4 with its requirements for the different proof was still open. However, the scope for the legislator is never open because he is usually bound to the general understanding of the text at the moment of votation.

6. The campaign

Among the initiants were about 40 organizations - including WWF and Greenpeace Switzerland - with a total of about 800‘000 members; owning 25 printed media with a total edition of about 500‘000. The cumulated budget was estimated to about 60 mio sFr.; how much of it has been spent for the campaign is not known but it could well be 10 to 15 mio sFr. The initiative was supported by the green party and the social democrats.

On the side of the opponents there were the pharmaceutical industry (Novartis and Roche); the food industry (Néstle) was only marginally involved as the initiative did not treat gene food. Very early in the campaign the Swiss Association for Experimental Biology (USGEB) initiated the position *Gene Technology & Society* at the Swiss Federal Technical Institute ETHZ in Zurich with the goal to initiate and coordinate activities among the researchers at the ten Swiss universities. The budget of the industry has been cited from the financial newspaper „*cash*“ as 35 mio sFr., however, this has been denied strictly by the industry and can therefore only be estimated to be between 25 -35 mio sFr.

Initiants as well as industry had professional public relation offices involved which were leading the according campaigns for them. The newspapers covered the subject very intensively, as a matter of fact many citizens felt quite overwhelmed at the end of the campaign and many scientists could hardly hear the subject any more. The campaign was polarized very strongly and even sentences were demanded against unfair perceived advertisements and reports. The thematic was unusually intensively treated also in the editorial part of the newspapers and the period started extremely early with its highest intensity between August 1997 through April 1998. The discussion was kept on a high emotional level especially in the letters to the editors. While a majority of the reports in spring 1997 were in favor for the initiative the situation changed with time and at the end of the campaign the influence was perceived mostly in the other direction.

Printed media have been used very intensively (78%), but not so TV and radio (61 and 46% respectively). In no former votation campaign the official information booklet of the federal government, the letters to the editors, the advertisements and the posters have reached such a high recognition in open public as in this campaign.

7. The votation result and its analysis

The initiative was rejected with 66.7% by a voting participation of 41.3%. Polarization among political parties was extreme: namely 38% between social democrats and radical democrats; 29% between the red-green block and the commoner block and 27% between members and non-members of environmental groups. It is surprising however that the 50% limit of YES votes has even in the rows of initiants hardly been exceeded. Even the social democrats which supported the initiative reached only 46% YES votes. Only the group of members of environmental organizations reached 55%. The members of labor unions voted with 63% against the initiative and can therefore not be distinguished against other groups. Especially collective were the radical democrats with 92% NO voters.

Interestingly, the group of young voters (18-29 years) and the 68-generation which are generally much more sceptical than the overall citizen can not be distinguished from the average. Also the religious group had not significant more YES votes. There is only a correlation with the monthly income: families with less than sFr. 3000.- income per month show a YES fraction of 49% while families with an income of more than sFr. 9000.- have a YES fraction of only 22%. However, in general it can be said that the influence of social parameters are much less significant than the political variables.

What are the motives for voting YES or NO? For the supporters of the initiative the motives can mostly be explained with ideas of protection of nature and environment in general: 57% feel that manipulation of nature is dangerous and more protection is necessary. 38% believe gene technology is risky and 15% have ethical arguments against gene technology. Obviously this arguments have been overridden by other arguments, what happened?

It seems that the initiative has not so much been perceived as a „protection-initiative" but more as a „prohibition-initiative". This is certainly at least partly due to the strategy of the opponents of the initiative which showed that in case of an acceptance of the initiative a lot of basic research would be forbidden. The industry has even consequently titled the initiative in its advertisements and brochures as „Gene Prohibition Initiative", a fact which has been heavily criticized by the initiants which described it as polemic maneuver, however, it seems that the strategy has been successful. Also the scientific community demonstrated the limited power of life sciences without the powerful gene technology instruments. While the opponents of the initiative could keep the main-subject strictly on bio-medical research and their benefits for the future, the initiants showed no consistent strategy but switched aimlessly between all possible and impossible subjects, claiming the initiative would also protect the public from gene food and trying to connect all related and unrelated events with gene technology and their initiative. For example the statement was made that with paragraph 4 also gene food could be forbidden in Switzerland because there exist alternatives, which was in clear controversy with the governmental statement which made clear that paragraph 4 could not be extended to imported gene food produced outside Switzerland.

Another important point might have been that by voting NO one could not be regarded simply as a NO-sayer but rather someone who defends positive values like progress and an open mind also towards Europe.

As a conclusion, the mere thought to protect nature and the environment could never have succeeded because despite the fact that 58% believe that deliberate release of genetically modified organisms is a risk, 44% of those have nevertheless voted against the initiative! Even of those ranking nature and environmental protection as top priority, only 51% voted YES. Therefore nature and environmental protection was only of limited relevance - also the economic aspect was of importance.

8. Public communication about new technologies

Information is increasingly used as an administrative instrument and the right of information of the public is extended into new areas of risks in view of social interactions and contexts. Modern communication concepts have their basis on authoritative positions: managers in business and administration provide information at their own discretion to lay persons. But such kind of procedures could as well be found to be manipulative in order to enhance public acceptance. In addition it has recently be shown that more information does not necessarily lead to more acceptance. Although more information makes it more likely that a definite opinion will come up; this opinion, however, can be positive as well as negative.

This has severe consequences in public communication. It can probably be assumed that the reason for a controversy is not necessarily opposition but the way experts communicate with the lay person. For example it is often found that experts use only rational arguments on the basis of potential risks for men and nature. However, if the public is opposing a new technology on the basis of ethical aspects, a communication which exclusively discusses technical aspects in terms of prevention of damages, is rather inefficient.

The generally accepted communication concept of a two-way communication, composed of message, transmitter, channel and receiver is only a very incomplete model and misses the point that credibility is not so much a function of the quality of the message itself as much more a function of the quality of the relationship.

Modern concepts of mediation, public participation and public forums are also problematic. Here the requirements and expectations on lay persons are unrealistic and sometimes naive. It seems that politicians and administrative officials tend to move from one extreme to another: from the situation where only experts are taken seriously there is now a tendency to accept the judgment of a lay person with absolutely no experience and knowledge at the same level as that of the expert. Experts are needed and usually well accepted as experts by the lay public, problematic however is the decision process. As long as the public get the impression that only experts - which are very often regarded as biased persons, either belonging to the industry lobby or to the science lobby - are important in the decision finding process, they oppose to the decision. Why is this so?

There is at least one common element in all risk concepts, be it scientific or not: the distinction between reality and possibility. Risk is associated with the possibility that an undesirable or desirable state of reality may occur. Therefore experts are not only bringing into the decision process their knowledge about possibilities but also their images about reality and desirable realities. This might be the main problem because in matters of public concerns and values, public preferences and desired life-styles, everybody is its own expert. In a public forum as described before, the lay person is asked to extrapolate from his own image of reality, by his own concept to interpret probabilities to a future undesirable or desirable new reality. It is only natural that such a process is not very efficient in meeting the expectations of a pluralistic society.

If it is acknowledged that experts have some valuable data and all - inclusive the experts - have their own individual and social values and ways to measure utility, these sources of knowledge should be integrated, rather than separated.

9. First Proposal: Biosafety Competence Center

Our proposal is a consultants model. If the responsible institution treats the whole decision finding process with sufficient public involvement, political protest and conflicts should be minimized. Important is the fact that experts are to be consulted on all stages as well as input from a broad public in order to guarantee procedural fairness. Adequate risk communication through the whole process is a prerequisite for this kind of policy implementation.

On a practical basis competent centers are needed which fulfill the following criteria:

- The center has professional skills in risk assessment methodology.
- The center has sociological and cultural competence in order to manage and lead the decision making process.
- The center has competence in good risk communication practices in order to improve social learning of risk and mediate institutional and regulative responses to risk.
- The center has either executive competence itself or alternatively a sufficient high position of trust within the responsible regulative administration.

In the near future technical risk analysis must broaden its scope, methodology of social science must be used to inform policy makers about public concerns, use better methods of mutual communication and provide better models of discourse to integrate social and cultural needs in the decision making process.

10. Second Proposal: Liaison between Science and School

Assuming a doubling of our knowledge every five to ten years and taking into account that scholars and high school students influence their parents competently it is only logic to concentrate on high school teachers and secondary school teachers. The english Biotechnology and Biological Sciences Research Council offers a wide ranging educational liaison service which is exemplary and could serve as a model for Switzerland.

The planned Swiss service should include:

- Establishment of a liaison Scientist - Teacher. This includes the maintaining of a personnel contact between all high school teacher with an active scientist at a university.
- Organization of a national science week.
- Organization of workshops and science clubs for all levels of school and the public.
- Stimulate the connections between graduated students and scholars on all levels.

11. Third Proposal: Public Information System on the Internet

All available information about facts are subject to ideological representations, even the scientific representation is not „neutral" as even the selection of a research object could be regarded as valuation. The representation of a biorisk by Greenpeace is in general not identical with the representation by the industry and this is in general not the same like the scientific representation. The different world views can be described as: scientific, optimistic, sceptical, popular and official. Of course it would be possible to find other definitions, however, we assume that all representations can be assigned to one of those five.

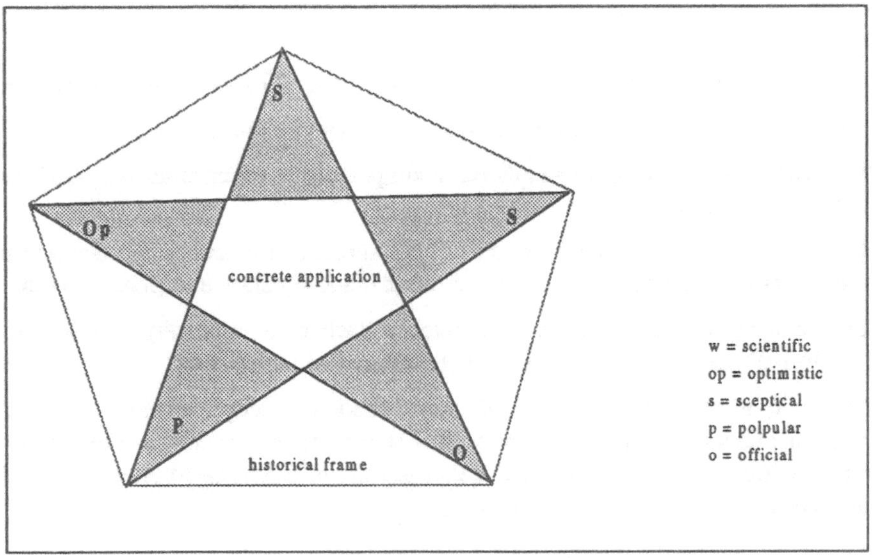

Figure 1: Each concrete application must be represented by all relevant views as there are: scientific, optmimistic, sceptical, popular and official.

From this, the following thesis are concluded:

- All representations include some valuation, there exist no „neutral" representation.
- Every new institution would make its own valuation or take over existing world views except it would become just another „scientific" institution.
- Only way out of this dilemma is to represent the „facts" in a pattern containing all five views.
- A synopsis is not possible, however, obvious mistakes can be pointed out and the five representation should be put in a historical frame in order to allow a reflected view.

The following graphics demonstrates the principle for the functioning of an internet site which will be maintained for the open public.

12. Conclusion

Only 26% of all citizens said NO to the initiative and only 13% said YES, 59% did not express their opinion! What does this mean?

First of all it can be concluded that the discourse on gene technology is not yet over, there is still a strong demand for information, education and discussion. Until a large proportion of Swiss citizens will accept gene technology in their daily life and gain - or regain - credibility in experts, officials and industry a lot of work has yet to be done.

During the campaign government, scientists and industry have made many promises to improve the discourse on gene technology and it has been recognized that scientists had for a long period failed to inform adequately and discuss their motivation and the fear of citizens. Government has enacted many corrections to existing laws, implemented a national biosafety committee and a national ethics committee, however, it is yet to demonstrate that these institutions are not mere strategical instruments but will play an important role in future.

Scientists have the problem of demonstrating the relevance for science to individual well-being, human health and national wealth. Therefore, they are forced to become more active and allocate a certain amount of their time to inform the citizen and discuss the relevant issues.

It seems as the rare resource „attention" from open public has to compete with other rare resources like time and money. The challenge might be harder than we realize at the moment!

13. References

(1) Sybille Hardmeier, Daniel Scheiwiller: Analyse der eidgenössischen Abstimmung vom 7. Juni 1998

(2) Making that link; Biotechnology and biological science research council (UK) 1996

(3) Andreas Thomann; Was bewegt die Schweiz? CS Bulletin 1/1998

(4) Manuel Eisner; Gentechnologie und gesellschaftliche Moral, Bio World 1/98

SUBJECT INDEX

GPSR Compliance

The European Union's (EU) General Product Safety Regulation (GPSR) is a set of rules that requires consumer products to be safe and our obligations to ensure this.

If you have any concerns about our products, you can contact us on ProductSafety@springernature.com

In case Publisher is established outside the EU, the EU authorized representative is:

Springer Nature Customer Service Center GmbH
Europaplatz 3
69115 Heidelberg, Germany

Batch number: 10584568

Printed by Printforce, the Netherlands